Hydrodynamic stability theory

MECHANICS : ANALYSIS

Editors: V.J.Mizel and G.Æ. Oravas

M.A. Krasnosel'skii, P.P. Zabreyko, E.I. Pustylnik and P.E. Sobolevski, Integral Operators in Spaces of Summable Functions. 1976.
ISBN 90-286-0294-1.

V.V. Ivanov, The Theory of Approximate Methods and Their Application to the Numerical Solution of Singular Integral Equations. 1976.
ISBN 90-286-0036-1.

A. Kufner, J. Oldrich and F.Cl. Svatopluk (eds), Function Spaces. 1977.
ISBN 90-286-0015-9.

S.G. Mikhlin, Approximation on a Rectangular Grid. With Application to Finite Element Methods and Other Problems. 1979.
ISBN 90-286-0008-6.

D.G.B. Edelen, Isovector Methods for Equations of Balance. With Programs for Computer Assistance in Operator Calculations and an Exposition of Practical Topics of the Exterior Calculus. 1980.
ISBN 90-286-0420-0.

R.S. Anderssen, F.R. de Hoog and M.A. Lukas (eds), The Application and Numerical Solution of Integral Equations. 1980.
ISBN 90-286-0450-2.

R.Z. Has'minski, Stochastic Stability of Differential Equations. 1980.
ISBN 90-286-0100-7.

A.I. Vol'pert and S.I. Hudjaev, Analysis in Classes of Discontinuous Functions and Equations of Mathematical Physics. 1985.
ISBN 90-247-3109-7.

A. Georgescu, Hydrodynamic Stability Theory. 1985.
ISBN 90-247-3120-8.

Hydrodynamic stability theory

By

Adelina Georgescu

Institute of Mathematics
Bucharest, Romania

1985 **MARTINUS NIJHOFF PUBLISHERS**
a member of the KLUWER ACADEMIC PUBLISHERS GROUP
DORDRECHT / BOSTON / LANCASTER

Distributors

for the United States and Canada: Kluwer Boston, Inc., 190 Old Derby Street, Hingham, MA 02043, USA
for the UK and Ireland: Kluwer Academic Publishers, MTP Press Limited, Falcon House, Queen Square, Lancaster LA1 1RN, UK
for Hungary, Albania, Bulgaria, China, Cuba, Czechoslovakia, German Democratic Republic, Democratic People's Republic of Korea, Mongolia, Poland, Romania, Soviet Union, Democratic Republic of Vietnam and Yugoslavia: Editura Ştiinţifică şi Enciclopedică, Piaţa Scînteii 1, Bucharest 33, Romania
for all other countries: Kluwer Academic Publishers Group, Distribution Center, P.O. Box 322, 3300 AH Dordrecht, The Netherlands

Library of Congress Cataloging-in-Publication Data

Georgescu, Adelina.
Hydrodynamic stability theory.

(Mechanics—analysis; 9)
Revised translation of: Teoria stabilităţii hidrodinamice.
Includes bibliographies and index.
1. Hydrodynamics. 2. Navier-Stokes equations—Numerical solutions. 3. Stability. I. Title. II. Series.
QA911.G4413 1985 532'.5 85-18903
ISBN 90-247-3120-8

Book information

Revised, updated translation of the Romanian edition "Teoria stabilităţii hidrodinamice", first published by Editura Ştiinţifică şi Enciclopedică, Bucharest, 1976. Translated by dr. Adelina Georgescu. Translation edited by Professor David Sattinger.

PRINTED IN ROMANIA

CONTENTS

FOREWORD

The great number of varied approaches to hydrodynamic stability theory appear as a bulk of results whose classification and discussion are well-known in the literature. Several books deal with one aspect of this theory alone (e.g. the linear case, the influence of temperature and magnetic field, large classes of globally stable fluid motions etc.). The aim of this book is to provide a complete mathematical treatment of hydrodynamic stability theory by combining the early results of engineers and applied mathematicians with the recent achievements of pure mathematicians. In order to ensure a more operational frame to this theory I have briefly outlined the main results concerning the stability of the simplest types of flow.

I have attempted several definitions of the stability of fluid flows with due consideration of the connections between them. On the other hand, as the large number of initial and boundary value problems in hydrodynamic stability theory requires appropriate treatments, most of this book is devoted to the main concepts and methods used in hydrodynamic stability theory. Open problems are expressed in both mathematical and physical terms.

In order to understand hydrodynamic stability theory and the reality underlying the Navier-Stokes model the reader is of necessity assumed to be highly conversant with mathematics (the calculus of variations, differential equations, spectral theory of linear operators, nonlinear functional analysis, differential topology etc.). Additionally, the mathematical formulation of the problems of hydrodynamic stability theory makes use of highly abstract concepts from the above theories. Practice has taught us that whereas undergraduates or graduates in mathematics can deal with these concepts at a high level of abstraction, they are unable to take good account of these abstract formulations in tackling particular cases from mechanics and eventually to grasp the phenomena through equations. Besides, most of the talks I had with experts in boundary layer theory, turbulence, chemistry, only reinforced my belief that most of these specialists cannot abstract the physical meanning from the beautiful results of pure mathematicians. I have attempted to overcome these difficulties by an appropriate presentation and explanation of abstract problems and by associating physical facts to the mathematical concepts discussed.

The main topics discussed in this book refer to the mathematical linear and nonlinear stability of fluid flows. In order to manage a unified treatment of this theory, I have discussed the linear case only after elucidating its connection with the nonlinear one. Besides I have made use of both classical and modern methods of treatment.

The first three chapters provide a general account of hydrodynamic stability theory. Thus, after reducing the nonlinear stability problem to a nonlinear spectral one (i.e. after the completeness of normal modes has been proven) results of nonlinear stability are derived from linear theory in virtue of the linearization principle. In that respect Chapter 1 is a brief exposition of the classical linear theory. Chapter 2 analyses the principle of exchange of stabilities, which relates stability with bifurcation for the solutions of the Navier-Stokes equations. Stress is laid on the generalized theories developed by O. A. Ladyzhenskaya, G. Iooss, C. Foias, G. Prodi and to the energetic methods of J. Serrin, J. L. Synge and D. D. Joseph. This generalized frame is a suitable background for proving the principle of exchange of stabilities and the linearization principle. Chapter 3 is a treatment of the bifurcation of solutions of the Navier-Stokes equations. The stabillity of secondary flows (including basic results of D. D. Joseph, D. H. Sattinger, W. Velte, V. Yudovich) is also discussed. Both stability and bifurcation are used to explain the origin of turbulence. Three phenomenological theories on the nature of turbulence (Leray, Landau-Hopf, Ruelle-Takens) and Sattinger's theory of symmetry breaking and pattern formation are the topic of Chapter 4. The contribution of stability and bifurcation in selecting from the solutions of the Navier-Stokes equations, that solution which corresponds to the real motion is duly emphasized. The last chapter discusses Gheorghitza's linear theory of stability in the presence of a porous medium.

Compared with the first edition (1976) this book adds a chapter on the nature of turbulence which presents the results of stability and bifurcation theory obtained over the last decade. Appendices 3—8 give several recent methods for solving various problems in hydrodynamic stability theory.

I am grateful to Professor St. I. Gheorghitza, unfortunately no longer with us, for his authoritative guidance of my studies in fluid mechanics and for his encouragement in writing this book. Particular thanks and acknowledgements are due to Dr. Mihnea Moroianu who made valuable comments and remarks and gave much helpful criticism to most of the mathematics involved in this book. Last but not least, I would like to express my gratitude to Professor David Sattinger who has substantially contributed to the publication of this edition.

INTRODUCTION

The role of hydrodynamic stability theory (h.s.th.) in fluid mechanics is paramount, especially when we have to deal with the problem of minimum consumption of energy. This theory deserves special mention in the aerodynamics of profiles in supersonic regime, the construction of automation elements by fluid jets and the technique of emulsions. The rapid development of this theory in recent decades has resulted in a huge number of studies that require much effort to cover. The aim of this book is to provide a unified presentation of the fundamentals in h.s.th. and its achievements.

From the mathematical standpoint a solution $\bar{u}$ of the Navier-Stokes equations can exist for any value of the Reynolds number Re, but it corresponds to the observed motion only for Re smaller than some critical value Re_{cr}. For $\mathrm{Re} > \mathrm{Re}_{cr}$, perturbations existing in the ambient medium are amplified destroying the motion $\bar{u}$, the subsequent fluid flow corresponding to some other solution u of Navier-Stokes equations. We say that for $0 \leqslant \mathrm{Re} \leqslant \mathrm{Re}_{cr}$, the motion $\bar{u}$ is stable, and for $\mathrm{Re} > \mathrm{Re}_{cr}$ it is unstable.

Hydrodynamic stability theory deals with the determination of the value Re_{cr}, therefore of the ranges where various solutions of the Navier-Stokes equations correspond to the actual flow. In addition, hydrodynamic stability theory provides the form of the most unstable disturbances.

This theory is connected with the problem of turbulence since one cause of the appearance of turbulence for any motion is instability, and for several motions, when $\mathrm{Re} = \mathrm{Re}_{cr}$, the laminar flow turns into a turbulent one.

Although several early concerns related to hydrodynamic stability go back as far as 1764, when Euler [30][1] raised the problem of convection currents deriving from the slight compressibility of a heavy fluid, a systematic study of h.s.th. was undertaken only in the mid past century as a consequence of the Hagen (1839) and Poiseuille (1842) experimental evidence of two regimes of fluid flows—laminar and turbulent. In 1848 Helmholtz [20] gave an accurate qualitative analysis of the instability of the surface of discontinuity of tangential velocity in stratified flows and Tyndall [4] studied the Helmholtz instability of a layer of vortices which surround a circular

[1] The references can be found at the end of Chap. 1.

jet of air. Ten years later, h.s.th. attracted the attention of Rayleigh [48] who laid (1880) the foundations of the linear theory and was the first to apply the method of small perturbations — known then as the time exponential method — to the study of inviscid fluids [49].

An important change in the dynamics of viscous fluid was brought about by the first attempt to establish a theory of turbulence. Thus, in 1883 O. Reynolds [55] made the first observations on transition and the h.s. criterion $\mathrm{Re} < \mathrm{Re}_{cr}$, relating this physical parameter to the passage from the laminar to the turbulent regime. Reynolds advanced the idea that turbulence is a consequence of instability (more precisely, of the amplification of disturbances) and this idea was taken over in all the subsequent studies on turbulence. From the equations known as Reynolds equations he derived (1895) the energy balance equation [56] to which he applied the energetic method based on the use of several positively-defined integrals. Thus another credit to Reynolds was the introduction of the second principal method in h.s., which is currently employed in nonlinear theory. Orr [44] studied the energy equations by means of variational methods and the energy method became conservative in papers concerning h.s. written early in this century; but, as was noticed by D. D. Joseph [25], [28], the results obtained in this way by Sharpe, Orr, Lorentz and others are erroneous since they neglect just the most unstable perturbations.

Early in this century studies on hydrodynamic stability were connected with the Bénard experiments on thermal convection in thin liquid layers. In 1907—1908 Orr and Sommerfeld derived the equation which lay at the basis of the linear theory of the stability of plane parallel motions. In 1910 Kelvin analysed the Helmholtz instability for water waves generated by wind and in 1916 Rayleigh [51] stated a stability criterion for the flow between cylinders. This criterion became completely validated (1938) at the hands of Synge [75]. The first experimental validation of h.s.th. is also related to Couette flow by the theoretical and experimental results of Taylor [76].

Boundary layer theory, whose theoretical foundations were laid by Prandtl in 1904, is closely related to h.s., owing to the practical importance of transition in this motion. In the simplest cases of boundary layer theory, the problems of stability have been investigated for a long time by approximating the Blasius velocity profile of the motion by segments of straight lines (L. Prandtl (1921), O. Tietjens (1925) etc.). In this way the very influence of the curvature of the profile is lost and thus the role played by the curvature in the stability of this motion cannot be duly considered. That is why the stability curve for this case was given as late as 1929 in a basic paper [80] of Tollmien, who approximated the Blasius profile by a straight line joined by a parabola or by the curve corresponding to a 4th order polynominal. Subsequently, it was found that the expressions given by Tollmien for the solutions of the Orr-Sommerfeld equation are suitable in numerical calculations [53], [3]. Nowadays they are still used as a tool in papers dealing with linear stability.

Another remarkable paper on h.s.th. is due to Heisenberg [19] (1924) who indicated the instability (for some Reynolds numbers) of plane Poiseuille motion by asymptotic methods and gave the form of the stability curve. Shortly after the publication (1924) of Heisenberg's paper, E. Noether gave a proof of the stability of this motion for any Reynolds number. This fact aroused much dispute and a considerable number of published articles [58], [68], [34] confirmed one or the other of these two opposite theses. Finally the rigorous proof given by A. L. Krylov (1963) confirmed the validity of Heisenberg's results.

The first paper (1926) concerning the analogy between the stability of rotating flows and of thermal convection became instrumental in the study of these two motions (see, for instance, [18], [61]).

In 1930 Odqvist published his well-known memoirs on the integrodifferential form of Navier-Stokes equations. The next decade witnessed intensified developments in fluid mechanics, some of which are even today far from being over. Thus, in 1932 H. Schlichting undertook the investigation of boundary layer flows and his results lay at the basis of most researches [59], [9], [40] carried out during a quarter century. In 1933—1934 three memoirs due to J. Leray introduced generalized solutions of Navier-Stokes equations and provided the first global existence theorem for these equations and elaborated the method called by him the Arzelà-Schmidt method. That same year (1933), H. B. Squire proved the famous theorem which bears now his name. But the essential contribution (1938) to the development of h.s. is due to Synge [74] who related the Reynolds idea with Serrin's in the energetic method. In a well-known monograph [17] S. Goldstein synthetized many results in h.s. (1938). In 1940 A. Pellew and R. V. Southwell proved the validity of the principle of exchange of stabilities for Bénard convection; the discussion of this principle became the object of many articles which dealt with other types of fluid flows.

Since 1930, many papers on h.s. have taken good account of E. Langer's results concerning asymptotic solutions of differential equations; some of these results have been applied in the h.s.th. by Langer himself [35], [36].

L. D. Landau proposed a theoretical model for transition (1944) based on supercritical branching of the solutions of the Navier-Stokes equations. This model was substantiated mathematically (1952) by E. Hopf for systems of nonlinear equations close to Navier-Stokes equations.

C.C. Lin, a famous specialist in h.s.th., whose activity in this field extends to the present, published his first paper on stability of fluid motions in 1944. In 1955 he published a monograph [38] on h.s.

in which the mathematical formulation of the problems is essentially different from the conservative treatment.

In 1947, Tollmien's theoretical results were experimentally verified by G. B. Schubauer and H. K. Skramstad [60] in a wind tunnel of low turbulence. That same year, W. Wasow took up the study of asymptotic solutions of the Orr-Sommerfeld equation, especially for the case of plane Couette flow. The importance of these studies grew in 1971 when V. A. Romanov used them in the proof of the stability of this flow for every Reynolds number (for small initial energy of the perturbation).

In 1954 S. F. Shen gave a numerical integration of the Orr-Sommerfeld equation in the case of boundary layer, which reinforced the validity of Tollmien's results [65]. The same year witnessed the publication of D. Grohne's article on the spectrum of eigenvalue problem for the Orr-Sommerfeld equation.

The intermittent character of the transition of motions in pipes was identified for the first time (1956) by J. C. Rotta [57].

A very important contribution to the development of a rigorous theory of the asymptotic method in h.s. [53], [54] is due to W.H. Reid who, since the years 1957 used this method in examining numerous cases of fluid flows. His studies are well-known for their accuracy.

The generalized solutions of Leray were analysed only in 1951 by E. Hopf, and S. G. Krein in 1953 and 1957. At the hands of A. A. Kiselev and O. A. Ladyzhenskaya (1957) these generalized solutions led to the modernization of the mathematical tools of investigation of fluid flows.

In 1958 J. T. Stuart developed an energetic method related to Reynolds'idea and to the Landau-Hopf conjecture. This method was frequently used in the investigation of transition and therefore of h.s. J. Serrin reconsidered (1959) the energetic method [62] and his study became fundamental for the subsequent development of the theory of stability in the mean.

The importance of the initial value problem in h.s. was discussed in several papers by K. M. Case (beginning in 1960). That same year V. I. Yudovich began his important studies on generalized solutions of the Navier-Stokes equations.

In 1961 V. S. Sorokin published the first study on the branching of solutions of Navier-Stokes equations, using the Liapunov-Schmidt method.

In 1961 the well-known monograph of S. Chandrasekhar [5] concerning classical hydrodynamic and hydromagnetic stability was published.

The first part of the linearization principe in h.s. was proved by G. Prodi in 1962. Other worthy contributions in the field belong to the synthetical work (1962) by A. S. Monin and A. M. Yaglom [41]. Later completions of [41] were given in a chapter of the monograph [42] on turbulence by the same authors. We quote also the monographs [66] and [74].

In 1964 and 1966, W. Velte published two fundamental papers in the theory of bifurcation and stability of the secondary solutions for the Navier-Stokes equations.

The analysis of h.s.th. by the energetic methods was undertaken in 1965 by D. D. Joseph, whose intensive activity has lead to the theory of the global stability of fluid flows [30].

A substantial contribution to the stability and branching problem for solutions of the Navier-Stokes equations was given by G. Iooss, D. H. Sattinger and I. Kirchgässner.

Three trends become apparent (1970—1980) with regard to h.s. and bifurcation. 1. The Ruelle-Takens theory and, related with it, the theory of the Lorentz strange attractor; 2. The generic finiteness of the set of the stationary (and last time — nonstationary) solutions of the Navier-Stokes equations; 3. Sattinger's theory on bifurcation, symmetry breaking and pattern formation.

In addition to the works dealing with h.s. in a generalized frame (those published after 1964 are discussed in Chaps. 2 and 3 herein) papers on classical h.s.th. still continue to appear.

Numerous works on h.s. treat problems of remarkable practical importance: convection, convection in porous media and instability of fluid flows in the presence of porous media [16], [43], [86]. The last among these problems, initiated in 1960 by St. I. Gheorghitza, is the topic of Chap. 5.

Chapter 1

CLASSICAL HYDRODYNAMIC STABILITY

§ 1.1. SETTING OF THE PROBLEM

1.1.1. Stability in the small

Let $\Omega \subset R^3$ be a domain (i.e. an open connected set) in which the flow of an incompressible viscous Newtonian fluid, acted upon by the body force $\mathbf{f}(\mathbf{x}, t)$ takes place. The velocity $\mathbf{u}(\mathbf{x}, t) = (u_1, u_2, u_3)$ and pressure $p(\mathbf{x}, t)$ are the solutions of the mathematical problem of hydrodynamics [23]

$$\frac{\partial \mathbf{u}}{\partial t} + (\mathbf{u} \cdot \operatorname{grad})\, \mathbf{u} = \mathbf{f} - \frac{1}{\rho} \operatorname{grad} p + \nu \Delta \mathbf{u}, \tag{1.1.1}$$

$$\operatorname{div} \mathbf{u} = 0, \tag{1.1.2}$$

$$\mathbf{u}(\mathbf{x}, 0) = \mathbf{u}_0(\mathbf{x}), \tag{1.1.3}$$

$$\mathbf{u}\,|_{\partial\Omega} = \mathbf{w}. \tag{1.1.4}$$

The vector equation (1.1.1) is called the Navier-Stokes equation, (1.1.2) is the continuity equation, (1.1.3) represents the initial condition, and (1.1.4) is the condition on the boundary $\partial\Omega$ of Ω[1]; $\mathbf{x} = (x_1, x_2, x_3)$ is the position vector of an arbitrary point in the domain of motion Ω, t denotes the time, the constant ρ is the density and ν represents the kinematic viscosity which is assumed to be constant.

Problem (1.1.1)—(1.1.4) can vary owing to a variation of the initial conditions or of boundary conditions or of the body force. In the sequel, we shall assume that *only the initial conditions are those which are varied*, and this assumption reduces the problem to the study of the perturbations acting only at the initial moment (in reality they are acting all the time). Thus, consider a motion, called the basic motion corresponding to a solution $(\bar{\mathbf{u}}, \bar{p})$ of the problem

$$\frac{\partial \bar{\mathbf{u}}}{\partial t} + (\bar{\mathbf{u}} \cdot \operatorname{grad})\, \bar{\mathbf{u}} = \mathbf{f} - \frac{1}{\rho} \operatorname{grad} \bar{p} + \nu \Delta \bar{\mathbf{u}}, \tag{1.1.1$'$}$$

$$\operatorname{div} \bar{\mathbf{u}} = 0, \tag{1.1.2$'$}$$

$$\bar{\mathbf{u}}(\mathbf{x}, 0) = \bar{\mathbf{u}}_0(\mathbf{x}), \tag{1.1.3$'$}$$

$$\bar{\mathbf{u}}\,|_{\partial\Omega} = \mathbf{w} \tag{1.1.4$'$}$$

[1] When Ω is unbounded we impose also a condition as $|\mathbf{x}| \to \infty$.

and assume that at the initial moment the basic motion is subjected to the perturbation $(\mathbf{v}_0, q_0)$. This induces a perturbation $(\mathbf{v}, q)$ for all time $t > 0$, such that for the same body forces $\mathbf{f}$ and for the same boundary conditions $\mathbf{w}$, the fluid has the perturbed flow $(\mathbf{u}, p)$ corresponding to the initial condition $\mathbf{u}_0 = \bar{\mathbf{u}}_0 + \mathbf{v}_0$, where

$$\mathbf{u} = \bar{\mathbf{u}} + \mathbf{v}, \; p = \bar{p} + q. \tag{1.1.5}$$

In most studies on hydrodynamics the pressure is eliminated by application of the curl operator to equation (1.1.1); and the study of the solution $(\mathbf{u}, p)$ is reduced to that of the velocity $\mathbf{u}$. In the following, the velocity will be referred to as the solution of the hydrodynamics problem.

We say that the basic flow $\bar{\mathbf{u}}$ *is stable to perturbations of the initial conditions* if every other motion $\mathbf{u}$ which at the initial time $t = 0$ is sufficiently close to $\bar{\mathbf{u}}_0$, subsequently remains as close to $\bar{\mathbf{u}}$ as desired; if, in addition, the motion $\mathbf{u}$ approaches the basic motion as $t \to \infty$ then we say that $\bar{\mathbf{u}}$ is *asymptotically stable* (in Liapunov's sense). In other words, a basic flow is asymptotically stable if disturbances are damped out and it is asymptotically unstable if disturbances are amplified. That is why a real motion can correspond only to a stable solution; whereas an unstable motion will never be observed in reality, since it will be destroyed by the perturbations existing at every time in the ambient medium.

The mathematical formulation of the stability definition depends on the mathematical model attached to the motion. When the fluid flow corresponds to the classical solution of the problem (1.1.1)—(1.1.4), the mathematical formulation is as follows.

Definition 1.1.1 The solution $\bar{\mathbf{u}}$ is said to be *stable in the small* if for every $\varepsilon > 0$ there exists $\eta(\varepsilon) > 0$ such that $\sup\limits_{x,t} |\mathbf{u}(\mathbf{x}, t) - \bar{\mathbf{u}}(\mathbf{x}, t)| < \varepsilon$ for $\sup\limits_{x} |\mathbf{u}_0(\mathbf{x}) - \bar{\mathbf{u}}_0(\mathbf{x})| < \eta(\varepsilon)$; $\bar{\mathbf{u}}$ is said to be *asymptotically stable in the small* if it is stable in the small and, in addition, $\lim\limits_{t\to\infty} |\mathbf{u}(\mathbf{x}, t) - \bar{\mathbf{u}}(\mathbf{x}, t)| = 0$, the limit being uniform with respect to $\mathbf{x}$.

A motion is called *unstable* if it is not stable.

Subtracting equation (1.1.1)′ from (1.1.1), it follows that $\mathbf{v}$ is the solution of the following initial-value problem

$$\frac{\partial \mathbf{v}}{\partial t} + (\bar{\mathbf{u}} \cdot \operatorname{grad})\, \mathbf{v} + (\mathbf{v} \cdot \operatorname{grad})\, \bar{\mathbf{u}} + (\mathbf{v} \cdot \operatorname{grad})\, \mathbf{v} = -\frac{1}{\rho} \operatorname{grad} q + \nu \Delta \mathbf{v}, \tag{1.1.6}$$

$$\mathbf{v}(\mathbf{x}, 0) = \mathbf{v}_0, \tag{1.1.7}$$

in the class of solenoidal functions ($\operatorname{div} \mathbf{v} = 0$), which vanish on $\partial\Omega$ and have all the derivatives in equation (1.1.6). Therefore the null solution of the problem (1.1.6), (1.1.7) corresponds to the basic motion, Definition 1.1.1 being equivalent to

Definition 1.1.2 The basic flow $\bar{\mathbf{u}}$ is stable to disturbances $\mathbf{v}_0$ of the initial conditions if for every $\varepsilon > 0$, there exists $\eta(\varepsilon) > 0$, such that $\sup_{x,t} |\mathbf{v}(\mathbf{x}, t)| < \varepsilon$ for $\sup_x |\mathbf{v}_0(\mathbf{x})| < \eta$; if, in addition, $\lim_{t\to\infty} \mathbf{v}(\mathbf{x}) = 0$, then $\bar{\mathbf{u}}$ is asymptotically stable.

Hydrodynamic stability deals with the determination of those conditions and limits giving rise to an asymptotically stable or unstable basic flow, without knowledge of the disturbances. In other words, this theory does not study the time-evolution of the initial disturbances but their asymptotic behaviour, and it indicates the value of the physical parameters at which the amplification of the perturbations starts. For the fluid flows described by the problem (1.1.1)—(1.1.4) we have a single parameter which shall be pointed out by passing from physical coordinates to the dimensionless ones, by means of the following characteristic quantities V — the velocity, l — the length, Vl^{-1} — time and ρV^2 — the pressure. Then we can denote the dimensionless quantities by the same symbols which have stood for the corresponding dimensional quantities, and hence the Navier-Stokes equation becomes

$$\text{(1.1.8)} \qquad \frac{\partial \mathbf{u}}{\partial t} + (\mathbf{u} \cdot \text{grad})\, \mathbf{u} = \mathbf{f} - \text{grad}\, p + \frac{1}{\text{Re}} \Delta \mathbf{u},$$

where Re is the positive dimensionless Reynolds number, equal to the ratio between inertia and viscosity forces

$$\text{(1.1.9)} \qquad \text{Re} = \frac{Vl}{\nu}.$$

In dimensionless coordinates, the mathematical problem of hydrodynamics is expressed by relations (1.1.8) and (1.1.2)—(1.1.4), the only physical parameter being the Reynolds number. Therefore we can point out the stability ranges for a basic motion bounded by particular values of this number. Now, we shall briefly indicate the connection between hydrodynamic stability, on one hand, and the existence, uniqueness, branching and the regularity of the solution $\bar{\mathbf{u}}$ of the Navier-Stokes equations on the other hand in terms of the Reynolds number. For small Re, the above-mentioned problem admits the basic solution $\bar{\mathbf{u}}$ which is unique and stable. Let Re_G be the Reynolds number at which the basic flow loses its stability. The resulting instability may be of two types *pure instability* and *instability by steps*. The perturbations acting on basic motions possessing pure instability for $\text{Re} \geqslant \text{Re}_G$ are amplified as $t \to \infty$, and the motions change fundamentally in character, passing to turbulence. In the case of instability by steps, for $\text{Re} = \text{Re}_G$, as $t \to \infty$ the perturbations tend toward a new laminar motion $\bar{\mathbf{u}}_1$, called the *secondary flow*. This motion is stable up to another Re'_G at which $\bar{\mathbf{u}}_1$ becomes unstable (pure or by steps) and so on.

Instability by steps is closely related to the problem of the bifurcation of solutions of the Navier-Stokes equations, Re_G being a point where stability is lost and a branching point as well; and thus it is related to the appearance of a continuum of motions following one another as Re increases. Assuming that the basic solution exists for every Re, it follows that Re_G is also a point of loss of uniqueness of this solution.

The joint study of stability and branching removes the paradoxes of some basic flows (for instance Couette flow between rotating cylinders) which are observed in experiments only for $\mathrm{Re} < \mathrm{Re}_G$, although the corresponding solution of the problem (1.1.8), (1.1.2)—(1.1.4) exists for any Re. In these cases, the basic flow (which is unique only for $\mathrm{Re} < \mathrm{Re}_G$) was derived earlier and, since it exists for any Re, nobody questioned about its uniqueness up to now. But for $\mathrm{Re} > \mathrm{Re}_G$ there exists also another solution calculated according to branching theory. As this solution is stable, it is the one which is observed in reality.

A rigorous study of hydrodynamic stability theory is closely related to the study of the global existence (for every $t > 0$) of the basic solution and of perturbations; the latter are observed in reality or can be generated artificially. In general, the global existence for solutions corresponding to the perturbations and to basic motions cannot be proved in the classical framework but only for generalized solutions (Chap. 2) belonging to some Hilbert spaces H, obtained by completion of some spaces of continuous functions. For these solutions we can define the *stability in the large* by replacing $\sup\limits_{x,\,t} |\mathbf{v}(\mathbf{x}, t)|$ in Definition 1.1.2 by the norm of H. For instance, for a given choice of H, a motion is called stable if $\int_\Omega |\mathbf{v}|^2 \, \mathrm{d}x \to 0$ as $t \to \infty$, for another choice, if $\int_\Omega |\operatorname{grad} \mathbf{v}|^2 \, \mathrm{d}x \to 0$ when $t \to \infty$. In the classical case, a kind of stability in the large is the so-called *stability in the mean* (corresponding to the former example), connected with the energy of the perturbations. In the second example, owing to the occurrence of the gradient of the velocity in the definition of the stability in the large, it follows that the second condition is stronger than the stability in the small which was concerned only with the velocity. The generalized solutions correspond to motions whose velocities and gradients of the velocities are bounded in the mean, but may nevertheless be unbounded pointwise. Thus, the generalized solutions are appropriate to the study of the last transition stages, whilst the classical model corresponds to laminar flows or to the first stages of transition. Since the classical solutions are particular case of the generalized ones, it follows that these latter describe both the laminar regime and the transitional regime (the only regimes discussed in this book); the third regime of motion—turbulence—requires a statistical model.

In practice, the Navier-Stokes equations are of interest only if their solutions correspond to a real motion, and therefore, in particular, if they are stable, whence the necessity of studying the stability of the classical and generalized basic or secondary solutions of these equations. The available numerical results lead to the conclusion that the Navier-Stokes equations are suited for the description of the usual fluid flows, provided that they belong to those function classes that are imposed by experiments.

1.1.2. Stability in the mean

As in the preceding section, by perturbation we mean a vector function $\mathbf{v}$ equal to the difference between the velocity of the perturbed motion and the basic one. In dimensionless coordinates it satisfies the following problem

$$\frac{\partial \mathbf{v}}{\partial t} + (\overline{\mathbf{u}} \cdot \text{grad})\, \mathbf{v} + (\mathbf{v} \cdot \text{grad})\, \overline{\mathbf{u}} + (\mathbf{v} \cdot \text{grad})\, \mathbf{v} =$$

$$= -\text{grad}\ q + \frac{1}{\text{Re}}\, \Delta \mathbf{v}, \tag{1.1.6$'$}$$

$$\mathbf{v}(\mathbf{x}, 0) = \mathbf{v}_0 \tag{1.1.7$'$}$$

in the class (called *difference motions*) consisting of solenoidal vectors vanishing on $\partial\Omega$, that is

$$\text{div}\ \mathbf{v} = 0, \tag{1.1.10}$$

$$\mathbf{v}\,|_{\partial\Omega} = 0. \tag{1.1.11}$$

Let us now derive the equation for the perturbation energy $\mathbf{v}$, assuming that the domain of motion Ω is bounded. If Ω is unbounded, we assume that $\mathbf{v}$ is periodic with respect to $\mathbf{x}$. Then Ω can be replaced by the periodicity cell (denoted also by Ω). Multiplying equation (1.1.6)′ by $\mathbf{v}$, and integrating the resulting equation over Ω, we obtain

$$\frac{\mathrm{d}}{\mathrm{d}t}\int_\Omega \frac{|\,\mathbf{v}\,|^2}{2}\,\mathrm{d}x + \int_\Omega (\overline{\mathbf{u}} \cdot \text{grad})\, \mathbf{v} \cdot \mathbf{v}\,\mathrm{d}x + \int_\Omega (\mathbf{v} \cdot \text{grad})\, \overline{\mathbf{u}} \cdot \mathbf{v}\,\mathrm{d}x +$$

$$+\int_\Omega (\mathbf{v} \cdot \text{grad})\, \mathbf{v} \cdot \mathbf{v}\,\mathrm{d}x = -\int_\Omega \mathbf{v}\ \text{grad}\ q\,\mathrm{d}x +$$

$$+\frac{1}{\text{Re}}\int_\Omega \Delta \mathbf{v} \cdot \mathbf{v}\,\mathrm{d}x. \tag{1.1.12}$$

For solenoidal vectors $\mathbf{v}_1$, the identity $\operatorname{div}(\mathbf{v}_1\varphi) = \mathbf{v}_1 \operatorname{grad}\varphi + \varphi \operatorname{div}\mathbf{v}_1$ comes to $\operatorname{div}(\mathbf{v}_1\varphi) = \mathbf{v}_1 \operatorname{grad}\varphi$ which, by integrating on Ω and applying the flux-divergence formula, becomes

$$\int_\Omega \mathbf{v}_1 \operatorname{grad}\varphi \, dx = \int_\Omega \operatorname{div}(\mathbf{v}_1\varphi)\, dx = \int_{\partial\Omega} \mathbf{v}_1 \varphi \mathbf{n}\, d\sigma.$$

Moreover, if $\mathbf{v}_1|_{\partial\Omega} = 0$, then from the above equality it follows that every difference motion satisfies the relation

$$\int_\Omega \mathbf{v}_1 \operatorname{grad}\varphi\, dx = 0. \tag{1.1.13}$$

Clearly, $(\bar{\mathbf{u}}\cdot\operatorname{grad})\,\mathbf{v}\cdot\mathbf{v} = \bar{\mathbf{u}}\cdot\operatorname{grad}\dfrac{|\mathbf{v}|^2}{2}$ and $(\mathbf{v}\cdot\operatorname{grad})\,\mathbf{v}\cdot\mathbf{v} = \mathbf{v}\cdot\operatorname{grad}\dfrac{|\mathbf{v}|^2}{2}$.

Hence, taking into account formula (1.1.13), it follows that

$$\int_\Omega (\bar{\mathbf{u}}\cdot\operatorname{grad})\,\mathbf{v}\cdot\mathbf{v}\, dx = \int_\Omega (\mathbf{v}\cdot\operatorname{grad})\,\mathbf{v}\cdot\mathbf{v}\, dx =$$

$$= \int_\Omega \mathbf{v}\cdot\operatorname{grad} q\, dx = 0. \tag{1.1.14}$$

Integrating by parts, for vectors $\mathbf{v}$ vanishing on $\partial\Omega$, we obtain

$$\int_\Omega \Delta\mathbf{v}\cdot\mathbf{v}\, dx = -\int_\Omega |\operatorname{grad}\mathbf{v}|^2\, dx,$$

where $|\operatorname{grad}\mathbf{v}|^2 = \sum_{1\leqslant i,j\leqslant 3} \dfrac{\partial v_i}{\partial x_j}\dfrac{\partial v_i}{\partial x_j}$. Therefore, taking into account (1.1.14), equality (1.1.12) turns into the Reynolds-Orr energy equation

$$\frac{d}{dt}\int_\Omega \frac{|\mathbf{v}|^2}{2}\, dx = -\int_\Omega (\mathbf{v}\cdot\operatorname{grad})\,\bar{\mathbf{u}}\cdot\mathbf{v}\, dx -$$

$$-\frac{1}{\mathrm{Re}}\int_\Omega |\operatorname{grad}\mathbf{v}|^2\, dx. \tag{1.1.15}$$

For solenoidal vectors $\mathbf{v}$, we have the identity $(\mathbf{v}\cdot\operatorname{grad})\,\bar{\mathbf{u}}\cdot\mathbf{v} = \operatorname{div}[(\bar{\mathbf{u}}\cdot\mathbf{v})\cdot\mathbf{v}] - (\mathbf{v}\cdot\operatorname{grad})\,\mathbf{v}\cdot\bar{\mathbf{u}}$, which integrated on Ω, after applying the flux-divergence formula and Condition (1.1.11) yields

$$\int_\Omega (\mathbf{v}\cdot\operatorname{grad})\,\bar{\mathbf{u}}\cdot\mathbf{v}\, dx = -\int_\Omega (\mathbf{v}\cdot\operatorname{grad})\,\mathbf{v}\cdot\bar{\mathbf{u}}\, dx.$$

Hence, an equivalent form of the energy equality is

$$\frac{d}{dt}\int_\Omega \frac{|\mathbf{v}|^2}{2}\, dx = \int_\Omega (\mathbf{v}\cdot\operatorname{grad})\,\mathbf{v}\cdot\bar{\mathbf{u}}\, dx - \frac{1}{\mathrm{Re}}\int_\Omega |\operatorname{grad}\mathbf{v}|^2\, dx. \tag{1.1.15$'$}$$

Let $K(t) = \int_\Omega \frac{1}{2} |\mathbf{v}|^2 \mathrm{d}x$ be the mean energy of the perturbation. The first term on the right-hand side of (1.1.12) represents the quantity of energy which passes from the basic motion to the perturbation and the second term is the dissipation energy due to the viscosity (which is always positive). Then the energetic balance can be expressed as: the increase in time of the mean energy of the perturbations takes place on account of the energy transfer from the basic motion to the perturbation and is diminished by dissipation due to viscosity.

Definition 1.1.3 The basic motion $\bar{\mathbf{u}}$ *is stable in the mean* to the perturbations of the initial conditions, if the energy $K(t)$ of every perturbation $\mathbf{v}$, solution of the problem (1.1.6)', (1.1.7)' remains bounded for every $t \geqslant 0$[2]; if in addition $K(t) \to 0$ for $t \to \infty$, we say that $\bar{\mathbf{u}}$ *is asymptotically stable in the mean.*

$K(t) = 0$ implies that $\mathbf{v} = 0$ almost everywhere while if $K(t)$ is bounded for every $t > 0$, then $\mathbf{v}$ remains bounded almost everywhere. Hence stability in the small is stronger then the stability in the mean. On the other hand, stability in the mean can be defined not only for perturbations expressed by continuous functions but also for perturbations expressed by Riemann integrable functions which can therefore admit points of discontinuity of zero Lebesgue measure on the positive time-axis [63]. In Chap. 2 a similar physical argument is shown to justify the use of Lebesgue integrable perturbations having a set of points of discontinuity of nonvanishing Lebesgue measure. This last property indicates the use of generalized solutions, and thus the concept of stability in the large for the transition regime, where large jumps for velocity and its gradient occur. The stability in the mean, requiring the Riemann integrability of the velocity function, can be applied to the laminar flows only.

From Definition 1.1.3, it follows that a *sufficient condition for stability* in the mean is

$$\frac{\mathrm{d}K(t)}{\mathrm{d}t} < 0, \text{ for all } t > 0, \tag{1.1.16}$$

and every $\mathbf{v}$ in the set of perturbations. A stronger condition, which is also the only one for which there exist numerical criteria, is that the relation (1.1.16) holds in the set of all difference motions.

The above method, by which the study of hydrodynamic stability is reduced to the study of the increasing difference motions energy, is called the *energy method*. It was introduced in hydrodynamic stability by Reynolds [56] and Orr [44].

---

[2] Sometimes a stronger condition than Definition 1.1.2 is used which can be obtain ed by replacing sup $|\mathbf{v}|$ by the norm $K(t)$.

Using difference motions instead of perturbations (the single ones possessing a hydrodynamic sense in the class of difference motions), the energy method yields only bounds for the stability but not its exact limits; hence, instability conditions cannot be obtained by this method [74]. On the other hand, using a class larger than that of perturbations, one has the advantage of obtaining results which are valid for *perturbations of arbitrary form and amount.*

Since for $\mathbf{v} \not\equiv 0, \int_\Omega |\operatorname{grad} \mathbf{v}|^2 \, dx$ is positive, it follows from the energy equation that the viscous dissipation tends to diminish the increase of the energy of perturbations and hence, by the above criterion, it plays a stabilizing role. For large gradients of $\bar{\mathbf{u}}$, the first term from the right-hand side in (1.1.15) tends to enlarge the increase of the energy of the perturbations and thus to induce instability. The form (1.1.15)′ of the energy equation allows us to extend this conclusion also for large $\bar{\mathbf{u}}$. On the contrary, small $\bar{\mathbf{u}}$ or $|\operatorname{grad} \bar{\mathbf{u}}|$ will have a stabilizing effect. Finally, for *small Re every motion is stable in the mean* [63] since $\frac{1}{\mathrm{Re}}$ is large and the leading term is that expressing the viscous dissipation.

Generally speaking, in order to know whether $\bar{\mathbf{u}}$ is stable in the mean or not, we must know the ratio of the two terms in the right-hand side of (1.1.15) or (1.1.15)′; so, if

$$\frac{-\int_\Omega (\mathbf{v} \cdot \operatorname{grad})\, \bar{\mathbf{u}} \cdot \mathbf{v} \, dx}{\frac{1}{\mathrm{Re}} \int_\Omega |\operatorname{grad} \mathbf{v}|^2 \, dx} < 1, \tag{1.1.17}$$

then, by (1.1.15), we have $\frac{dK}{dt} < 0$. Hence, $\bar{\mathbf{u}}$ is stable in the mean. Consequently, from relation (1.1.15) we may deduce the following stability criterion

$$\mathrm{Re} < \widetilde{\mathrm{Re}}. \tag{1.1.18}$$

In (1.1.18), $\frac{1}{\widetilde{\mathrm{Re}}}$ is the extremum of the variational problem with constraints

$$\frac{1}{\widetilde{\mathrm{Re}}} = \max \frac{-\int_\Omega (\mathbf{v} \cdot \operatorname{grad})\, \bar{\mathbf{u}} \cdot \mathbf{v} \, dx}{\int_\Omega |\operatorname{grad} \mathbf{v}|^2 \, dx}, \tag{1.1.19}$$

in the class of difference motions. We shall indicate now another way of determining $\widetilde{\mathrm{Re}}$. First, we notice that the isoperimetric variational problem (1.1.19), (1.1.10), (1.1.11) can be written also as

$$(1.1.19)' \qquad \frac{1}{\widetilde{\mathrm{Re}}} = \max \left\{ -\int_\Omega (\mathbf{v} \cdot \operatorname{grad})\, \bar{\mathbf{u}} \cdot \mathbf{v}\, dx \right\},$$

where $\mathbf{v}$ must satisfy the constraints

$$(1.1.20) \qquad \int_\Omega |\operatorname{grad} \mathbf{v}|^2\, dx = 1, \ \operatorname{div} \mathbf{v} = 0$$

and the boundary condition (1.1.11). Using Lagrange's multipliers $\frac{1}{\mathrm{Re}^*}$ and $\lambda = \lambda(\mathbf{x}, t)$, from (1.1.19)′ and (1.1.20) we get

$$\delta \int_\Omega \left[(\mathbf{v} \operatorname{grad})\, \bar{\mathbf{u}} \cdot \mathbf{v} + \frac{1}{\mathrm{Re}^*} |\operatorname{grad} \mathbf{v}|^2 - 2\lambda \operatorname{div} \mathbf{v} \right] dx = 0,$$

whose Euler-Lagrange equation is

$$(1.1.21) \qquad (\mathbf{v} \cdot \operatorname{grad})\, \bar{\mathbf{u}} = -\operatorname{grad} \lambda + \frac{1}{\mathrm{Re}^*} \Delta \mathbf{v}.$$

In this way the variational problem (1.1.19)′, (1.1.20) is reduced to the eigenvalue problem (1.1.21), (1.1.20), (1.1.11) having $\frac{1}{\widetilde{\mathrm{Re}}}$ as eigenvalue, as is obvious from the following.

Theorem 1.1.1 (*Serrin* [62]). *If the variational problem* (1.1.19)′ (1.1.20), (1.1.11) *admits solutions, then* $\frac{1}{\widetilde{\mathrm{Re}}}$ *is the largest eigenvalue of the eigenvalue problem* (1.1.21), (1.1.20), (1.1.11) *and the basic flow* $\bar{\mathbf{u}}$ *is stable for* $\frac{1}{\mathrm{Re}} > \frac{1}{\widetilde{\mathrm{Re}}}$.

Proof. Taking into account (1.1.20) and (1.1.11) together with the flux-divergence formula, multiplying (1.1.21) by $\mathbf{v}$ and integrating on Ω, we see that every eigenvector $\mathbf{v}$ of Equation (1.1.21) satisfies the relation

$$-\int_\Omega (\mathbf{v} \cdot \operatorname{grad})\, \bar{\mathbf{u}} \cdot \mathbf{v}\, dx = -\int_\Omega \left(-\operatorname{grad} \lambda + \frac{1}{\mathrm{Re}^*} \Delta \mathbf{v} \right) \mathbf{v}\, dx =$$

$$= \frac{1}{\mathrm{Re}^*} \int_\Omega |\operatorname{grad} \mathbf{v}|^2\, dx = \frac{1}{\mathrm{Re}^*}.$$

On the other hand, every extremal vector $\tilde{\mathbf{v}}$ of (1.1.19)′ is by construction, an eigenvector of equation (1.1.21). Therefore the eigenvalue $\frac{1}{\widetilde{\text{Re}}}$ corresponding to the eigenvector $\widetilde{\mathbf{v}}$ is the maximum of (1.1.19)′. In addition, $\frac{1}{\widetilde{\text{Re}}}$ is the largest eigenvalue of (1.1.21) since if a larger eivenvalue would exist, then for its corresponding eigenvector the integral in (1.1.19)′ would take a value larger than $1/\widetilde{\text{Re}}$. The theorem is proved.

Serrin's theorem leaves open the existence problem of an extremum vector for (1.1.19)′. This problem must be analysed separately for every basic motion. A complete and detailed analysis can be found, for instance, in [28].

By means of the energy method, the problem of determining a reliable limit for stability has been reduced to that of the determination of the largest eigenvalue $1/\widetilde{\text{Re}}$ of (1.1.21). This equation is very similar to the Navier-Stokes equation and therefore, very difficult to solve. Nevertheless, it has been solved in particular cases [28], [29], [30], each time yielding the corresponding eigenfunction $\widetilde{\mathbf{v}}$ which represents the most unstable difference motion. Generally $\widetilde{\mathbf{v}}$ is not a perturbation (i.e. a solution of the problem (1.1.6)′, (1.1.17)′) and therefore it may happen that the basic motion $\bar{\mathbf{u}}$ certainly stable for $\text{Re} < \widetilde{\text{Re}}$, is still stable at other values of Re greater than $\widetilde{\text{Re}}$. But if $\widetilde{\mathbf{v}}$ satisfies the problem (1.1.6)′, (1.1.7)′, then $\widetilde{\mathbf{v}}$ is the most unstable perturbation and $\widetilde{\text{Re}}$ is an *exact bound of stability* such that for $\text{Re} > \widetilde{\text{Re}}$ every perturbation is amplified. In particular this is the case for thermal convection in the Boussinesq approximation [24], the helicoidal flows [29] or the Hagen-Poiseuille motion [28] (for other types of flow, see [30]).

$\widetilde{\text{Re}}$, Re_G are called the *energy bound* of stability and the *global bound of stability*, respectively. Re_G possesses the property that, for $\text{Re} < \text{Re}_G$, every perturbation is damped out and for $\text{Re} > \text{Re}_G$ there exists at least an amplified perturbation. Thus the energy method becomes more appropriate as $\widetilde{\text{Re}}$ comes closer to Re_G.

Since the energy method takes into account only the mean characteristics of the motion it gives satisfactory results only for particular classes of motions. Numerous results in the literature show that this method is appropriate especially for *rotation and convective motions*. For plane flows, the results of this method are very weak, as the stability of these motions depends on small subdomains, known as friction layers of the domain of motion. In these cases local methods (e.g. the asymptotic analysis) must be used.

1.1.3. Linear stability

Small perturbations acting upon a fluid flow can be either infinitesimal or finite. Stability theory which takes into account perturbations of any amount is said to be *nonlinear* since it is governed by the nonlinear equations (1.1.6). *The linear theory of hydrodynamic stability* deals with stability against infinitesimal disturbances, since the powers of higher order than one of the quantities which characterize these perturbations are neglected. Then, by neglecting the nonlinear terms in (1.1.6) written in dimensionless coordinates, we obtain the linear initial-value problem

$$(1.1.6)'' \qquad \frac{\partial \mathbf{v}}{\partial t} + (\bar{\mathbf{u}} \cdot \operatorname{grad})\, \mathbf{v} + (\mathbf{v} \cdot \operatorname{grad})\, \bar{\mathbf{u}} = -\operatorname{grad} q + \frac{1}{\mathrm{Re}} \Delta \mathbf{v},$$

$$(1.1.7)'' \qquad \mathbf{v}(\mathbf{x}, 0) = \mathbf{v}_0,$$

in the class of difference motions, which represents the mathematical problem of linear hydrodynamic stability.

The definition of linear stability in the small and that of linear asymptotic stability in the small are derived from Definition 1.1.2 by requiring $\mathbf{v}$ to satisfy equation (1.1.6)''. But, often, another definition is frequently used in the linear theory of hydrodynamic stability. This definition differs from the asymptotic one by additional hypotheses regarding the form of the perturbations. More precisely it is assumed that any perturbation $\mathbf{v}$ can be obtained (see § 2.3) by superposition of some perturbations $\mathbf{v}_0(\mathbf{x})\, e^{-\sigma t}$, called *normal modes*, where the complex number σ is an eigenvalue and $\mathbf{v}_0$ is an eigenfunction of the following problem derived from (1.1.6)'' and (1.1.7)''

$$(1.1.22) \qquad \begin{cases} -\sigma \mathbf{v}_0 + (\bar{\mathbf{u}} \operatorname{grad})\, \mathbf{v}_0 + (\mathbf{v}_0 \operatorname{grad})\, \bar{\mathbf{u}} = -\operatorname{grad} q_0 + \dfrac{1}{\mathrm{Re}} \Delta \mathbf{v}_0, \\ \mathbf{v}_0 \,|_{\partial\Omega} = 0, \end{cases}$$

($\mathbf{v}_0$ being solenoidal).

Definition 1.1.4 The fluid flow $\bar{\mathbf{u}}$ is called *linearly asymptotically stable*, if the problem (1.1.22) has no eigenvalue with negative real part. It is called *asymptotically unstable*, if there exists at least one eigenvalue with negative real part and it is *neutrally* or *marginally stable* if there exists at least an eigenvalue with vanishing real part, the rest of the eigenvalues having positive or vanishing real parts.

Throughout this section and § 1.2 instead of linearly asymptotically stable motion, we shall say *linearly stable motion*.

The eigenvalues σ_i of the problem (1.1.22) depend on Re. In § 1.3 we shall see that every basic flow is stable for Reynolds numbers that are small enough. That is why the smallest Reynolds number for which there exists at least one eigenvalue having a negative real part is called the *bound of linear stability*. This bound is denoted by $\mathrm{Re}_{\mathrm{cr}}$ and is also called the *critical Reynolds number*. Denote $\xi_0(\mathrm{Re}) = \inf \mathcal{R}e(\sigma_i)$ where σ_i are the eigenvalues corresponding to some value of Re, and hence $\xi_0(\mathrm{Re}_{cr}) = 0$. Thus the bound of linear stability is saught among neutral perturbations (with $\mathcal{R}e\{\sigma_i\} = 0$).

In the case of unbounded domains (in one or both directions x_1 and x_2) it is assumed that the perturbations are periodic and bounded at infinity along these directions, therefore they are of the form

$$\mathbf{v} = \mathbf{v}_0(x_2, x_3) \exp\{\mathrm{i}\,\alpha\, x_1 - \sigma t\} \tag{1.1.23}$$

respectively

$$\mathbf{v} = \mathbf{v}_0(x_3) \exp\{\mathrm{i}\,\alpha\, x_1 + \mathrm{i}\,\beta\, x_2 - \sigma t\}, \tag{1.1.23$'$}$$

where α and β are real numbers. In this case σ depends on α, β and Re.

The main object of the linear theory of hydrodynamic stability is the delimitation in the plane (α, Re) (respectively in the space (α, β, Re)) of domains where the basic motion is lineary stable, from the domains where it is lineary unstable. The domains of stability or instability are separated by neutral curves (respectively surfaces). In particular, it is important to find the bound Re_{cr} such that for values smaller than Re_{cr} infinitesimal perturbations on the form considered are damped out and for values larger than Re_{cr} these perturbations are amplified. We remark that Re_{cr} represents also the bound of stability with respect to infinitesimal perturbations.

Although the stability problem can be set for every fluid motion, studies on hydrodynamic stability mainly concern the following classes of flows: plane parallel flows in channels, Couette-plane flow, Couette flow between rotating cylinders, Hagen-Poiseuille flow in tubes, almost parallel flows (boundary layers on the half-bounded flat plate, the jet, the wake, and motions in diverging channels).

Plane parallel flows are by definition those motions whose characteristics do not depend on one of the space coordinates, for instance x_3, with the velocity component in the Ox_3-direction vanishing. Put x, y, z, instead of x_1, x_2, x_3 and u, v, w instead of v_1, v_2, v_3 and consider a stationary basic motion $\overline{\mathbf{u}} = (U(y), 0, 0)$ in the domain Ω (called channel) bounded by the parallel planes $y = 0$ and $y = 1$ expanded up to infinity in the Ox and Oz directions. Therefore this motion is plane and it can be considered as taking place in the plane (x, y). The most simple steady flow in channels of depth l is the *plane Couette* flow. It is generated by a displacement of velocity V of the upper wall $y = 1$ (the lower wall $y = 0$ being at rest). The velocity and the Reynolds number of this flow are respectively

$$U(y) = y, \ \mathrm{Re} = \frac{Vl}{\nu}\cdot$$

The plane parallel flow which appears in a fixed channel of depth l under the action of a pressure gradient is called the plane Poiseuille flow and has the velocity and Reynolds number given respectively by the expressions

$$U(y) = -4y(y-1),\ \mathrm{Re} = \frac{Vl}{\nu},$$

where V is the velocity at the points of the symmetry plane of the channel.

The velocities of plane parallel Couette and Poiseuille flows are exact solutions of the Navier-Stokes equations, having the property that every linear combination of them is also an exact solution of these equations. This combination represents the velocity of a motion in a channel called *Couette-Poiseuille flow*. This motion is due to the relative displacement of the upper wall with respect to the lower one and a fall of pressure. Its velocity is not a symmetric function with respect to the line $y = \frac{1}{2}$, and is of the form

$$U(y) = ay - 4\,by(y-1), \tag{1.1.24}$$

with a and b real numbers.

Suppose now that on the basic flow $(U, 0, 0)$ one superposes plane infinitesimal perturbations in the form of normal modes

$$\mathbf{v} = \mathbf{v}_0(y)\, e^{i\alpha(x-ct)},\ \mathbf{v}_0(y) = (u(y), v(y)), \tag{1.1.23$''$}$$

called Tollmien-Schlichting transversal waves [3], where we put $(i\alpha)^{-1}\sigma = c = c_r + ic_i$. Using these notations, by Definition 1.1.4 we obtain stability if all $c_i < 0$, instability if there exist a $c_i > 0$ and the neutral case if at least one $c_i = 0$, the remaining c_i having vanishing or negative imaginary parts. c_i is called the *amplification factor*; c_r is the ratio between the *velocity of propagation of the wave* of perturbation and the characteristic velocity, and α is the wave number in the x-direction equal to $2\pi L^{-1}$, where L is the wave length of the Tollmien-Schlichting perturbation.

As $\mathbf{v}$ is selenoidal, we introduce the stream function $\Psi(x, y, t) = \varphi(y)\, e^{i\alpha(x-ct)}$ and obtain

$$u(x, y, t) = \frac{\partial \Psi}{\partial y} = \frac{d\varphi}{dy}\, e^{i\alpha(x-ct)},\ v(x, y, t) = -\frac{\partial \Psi}{\partial x} = -\,i\,\alpha\,\varphi\, e^{i\alpha(x-ct)},$$

whence

$$u(y) = \varphi'(y),\ v(y) = -\,i\,\alpha\,\varphi.$$

---

[3] For every finite t this wave is a cylindrical surface with the generatrices parallel to the Oz-axis and therefore perpendicular to the plane of the basic flow.

Then, setting $p(x, y, t) = p(y)e^{i\alpha(x-ct)}$, the eigenvalue problem (1.1.22) becomes

$$-i\alpha\varphi U' - i\alpha c\varphi' + U\alpha i\varphi' = -i\alpha p + \frac{1}{\mathrm{Re}}(-\alpha^2\varphi' + \varphi'''),$$

$$-\alpha^2 c\varphi + \alpha^2 U\varphi = -p' + \frac{1}{\mathrm{Re}}(i\alpha^3\varphi - i\alpha\varphi'').$$

By eliminating the pressure, we obtain the Orr-Sommerfeld equation

$$(1.1.25) \quad (U - c)(\varphi'' - \alpha^2\varphi) - \varphi U'' = \frac{1}{i\alpha\,\mathrm{Re}}(\varphi^{\mathrm{IV}} - 2\alpha^2\varphi'' + \alpha^4\varphi).$$

The boundary condition for problem (1.1.25) can be written as

$$(1.1.26) \quad \varphi(0) = \varphi'(0) = \varphi(1) = \varphi'(1) = 0.$$

Therefore, the linear stability problem for plane parallel motions (1.1.24) to perturbations of the form (1.1.23)'' is equivalent to the eigenvalue problem (1.1.25), (1.1.26) where c is the eigenvalue and φ is the eigenfunction.

In the linear theory of hydrodynamic stability it is assumed that problem (1.1.25), (1.1.26) also governs the stability of almost parallel flows (i.e. with U an arbitrary function of y). Among the almost parallel flows, *the boundary layer flow on a half-bounded plate* is most important for practical purposes, and, at the same time, it is most studied. The expression for this type of flow of the velocity has been deduced by Blasius as a series. For the purposes of hydrodynamic stability it is sufficient to take [80]

$$(1.1.27) \quad U(y) = 1.68\, y - 1.533\, y^4.$$

Let us now analyse linear stability with respect to perturbations with symmetry of rotation for the Couette flow between two coaxial cylinders of radii r_1 and $r_2(>r_1)$, having uniform rotations of angular velocities ω_1 and ω_2. By definition, *Couette-flow* is the stationary flow with symmetry of rotation, between two rotating cylinders observed for Reynolds numbers small enough. We take as characteristic quantities $\omega_1 r_1$ for velocity and r_1 for length; we denote by ξ, θ, ζ the dimensionless cylindrical coordinates; and we take the cylinders axis to be the Oz-axis. Then the Couette flow velocity is $\bar{\mathbf{u}} = (0, \bar{v}_\theta, 0)$, where

$$\bar{v}_\theta = a\xi + \frac{b}{\xi}\left(\text{for } 1 < \xi < \alpha\left(\alpha = \frac{r_2}{r_1}\right)\right), \quad a = \frac{\omega_2 r_2^2 - \omega_1 r_1^2}{r_2^2 - r_1^2},\ b = \frac{r_2^2(\omega_1 - \omega_2)}{\omega_1(r_2^2 - r_1^2)}.$$

Let the perturbations characterized by the quantities $q(\xi, \theta, \zeta, t)$ have the form of normal modes

$$q(\xi, \theta, \zeta, t) = q(\xi)\, e^{\sigma t + iN\theta + i\lambda\zeta} \tag{1.1.28}$$

(where N and λ are positive numbers) and denote by $(v_\xi, v_\theta, v_\zeta)$ the velocity of these perturbations. The perturbations which do not depend on θ (i.e. which have $N = 0$) are called *perturbations with symmetry of rotation,* and those which have $N = 0$ and $v_\theta = 0$ are called *axi-symmetric perturbations.*

Equations (1.1.6)'' can be written in cylindrical coordinates as follows

$$\frac{\partial v_\xi}{\partial t} + \frac{\bar{v}_\theta}{\xi} \cdot \frac{\partial v_\xi}{\partial \theta} - \frac{2\bar{v}_\theta}{\xi} \cdot v_\theta = -\frac{\partial p}{\partial \xi} + \frac{1}{\mathrm{Re}}\left(\Delta v_\xi - \frac{v_\xi}{\xi^2} - \frac{2}{\xi^2}\frac{\partial v_\theta}{\partial \theta}\right), \tag{1.1.29}$$

$$\frac{\partial v_\theta}{\partial t} + \frac{\bar{v}_\theta}{\xi}\frac{\partial v_\theta}{\partial \theta} + v_\xi \frac{d\bar{v}_\theta}{d\xi} + v_\xi \frac{\bar{v}_\theta}{\xi} = -\frac{1}{\xi}\frac{\partial p}{\partial \theta} +$$

$$+ \frac{1}{\mathrm{Re}}\left(\Delta v_\theta - \frac{v_\theta}{\xi^2} + \frac{2}{\xi^2}\frac{\partial v_\xi}{\partial \theta}\right), \tag{1.1.30}$$

$$\frac{\partial v_\zeta}{\partial t} + \frac{\bar{v}_\theta}{\xi} \cdot \frac{\partial v_\zeta}{\partial \theta} = -\frac{\partial p}{\partial \zeta} + \frac{1}{\mathrm{Re}}\Delta v_\zeta, \tag{1.1.31}$$

where $\Delta = \dfrac{\partial^2}{\partial \xi^2} + \dfrac{1}{\xi}\dfrac{\partial}{\partial \xi} + \dfrac{1}{\xi^2}\dfrac{\partial^2}{\partial \theta^2} + \dfrac{\partial^2}{\partial \zeta^2}\cdot$ The solenoidality condition of the perturbation may be written as

$$\frac{1}{\xi}\frac{\partial}{\partial \xi}(\xi v_\xi) + \frac{1}{\xi}\frac{\partial v_\theta}{\partial \theta} + \frac{\partial v_\zeta}{\partial \zeta} = 0, \tag{1.1.32}$$

and the boundary conditions become

$$v_\xi = v_\theta = v_\zeta = 0 \text{ for } \xi = 1 \text{ and } \xi = \alpha. \tag{1.1.33}$$

In the case of the perturbations with symmetry of rotation,

$$\mathbf{v} = \mathbf{v}_0(\xi)\, e^{\sigma t + i\lambda\zeta}. \tag{1.1.34}$$

Introducing the stream function $\psi(\xi, \zeta, t) = f(\xi) e^{\sigma t + i\lambda\zeta}$, for which

$$v_\xi(\xi, \zeta, t) = -\frac{1}{\xi}\frac{\partial}{\partial \zeta}(\xi\psi),\quad v_\zeta(\xi, \zeta, t) = \frac{1}{\xi}\frac{\partial}{\partial \xi}(\xi\psi),$$

setting $v_\theta = ig(\xi) e^{\sigma t + i\lambda\zeta}$, and eliminating the pressure between (1.1.29), (1.1.30) and (1.1.31), we get

$$(L - \lambda^2 - \sigma)(L - \lambda^2) f = -2\lambda\, \mathrm{Re}(a + b\xi^{-2})\, g, \tag{1.1.35}$$

$$(L - \lambda^2 - \sigma)\, g = -2\lambda\, \mathrm{Re}\, af, \tag{1.1.36}$$

where $L = \frac{d^2}{d\xi^2} + \frac{1}{\xi}\frac{d}{d\xi} - \frac{1}{\xi^2}$, and $\mathrm{Re} = \frac{\omega_1 r_1^2}{\nu}$. The boundary conditions (1.1.33) become

$$f = \frac{df}{d\xi} = g = 0 \quad \text{for } \xi = 1 \text{ and } \xi = \alpha. \tag{1.1.37}$$

Hence, the eigenvalue problem governing linear stability with respect to perturbations with symmetry of rotation of Couette flow is given by relations (1.1.35)—(1.1.37), where σ is the eigenvalue and f and g are the eigenfunctions.

Put $\beta = \frac{\omega_2}{\omega_1}$. Then $\beta > 0$ corresponds to the case of rotation in the same sense of the cylinders, $\beta = 0$ corresponds to the case when the outer cylinder is at rest, and $\beta < 0$ to the case of counterrotation. For $\beta = 0$, problem (1.1.35)—(1.1.37) is selfadjoint, which is not true in the general case.

If $\beta > \frac{1}{\alpha^2}$, the Synge's criterion (Section 1.3.2.) ensures stability to perturbations with symmetry of rotation. It remains therefore to study the stability of the Couette flow between cylinders which rotate in the same sense and for which $0 < \beta \leqslant \frac{1}{\alpha^2}$, and the case of counterrotation—the latter case seldom being treated in the literature.

Unlike the energy theory where *all perturbations* must be considered linear stability theory takes into account *perturbations of a certain form and dimension* (suggested by experiments).

So far we have illustrated only the setting of the mathematical problem of linear stability for particular motions and perturbations. To various types of perturbations there correspond various Re_{cr}; *the critical perturbations* are those which correspond to the smallest Re_{cr}.

In the studies concerning the linear stability of the Couette flow with respect to perturbations with symmetry of rotation, an important role has been played by the approximation that the gap between the cylinders is negligible. In that case problem (1.1.35)—(1.1.37) becomes again selfadjoint. In fact, consider the transformation

$$r = \frac{(\xi - 1)\, r_1}{r_2 - r_1}, \quad z = \frac{\zeta r_1}{r_2 - r_1},$$

which corresponds to a dimensionalization by means of a characteristic length $r_2 - r_1$, so that the small gap between cylinders becomes dilated so as to enable a more detailed analysis. Choose the quantity $\omega_1(r_2 - r_1)$ as characteristic velocity and let Re' be the corresponding

Reynolds number $\mathrm{Re}' = \frac{\omega_1(r_2 - r_1)^2}{\nu} = \mathrm{Re}\,\frac{r_1^2}{(r_2 - r_1)^2}$. Let $\lambda = \frac{kr_1}{r_2 - r_1}$, and let f' and g' be quantities made dimensionless by the new characteristic quantities corresponding to f and g. Neglecting those powers greater than 1 of the small parameter $\frac{r_2 - r_1}{r_1}$ in the problem (1.1.35)—(1.1.37), we obtain

(1.1.35)′ $$(\mathrm{D}^2 - k^2 - \sigma)(\mathrm{D}^2 - k^2) f' = 2\,k^2\,\mathrm{Re}'\omega g',$$

(1.1.36)′ $$(\mathrm{D}^2 - k^2 - \sigma)\, g' = -2\,\mathrm{Re}'\, af',$$

(1.1.37)′ $$f' = \mathrm{D}f' = g' = 0 \text{ for } r = 0 \text{ and } r = 1,$$

where $\mathrm{D} = \frac{\mathrm{d}}{\mathrm{d}r}$, and $\omega = a + b\left(1 + \frac{r_2 - r_1}{r_1}\, r\right)^{-2}$. Instead of ω take the value $\bar{\omega} = \frac{1}{2}\left(1 + \frac{\omega_2}{\omega_1}\right)$, which is just the value of ω for $\xi = \frac{1}{2}\left(1 + \frac{r_2}{r_1}\right)$. Eliminating g' between (1.1.35)′ and (1.1.36)′ and considering the case $\sigma = 0$, we get the selfadjoint equation

(1.1.38) $$(\mathrm{D}^2 - k^2)^3 f' = 4\,k^2\left(\frac{r_2 - r_1}{r_1}\right)^4 \mathrm{Re}^2\, a\,\bar{\omega}\, f'.$$

Denote the Taylor number by $T = -4\left(\frac{r_2 - r_1}{r_1}\right)^2 \mathrm{Re}^2 a\bar{\omega} > 0$. From the problem (1.1.38), (1.1.37)′ one can determine T in terms of k.

Choose $\omega = \bar{\omega}$, eliminate g between (1.1.35) and (1.1.36) and assume that in the neutral case we have $\sigma = 0$[4]. Hence

(1.1.39) $$(L - \lambda^2)^3 f = 4\lambda^2 \mathrm{Re}^2\, a\,\bar{\omega} f,$$

under conditions (1.1.37), (1.1.39) leads to a critical Reynolds number corresponding to the minimum value of $K = -4\,\mathrm{Re}^2 \cdot a\,\bar{\omega} > 0$. Therefore, in the critical case, $K = K_C\left(\frac{r_2}{r_1}\right)$ has a well-determined value, and it corresponds to the critical perturbations, having wave number $\lambda = \lambda_c\left(\frac{r_2}{r_1}\right)$. From (1.1.38) and (1.1.37)′, it follows that

[4] In other words, it is to assume that the principle of exchange of stabilities (§ 2.5) holds.

$T = \frac{(r_2 - r_1)^2}{r_1} K$. Therefore, in the critical case, the wave length of the critical disturbances is a multiple of the distance between cylinders. Taylor's calculations [76], according to which f' is expanded in series of Bessel orthogonal functions, lead to an infinite system of linear equations which gives $T_{cr} = 1\,709$. This value is very close to the value 1708.1 later obtained by Di Prima [8] using a very simple variational principle. The case of counterrotation was treated by asymptotic methods by Mecksyn [39]. The stability characteristics of the Couette flow for various ratios of the cylinder angular velocities are represented in fig. 1.1.1.

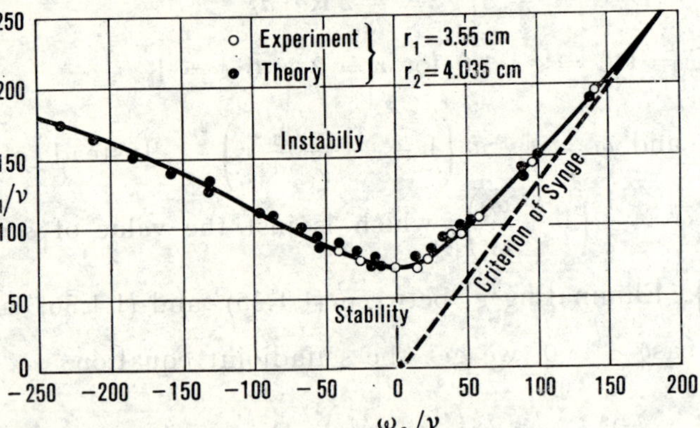

Fig. 1.1.1.

Recently, owing to the cellular character of the Couette flow instability, the stability of this motion and of the cellular Taylor flow which appears supercritically (for $T > T_{cr}$ and small $|T - T_{cr}|$) was studied in close connection with the branching problem of the steady solutions of the Navier-Stokes equations (Ch. 3).

Convective motions are another type of cellular secondary motions that are generated by the loss of stability of the equilibrium of a non-uniformly heated fluid layer. A closed analogy exists between the eigenvalue problems attached to the linear stability of rotation motions and those of the termal convection. The first to have proved this analogy was Jeffreys [31] when treating the particular case of the small gap between cylinders having the same sense of rotation. (For the study of the stability of convection motions see Chapter 3.)

The Hagen-Poiseuille flow in a circular tube is the steady motion observed at small Reynolds numbers, generated by a pressure fall along the tube. If the axis of the cylinder is directed in the ζ-direction, then its velocity in cylindrical coordinates ξ, θ, ζ is $(0, 0, 1 - \xi^2)$. An analysis analogous to that of the Couette flow shows that the stability of this flow to axi-symmetric perturbations is described by the eigenvalue problem

$$\{(L - \lambda^2) - [\sigma + i\lambda \mathrm{Re}(1 - \xi^2)]\}(L - \lambda^2)f = 0, \quad \text{(for } 0 \leqslant \xi < 1) \tag{1.1.40}$$

$$f = \frac{df}{d\xi} = 0, \quad \text{for } \xi = 1. \tag{1.1.41}$$

In the case of perturbations with symmetry of rotation, the stability is ensured for all Re [74]. The stability of this motion has been much studied [64], [1]; for the most recent results, see [30].

In spite of their simplicity, the eigenvalue problems derived in this section entail difficult calculations. The case of the plane Couette flow is analysed in detail in Section 2.4.3, where this flow is shown to be lineary stable for every Re; the linearization principle, according to which a motion lineary stable to small perturbations at the initial time is also nonlineary stable, holds for this flow. But perturbations finite at the initial time give rise to motions which for $t \to \infty$ tend to another periodic motion, whose velocity is a secondary solution branching from the basic plane Couette flow. Since for small Reynolds numbers every flow is lineary stable (§ 1.3), in order to solve the eigenvalue problem of the linear stability of some motion, we need asymptotic methods with $\frac{1}{\mathrm{Re}}$ as a small parameter.

In a paper that generated much controversies [19], Heisenberg applied an asymptotic method for the case of the plane Poiseuille flow; its neutral stability curve, known as Heisenberg's tongue, is shown in fig. 1.1.2. Recent rigorous calculations (Section 2.4.2)

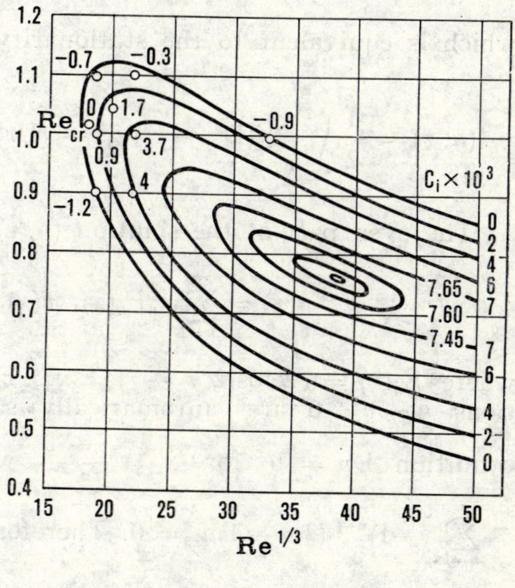

Fig. 1.1.2.

have validated Heisenberg's results and also the fact that plane Poiseuille flow is unstable for $\mathrm{Re} > \mathrm{Re}_{cr}$, where $\mathrm{Re}_{cr} \approx 5\,800$ [78], [37]. Experimentally it was observed that this motion loses its stability [7] for $\mathrm{Re} \approx 1\,000$; therefore, as in the case of the plane Couette flow, the initially finite perturbations destabilize the motion at subcritical Reynolds numbers.

The asymptotic methods are analysed in §1.2. We conclude this section by presenting Di Prima's variational method used in the study of the Couette flow stability for the case of Taylor's narrow gap approximation. To this purpose, denote $f' = v$, $r - \frac{1}{2} = x$ and $\mathrm{D} = \frac{\mathrm{d}}{\mathrm{d}x}$. Write problem (1.1.38). (1.1.37)′ in the following form

$$(1.1.42) \qquad (\mathrm{D}^2 - k^2)^3 v = -k^2 T v, \quad -\frac{1}{2} < x < \frac{1}{2},$$

$$(1.1.43) \qquad v = (\mathrm{D}^2 - k^2) v = \mathrm{D}(\mathrm{D}^2 - k^2) v = 0 \text{ for } x = \pm \frac{1}{2}.$$

Introducing the parameter $Z = \sqrt{T}$ and the function u such that $\mathrm{i}kZu = (\mathrm{D}^2 - k^2) v$, this problem becomes

$$(1.1.42)' \qquad \begin{cases} (\mathrm{D}^2 - k^2)^2 u = \mathrm{i}\, kZv, \\ (\mathrm{D}^2 - k^2) v = \mathrm{i}\, kZu, \end{cases}$$

$$(1.1.43)' \qquad u = \mathrm{D}u = v = 0 \quad \text{for } x = \pm \frac{1}{2},$$

which is equivalent to the stationarity of the functional

$$I(u, v) = \frac{1}{2} \int_{-\frac{1}{2}}^{\frac{1}{2}} \{[(\mathrm{D}^2 - k^2)u]^2 - (\mathrm{D}v)^2 - k^2 v^2\} \mathrm{d}x - \mathrm{i}kZ \int_{-\frac{1}{2}}^{\frac{1}{2}} uv \mathrm{d}x.$$

The even part of the solution (u, v) can be written in the form

$$u(x) = \sum_{n=1}^{\infty} a_n E_n(x), \quad v(x) = \sum_{n=1}^{\infty} b_n E_n(x),$$

where $E_n(x) = \sqrt{2} \cos(2n - 1) \pi x$. For this choice, the conditions $u = v = 0$ are automatically satisfied. Imposing now the condition $\mathrm{D}u = 0$ for $x = \pm \frac{1}{2}$, we obtain the constraint $\Gamma = \sum_{n=1}^{\infty} (-1)^{n+1} (2n - 1) a_n = 0$. Therefore, introducing the Lagrange multiplier μ, the variational problem is reduced to that of finding the stationary point of the functional $I - \mu\Gamma$. Since, for the above choice of u and v, I can be written

$$I = \sum_{n=1}^{\infty}\left\{\frac{1}{2}(A_n^2 a_n^2 - A_n b_n^2) - \mathrm{i}\, kZ a_n b_n\right\},$$

where $A_n = (2n-1)^2\pi^2 + k^2$, this stationary point must satisfy the conditions

$$\frac{\partial(I-\mu\Gamma)}{\partial a_n} = \frac{\partial(I-\mu\Gamma)}{\partial b_n} = 0$$

and the condition that $\Gamma = 0$; this yields the infinite system in a_n, b_n and μ

$$A_n^2 a_n - \mathrm{i}\, kZ b_n - \mu(-1)^{n+1}(2n-1) = 0, \tag{1.1.44}$$

$$A_n b_n + \mathrm{i}\, kZ a_n = 0, \tag{1.1.45}$$

$$\sum_{n=1}^{\infty}(-1)^{n+1}(2n-1)\, a_n = 0, \tag{1.1.46}$$

for $n = 1, 2, \ldots$ Multiplying (1.1.44) by A_n and adding to the result equality (1.1.45) multiplied by $\mathrm{i}\, kZ$, we obtain

$$(A_n^3 - k^2T)\, a_n - \mu(-1)^{n+1}(2n-1)\, A_n = 0.$$

Multiplying this equality by $\dfrac{(-1)^{n+1}(2n-1)}{-\mu(A_n^3 - k^2T)}$ and summing from 1 to ∞, we find

$$\sum_{n=1}^{\infty}\frac{(2n-1)^2 A_n}{A_n^3 - k^2T} = 0. \tag{1.1.47}$$

The convergence of this series is improved if instead of (1.1.47) we write the equivalent relation

$$\frac{\pi^2}{8} + \sum_{n=1}^{\infty}\left\{\frac{(2n-1)^2 A_n}{A_n^3 - k^2T} - \frac{1}{(2n-1)^2}\right\} = 0. \tag{1.1.47'}$$

From (1.1.47)′, by numerical integration, we obtain $T = T(k)$.

A choice of the solution (u, v) as old functions leads to weaker results.

Di Prima's method is also advantageous for non-Newtonian fluid flows [15].

1.1.4. **Rayleigh and Squire's theorems**

The first way to avoid the difficulties implied by the solution of the eigenvalue problem associated to the Orr-Sommerfeld equation is to suppose that the fluid is inviscid; putting $\nu = 0$ (or, equivalently,

$\mathrm{Re} = \infty$) in the Orr-Sommerfeld equation, we obtain the nonviscous Orr-Sommerfeld equation — called the *Rayleigh equation* —

$$(1.1.48) \qquad (\bar{u} - c)(\varphi'' - \alpha^2\varphi) - \bar{u}''\varphi = 0,$$

with the boundary conditions

$$(1.1.49) \qquad \varphi = 0 \text{ for } y = 0 \text{ and } y = 1.$$

Although the only information of physical significance about stability of the plane parallel flow $(\bar{u}, 0, 0)$ is that obtained by studying the solutions of the Orr-Sommerfeld equation, we assume as a first approximation that the same property holds for the solutions of the Rayleigh equation. For every real y, the only solutions of the Rayleigh equation which, as $\nu \to 0$, are limits of solutions of Orr-Sommerfeld equation, are [38] those corresponding to $c_i > 0$. So, in what follows, we shall restrict our discussion to this case alone.

Theorem 1.1.2 (*Rayleigh* [49]). *If there exist perturbations with* $c_i > 0$ $\bar{u}''(y)$ *must vanish for some* $y_s \in [0, 1]$.
Proof. The value y_c such that $\bar{u}(y) - c_r = 0$ is called the critical point. We note that for $y \in [0, 1]$, singular points for the Rayleigh equation can appear only in the neutral case at the critical points. Therefore, if $c_i > 0$, the Rayleigh equation has no singularities, and it can be written in the form

$$(1.1.48)' \qquad \varphi'' - \alpha^2\varphi - \frac{\bar{u}''}{\bar{u} - c}\varphi = 0.$$

Multiplying (1.1.48)' by $\bar{\varphi}$, integrating on [0, 1] and taking into account Condition (1.1.49), we obtain the equation

$$\int_0^1 \{|\varphi'|^2 + \alpha^2 |\varphi|^2\} \, dy + \int_0^1 \frac{\bar{u}''}{\bar{u} - c} |\varphi|^2 \, dy = 0,$$

in the complex field, the imaginary part of which is

$$\int_0^1 \frac{\bar{u}'' c_i |\varphi|^2}{|\bar{u} - c|^2} \, dy = 0.$$

The theorem follows immediately. We give the converse of Theorem 1.1.2 without proof

Theorem 1.1.3 (*Tollmien* [81]). *The plane parallel flows characterized by symmetric velocity profiles and the motions of the boundary layer type, for which* $\bar{u}$ *has some inflexion points in the domain* [0, 1] *of motion, are linearly instable.*

Theorem 1.1.4 (*Rayleigh* [49]). *If* $\bar{u}'' < 0$, $0 \leqslant \bar{u}$, $c_r > 0$, *then there exists at least a neutral wave perturbation for which the propagation velocity of the wave is equal to the basic velocity at a point* $y_1 \in$ $\in [0, 1]$.

Proof. Multiplying the Rayleigh equation (1.1.48) by $\bar{\varphi}$ and adding the obtained equality with its complex conjugate and then integrating by parts over [0, 1], we obtain

$$2\int_0^1 (\bar{u} - c_r)\,(|\,\varphi'\,|^2 + \alpha^2|\,\varphi\,|^2)\,\mathrm{d}y = -\int_0^1 \bar{u}''|\,\varphi\,|^2\,\mathrm{d}y.$$

Assuming now that $c_r \geqslant \bar{u}_{max}$, we reach a contradiction, since the right-hand side is positive. The theorem follows from the hypothesis that $\bar{u} \geqslant 0$, $c_r > 0$; a more complete analysis of the bounds for c_r for general velocity profiles shall be given in Theorem 1.3.3.

A substantial simplification of the mathematical problem of linear stability is obtained if three-dimensional perturbations may be replaced by two-dimensional ones. For some motions, as for instance plane parallel fluid flows, this possibility is ensured by the following

Theorem 1.1.5 *(Squire* [69]*). The plane parallel fluid flows become instable to two-dimensional Tollmien-Schlichting wave perturbations at values of the Reynolds number smaller than the value corresponding to three-dimensional perturbations.*

Proof. As we have seen, the linear stability problem for plane parallel flows $(U(y), 0, 0)$ to two-dimensional perturbations of the form (1.1.23) is described by relations (1.1.25), (1.1.26). Considering now three-dimensional perturbations of the form (1.1.23)′, from the relations (1.1.22) we obtain by an analogous approach

$$(1.1.50) \qquad (U - c)\,(\mathrm{D}^2 - \gamma^2)\,v - U''v = \frac{1}{\mathrm{i}\,\alpha\,\mathrm{Re}'}\,(\mathrm{D}^2 - \gamma^2)^2\,v,$$

$$(1.1.51) \qquad (U - c)\,(\beta u - v) - \frac{\mathrm{i}\beta}{\alpha}\,vU' = \frac{1}{\mathrm{i}\,\alpha\,\mathrm{Re}'}\,(\mathrm{D}^2 - \gamma^2)\,(\beta u - \gamma w),$$

$$(1.1.52) \qquad \mathrm{i}\alpha u + \mathrm{D}v + \mathrm{i}\beta w = 0,$$

$$(1.1.53) \qquad u = v = w = 0 \ \text{ for } \ y = 0,\ y = 1,$$

where $\mathrm{D} = \dfrac{\mathrm{d}}{\mathrm{d}y}$, $\gamma^2 = \alpha^2 + \beta^2$, and $\mathbf{v}_0(y) = (u, v, w)$. Taking into account the continuity equation (1.1.52), the boundary conditions for v are

$$(1.1.53)' \qquad v = \mathrm{D}v = 0, \quad \text{for} \quad y = 0, \quad y = 1.$$

Setting $\alpha\mathrm{Re}' = \gamma\mathrm{Re}$ and $v = \varphi$, (1.1.50) becomes

$$(1.1.50)' \qquad (U - c)\,(\mathrm{D}^2 - \alpha^2)\,\varphi - U''\varphi = \frac{1}{\mathrm{i}\,\gamma\,\mathrm{Re}}\,(\mathrm{D}^2 - \gamma^2)^2\,\varphi,$$

which, together with boundary conditions (1.1.53)′, represent the problem (1.1.25), (1.1.26) corresponding to two-dimensional perturbations. Thus, between the Reynolds numbers corresponding to two- and three-dimensional perturbations there exists the relation

$$\mathrm{Re}'_{cr} = \frac{\gamma\,\mathrm{Re}_{cr}}{\alpha} = \frac{\sqrt{\alpha^2+\beta^2}}{\alpha}\,\mathrm{Re}_{cr} > \mathrm{Re}_{cr}.$$

Whence the theorem.

Note that Squire's theorem ensures only that for $\mathrm{Re} < \mathrm{Re}_{cr}$ every three-dimensional perturbation is damped out; Jungclaus [32] proved that for appropriate choice of α and Re (with $\mathrm{Re} > \mathrm{Re}_{cr}$) there exists the possibility for the three-dimensional perturbations to be more unstable (i.e. to have a greater αc_i) than the two-dimensional ones.

Subsequently, Watson [84] determined a certain range for Re, inside which this property holds, and showed that for any α and Re, there exists a certain range for α such that the most unstable perturbations (corresponding to this last one) are the three-dimensional ones.

1.1.5. **Global stability**

For various values of Re, we examine the amplification of the energy $K(t)$ of the perturbations in terms of $K(0)$, stressing the physical significance of the results of the linear and nonlinear stability theories for every Re.

By Theorem 1.1.1, the energy relation (1.1.15) can be written as

$$\frac{\mathrm{d}K}{\mathrm{d}t} = \int_\Omega |\operatorname{grad}\mathbf{v}|^2\,\mathrm{d}x\left[\frac{-\int_\Omega (\mathbf{v}\cdot\operatorname{grad})\,\bar{\mathbf{u}}\cdot\mathbf{v}\,\mathrm{d}x}{\int_\Omega |\operatorname{grad}\mathbf{v}|^2\,\mathrm{d}x} - \frac{1}{\mathrm{Re}}\right] \leqslant$$

$$\leqslant \int_\Omega |\operatorname{grad}\mathbf{v}|^2\,\mathrm{d}x\left[\max\frac{-\int_\Omega (\mathbf{v}\cdot\operatorname{grad})\,\bar{\mathbf{u}}\cdot\mathbf{v}\,\mathrm{d}x}{\int_\Omega |\operatorname{grad}\mathbf{v}|^2\,\mathrm{d}x} - \frac{1}{\mathrm{Re}}\right] =$$

$$= \int_\Omega |\operatorname{grad}\mathbf{v}|^2\,\mathrm{d}x\left[\frac{1}{\widetilde{\mathrm{Re}}} - \frac{1}{\mathrm{Re}}\right].$$

For $\mathrm{Re} < \widetilde{\mathrm{Re}}$, taking into account the inequality $\int_\Omega |\operatorname{grad}\mathbf{v}|^2\,\mathrm{d}x \geqslant$ $\geqslant \tilde{\alpha}\int_\Omega |\mathbf{v}|^2\,\mathrm{d}x$, (Section 1.3.1), we have

$$\frac{\mathrm{d}K}{\mathrm{d}t} < 0 \text{ for every } t \geqslant 0 \tag{1.1.54}$$

and

$$\frac{\mathrm{d}K}{\mathrm{d}t} \leqslant \int_\Omega |\,\mathbf{v}\,|^2\,\mathrm{d}x \frac{\int_\Omega |\,\mathrm{grad}\,\mathbf{v}\,|^2\,\mathrm{d}x}{\int_\Omega |\,\mathbf{v}\,|^2\,\mathrm{d}x}\left[\frac{1}{\widetilde{\mathrm{Re}}} - \frac{1}{\mathrm{Re}}\right] \leqslant \frac{-2\tilde{\alpha}(\widetilde{\mathrm{Re}} - \mathrm{Re})}{\widetilde{\mathrm{Re}}\mathrm{Re}}\,K.$$

Whence

$$K(t) \leqslant K(0)\,\exp\left\{-2\tilde{\alpha}\,\frac{\widetilde{\mathrm{Re}} - \mathrm{Re}}{\widetilde{\mathrm{Re}}\mathrm{Re}}\,t\right\}.$$

Therefore,

$$\lim_{t\to\infty} \frac{K(t)}{K(0)} = 0. \tag{1.1.55}$$

We say that the basic flow $\bar{\mathbf{u}}$ is *attractive in the mean* if it satisfies relation (1.1.55). Note that a flow is asymptotically stable iff it is stable and attractive, these two properties being independent one of another. In hydrodynamic stability texts, a solution $\bar{\mathbf{u}}$ is frequently called asymptotically stable if it is attractive only; this is the reason why, throughout this section, instead of saying that $\bar{\mathbf{u}}$ is attractive in the mean we shall say that it is asymptotically stable in the mean, or just stable. The motion $\bar{\mathbf{u}}$ is *globally stable* or *unconditionally stable*, if relation (1.1.55) holds for every $K(0)$; $\bar{\mathbf{u}}$ is *conditionally stable* if relation (1.1.55) remains true only for $K(0) < \delta > 0$. The linearization principle (§ 2.4) ensures that the results of the linear theory of stability hold also for the nonlinear theory only if $K(0)$ is small. Therefore the linearly stable motion is conditionally stable. Since Re_{cr} of the linear theory represents a certain limit beyond which small amplified perturbations exist, it follows that $\mathrm{Re} > \mathrm{Re}_{cr}$ is a sufficient condition for global instability. On the other hand, for $\mathrm{Re} < \widetilde{\mathrm{Re}}$, the perturbations corresponding to every $K(0)$ are damped out; therefore, $\mathrm{Re} < \widetilde{\mathrm{Re}}$ represents a sufficient condition of global stability. By definition, the critical Reynolds number of global stability (Re_G) is the smallest Re_O, such that for $\mathrm{Re} < \mathrm{Re}_O$ the flow $\bar{\mathbf{u}}$ is globally stable. It follows then that $\mathrm{Re} < \mathrm{Re}_G$ represents a necessary and sufficient condition of global stability and

$$\widetilde{\mathrm{Re}} \leqslant \mathrm{Re}_G \leqslant \mathrm{Re}_{cr}.$$

For $\widetilde{\mathrm{Re}} < \mathrm{Re} < \mathrm{Re}_G$ the energy of the perturbations decreases to zero as $t \to \infty$; whereas for finite t, the energy $K(t)$ for some perturbations may first increase. In fact, let $\tilde{\mathbf{v}}$ be the perturbation corresponding to the maximum $(\widetilde{\mathrm{Re}})^{-1}$ of the expression $\int_\Omega \mathbf{v}\,(\text{grad})\,\overline{\mathbf{u}}\cdot\mathbf{v}\,\mathrm{d}x\cdot\left(\int_\Omega |\,\text{grad}\,\mathbf{v}\,|^2\,\mathrm{d}x\right)^{-1}$. Then, for $\mathrm{Re} > \widetilde{\mathrm{Re}}$, the energy of the perturbation $\mathbf{v}$ satisfies the relation

$$\left.\frac{\mathrm{d}K}{\mathrm{d}t}\right|_{t=0} = \int_\Omega |\,\text{grad}\,\mathbf{v}\,|^2\,\mathrm{d}x\cdot\left[\frac{1}{\widetilde{\mathrm{Re}}} - \frac{1}{\mathrm{Re}}\right] > 0.$$

When $\mathrm{Re}_G \leqslant \mathrm{Re} < \mathrm{Re}_{cr}$, there exist perturbations $\mathbf{v}'$ with positive energy at $t \to \infty$.

Since all the perturbations corresponding to some small $K(0)$ are damped out for $\mathrm{Re} < \mathrm{Re}_{cr}$, it follows that $\mathbf{v}'$ is a finite perturbation with large $K(0)$. On the other hand, by relation (1.1.54) for $\mathrm{Re} < \widetilde{\mathrm{Re}}$, the uniqueness of the perturbations $\mathbf{v}$ corresponding to a certain initial value $\mathbf{v}_0$ and the uniqueness of the steady basic motions (Section 1.3.1) follow easily [30]. Therefore, for $\mathrm{Re} < \widetilde{\mathrm{Re}}$ the steady basic flow $\overline{\mathbf{u}}$ is unique and unconditionally stable, all the perturbations damping out monotonically with respect to t; for $\widetilde{\mathrm{Re}} \leqslant \mathrm{Re} < \mathrm{Re}_G$, $\overline{\mathbf{u}}$ is stable (although some perturbations may be amplified at first) all perturbations damping out as $t \to \infty$. Finally, for $\mathrm{Re}_G \leqslant \mathrm{Re} \leqslant \mathrm{Re}_{cr}$ there exist finite perturbations, having positive energy as $t \to \infty$, their corresponding perturbed motions tending to a different motion $\overline{\mathbf{u}}_1$ as $t \to \infty$. Hence, the motion $\overline{\mathbf{u}}$ is linearly stable and nonlinearly stable of the "snap through" type (§ 2.5). For motions possessing this type of stability as, for instance, plane Couette and Poiseuille flows, Re_{cr} is a branching point at which a branch of secondary solutions appears; for $\mathrm{Re} = \mathrm{Re}_G$, from the secondary solution corresponding to Re_G a secondary branching for $\mathrm{Re} > \mathrm{Re}_G$ appears. This last branching contains just those stable motions to which tend all the perturbed motions corresponding to some finite perturbations, the remaining perturbed motions being still in the attraction zone of the flow $\overline{\mathbf{u}}$.

Re_G is determined experimentally when the basic flow changes its pattern. Mathematically it is determined if $\widetilde{\mathrm{Re}} = \mathrm{Re}_{cr}$.

1.1.6. Stability of the mean motion in transition regime

We shall now expose a method of the nonlinear theory of hydrodynamic stability which is implicit in the explanation of the appearance of turbulence. This method proceeds from an idea of Reynolds [56] according to which a transition motion $\mathbf{u}$ is decomposed into a mean motion $\overline{\mathbf{u}}$ and a fluctuation $\mathbf{v}$, this decomposition being identical with the one used so far in the laminar type of motion (where $\overline{\mathbf{u}}$ becomes the basic motion and $\mathbf{v}$ the perturbation). In this theory [70], [73], [71], [85], [10] the fluctuation is taken in the form of a power series of its amplitude; in particular, in the Stuart approximate method [70], which will be given below, the spatial form of the fluctuation is the same as that given by the linear neutral stability theory, its amplitude being an unknown function of time satisfying an equation of the type suggested by Landau (1944) (see § 3.6)

In this section we shall call *equilibrium motions* those motions having $\frac{\mathrm{d}K(t)}{\mathrm{d}t} = 0$ (i.e. corresponding to the case when the energy transferred from the mean motion to the fluctuation equilibrates the dissipation due to viscous forces). These motions are linearly unstable. The perturbations are first amplified exponentially with respect to time, as is prescribed by the linear theory. Then they reach such an amplitude that the mean momentum transfer by finite perturbations is considerably great and next the Rayleigh effort, which meassures this transfer, distorts the mean motion. This modifies the rate of change of the energy transfer from the mean motion to the fluctuations and therefore the rate of their increase. As the rate of increase of the fluctuations can be diminished by the mean flow, it follows that there exists a possibility (experimentally verified for the Taylor motion) that an equilibrium steady motion be stable for some $\mathrm{Re} > \mathrm{Re}_{cr}$. This proves the supercritical (nonlinear) stability; the subcritical instability can be explained in an analogous way. We shall illustrate these statements by considering the Poiseuille plane laminar flow having the velocity $(\bar{u}, 0, 0)$, $\bar{u} = 4y(1 - y)$ and the pressure $\bar{p} = -2x\mathrm{Re}^{-1} + \text{const}$. The corresponding transition motion has the velocity $(u, v, 0)$, where

$$u = \bar{u}(y, t) + u'(x, y, t), \quad v = v'(x, y, t), \tag{1.1.56}$$

$(u', v', 0)$ represents the fluctuation, and the bar indicates the averaging with respect to x. The energetic balance (Section 1.1.2) is

$$\frac{\mathrm{d}}{\mathrm{d}t}\iint_\Omega \frac{1}{2}(u'^2 + v'^2)\,\mathrm{d}x\mathrm{d}y = -\iint_\Omega \overline{u'v'}\,\frac{\partial\bar{u}}{\partial y}\,\mathrm{d}x\mathrm{d}y -$$

$$-\frac{1}{\mathrm{Re}}\iint_\Omega \left(\frac{\partial v'}{\partial x} - \frac{\partial u'}{\partial y}\right)^2 \mathrm{d}x\mathrm{d}y,$$

where Ω is a volume of fluid bounded by the planes $y = 0$ and $y = 1$, of length equal to the wave length and $\overline{u'v'}$ represents the Reynolds effort. This relation is the theoretical basis of the above considerations.

Introducing (1.1.56) into relations (1.1.1)—(1.1.4), we obtain the following equation of the mean motion

$$\frac{\partial\bar{u}}{\partial t} + \frac{\partial\overline{u'w'}}{\partial y} = -\frac{\partial\bar{p}}{\partial x} + \frac{1}{\mathrm{Re}}\frac{\partial^2\bar{u}}{\partial y^2}. \tag{1.1.57}$$

Since the pressure gradient may be assumed to be constant and the velocity $\bar{u}$ is independent of time in the equilibrium state, (1.1.57) is equivalent with

$$\text{(1.1.57)}' \qquad \frac{\partial \overline{u'v'}}{\partial y} = \frac{8}{\text{Re}} + \frac{1}{\text{Re}} \frac{\partial^2 \bar{u}}{\partial y^2}$$

under the boundary conditions $\bar{u} = 0$ at $y = 0$ and $y = 1$. Hence we find the distorted velocity profile of the plane Poiseuille motion in the form

$$\bar{u} = 4y(1 - y) + \text{Re} \int_0^y \overline{u'v'} \, \mathrm{d}y.$$

For the perturbed motion define now the stream function

$$\psi(x, y, t) = \varphi_0(y, t) + \varphi_1(y, t) \, \mathrm{e}^{\mathrm{i}\alpha(x - c_r t)} + \bar{\varphi}_1(y, t) \, \mathrm{e}^{-\mathrm{i}\alpha(x - c_r t)} +$$
$$+ \varphi_2(y, t) \, \mathrm{e}^{2\mathrm{i}\alpha(x - c_r t)} + \bar{\varphi}_2(y, t) \, \mathrm{e}^{-2\mathrm{i}\alpha(x - c_r t)} + \ldots,$$

(the bars stand for complex conjugation), and substitute in equations (1.1.1) and (1.1.2) $u = \frac{\partial \psi}{\partial y}$ and $v = -\frac{\partial \psi}{\partial x}$. Hence, after eliminating the pressure and separating the Fourier coefficients of the fluctuations, we obtain an infinite system of equations in $\bar{u}$, φ_1, φ_2, ... The first equation is (1.1.57) in which

$$\overline{u'v'} = \mathrm{i}\alpha\{\varphi_1' \bar{\varphi}_1 - \bar{\varphi}_1' \varphi_1 + 2(\varphi_2' \bar{\varphi}_2 - \bar{\varphi}_2' \varphi_2 + \ldots\}.$$

Neglecting the upper order harmonics $\varphi_2, \varphi_3, \ldots$, for $\bar{u}$ and φ_1 we obtain a system consisting of equation (1.1.57)′ and the Orr-Sommerfeld equation in which c_i is replaced by $\alpha \frac{\partial}{\partial t}$. If the motion is supercritic, the following additional hypothesis makes a physical sense: the fluctuation (u', v') differs from the corresponding linear perturbation only by a factor of amplitude $a(t)$. Then, from the relation of the energy balance, we deduce for a the Landau equation

$$\text{(1.1.58)} \qquad \frac{\mathrm{d}a^2}{\mathrm{d}t} = Aa^2 + Ba^4,$$

where A and B are known expressions of α, Re and φ_1 (φ_1 being given by the linear theory). The solution of this equation led Stuart to the result $\alpha \text{Re}_{cr} = 6\,150$, comparable with the value $\text{Re}_{cr} = 5\,780$, for $\alpha = 1.02$, obtained by Thomas [78] by numerical integration.

§ 1.2. ORR-SOMMERFELD EQUATION

1.2.1. Nonviscous Orr-Sommerfeld equation

In the absence of viscosity, putting $w(y) = \bar{u}(y) - c$ and assuming that $(y - y_c)^{-1} \cdot w(y)$ is analytic in the neighbourhood of $y = y_c$, the eigenvalue problem (1.1.25), (1.1.26) becomes

$$\varphi'' - (\alpha^2 + w''w^{-1})\,\varphi = 0, \tag{1.1.48$'$}$$

$$\varphi(0) = \varphi(1) = 0. \tag{1.1.49}$$

Unlike the Orr-Sommerfeld equation, in the neutral case $c_i = 0$ the Rayleigh equation (1.1.48)$'$, for y real, has singularities at the critical points y_c for which $\bar{u}(y_c) - c_r = 0$. In the following considerations, we assume the existence of a single critical point; the case of two critical points, corresponding for instance to axi-symmetric motions in channels (1.1.24) was first treated in 1966 [47]. The determinantal equation corresponding to y_c has the roots 1 and 0; consequently by the reduction method, two linearly independent solutions of equation (1.1.48)$'$ about y_c are

$$\varphi_1 = \sum_{k=1}^{\infty} a_k (y - y_c)^k, \tag{1.2.1}$$

$$\varphi_2 = \sum_{k=0}^{\infty} b_k (y - y_c)^k - \frac{w_c''}{w_c'}\,\varphi_1 \cdot \ln(y - y_c), \tag{1.2.2}$$

in which

$$a_{n+1} = \frac{1}{n(n+1)} \sum_{k=1}^{\infty} a_{n-k+1}\, q_k (n \geqslant 3),\ a_1 = 1,\ a_2 = \frac{w_c''}{2\,w_c'}, \tag{1.2.3}$$

$$b_n = \frac{1}{n(n-1)} \left[\sum_{k=1}^{\infty} b_{n-k} q_k + \right.$$

$$\left. + \frac{w_c''}{w_c'}\,(2n-1)\,a_n \right],\ (n \geqslant 2),\ b_0 = 1,\ b_1 = 0, \tag{1.2.4}$$

where denoting

$$x_k = \frac{w_c^{(k)}}{k!\,w_c'}\ (k \geqslant 1),$$

the quantities q_k are the coefficients of the following expansion [12]

$$-(y - y_c)^2\,(\alpha^2 + w''w^{-1}) = \sum_{k=0}^{\infty} q_k (y - y_c)^k.$$

If now instead of (1.1.48)' we consider equation (1.1.48), the coefficients of the expansion (1.2.1) and (1.2.2) have the expressions

$$a_n = x_n + \frac{\alpha^2}{n(n-1)} \sum_{k=1}^{n-2} a_k x_{n-k-1} +$$

$$(1.2.3)' \qquad + \frac{1}{n-1} \sum_{k=0}^{n-3} a_{k+2} x_{n-k-1}(n-2k-3),$$

$$b_n = \frac{x_{n+1}(n+1)}{n-1} + \frac{\alpha^2}{n(n-1)} \sum_{k=0}^{n-2} b_k x_{n-k-1} + \frac{1}{n-1} \sum_{k=0}^{n-3} b_{k+2} x_{n-k-1} \times$$

$$(1.2.4)' \quad \times (n-2k-3) - \frac{2x_2}{n(n-1)} \sum_{k=1}^{n} a_{n-k+1}\, x_k(2n-2k+1), \quad (n \geqslant 2)$$

and again $a_1 = 1$, $b_0 = 1$, $b_1 = 0$. The method used to deduce the expressions (1.2.3)' and (1.2.4)' is called the Frobenius method and it has been applied to the Rayleigh equation by Tollmien in the case of the boundary layer on the half bounded flat plate. Let us note the advantage for numerical calculations, presented by relations (1.2.3)', (1.2.4)' comparatively to (1.2.3), (1.2.4).

Note also the importance of the ratio $\frac{w_c''}{w_c'}$. Approximating the boundary layer profile by segments of straight lines (i.e. by first-order polynomials with respect to y), this ratio vanishes and the stability curve cannot be found. In order to calculate this ratio, Tollmien [80] approximated the Blasius profile by the fourth-order polynomial (1.1.27), the remaning estimates requiring only the approximation of this profile by a parabola joined with two straight-lines

$$(1.1.27)' \quad \bar{u}(y) = \begin{cases} 1.68y, & 0 \leqslant y \leqslant 0.175, \\ 1-(1.015-y)^2, & 0.175 \leqslant y \leqslant 1.015, \\ 1, & 1.015 \leqslant y < \infty. \end{cases}$$

The solution φ_1 is regular, and hence it can be taken as an approximate solution of the Orr-Sommerfeld equation, while φ_2 has a singularity at y_c and hence must be corrected (Section 1.2.2). After the correction, the solutions φ_1 and φ_2 of the Rayleigh equation approximate two of the four linearly-independent solutions of the Orr-Sommerfeld equation. These two solutions shall be called slow solutions, since they have small high-order derivatives, the other ones being rapid solutions.

Let us look for two linearly-independent solutions of the Rayleigh equation as power series of α^2

$$(1.2.5) \quad \varphi(y) = w\{q_0(y) + \alpha^2 q_1(y) + \alpha^4 q_2(y) + \ldots + \alpha^{2n} q_n(y) + \ldots\}.$$

We introduce this form into the Rayleigh equation, and hence

$$q_{01}(y) = 1,\ q_{02}(y) = \int w^{-2}\,\mathrm{d}y,$$

$$q_{n+1}(y) = \int w^{-2}\,\mathrm{d}y \int w^2 q_n(y)\,\mathrm{d}y,$$

the solutions of which are

$$\varphi_1 = w\left\{1 + \alpha^2 \int_{y_1}^{y} w^{-2}\,\mathrm{d}y \int_{y_1}^{y} w^2\,\mathrm{d}y + \ldots\right\}, \tag{1.2.6}$$

$$\varphi_2 = w\left\{\int_{y_1}^{y} w^{-2}\,\mathrm{d}y + \alpha^2 \int_{y_1}^{y} w^{-2}\,\mathrm{d}y \int_{y_1}^{y} w^2\,\mathrm{d}y \int_{y_1}^{y} w^{-2}\,\mathrm{d}y + \ldots\right\}. \tag{1.2.7}$$

These solutions have been obtained by Heisenberg [19] by the so-called Heisenberg method. The relationship between the solutions of Heisenberg type and those of Tollmien type shall be given in Section 1.2.3

1.2.2. Tollmien's solution for Blasius' boundary layer problem

The energy relation (1.1.15) leads, as will be seen in § 1.3, to stability for small Reynolds numbers. Hence instability is expected to appear for large Re.

Introduce the parameter

$$\varepsilon = (\alpha \mathrm{Re} w_c')^{-\frac{1}{3}}$$

put

$$\eta = \frac{y - y_c}{\varepsilon},$$

and look for the solution of the Orr-Sommerfeld equation of the form

$$\varphi = X_0 + \varepsilon X_1 + \ldots$$

After substituting φ in (1.1.25) we obtain for X_0 and X_1 the equations

$$\mathrm{i}\,\frac{\mathrm{d}^4 X_0}{\mathrm{d}\eta^4} + \eta\,\frac{\mathrm{d}^2 X_0}{\mathrm{d}\eta^2} = 0, \tag{1.2.8}$$

$$\mathrm{i}\,\frac{\mathrm{d}^4 X_1}{\mathrm{d}\eta^4} + \eta\,\frac{\mathrm{d}^2 X_1}{\mathrm{d}\eta^2} = \frac{w_c''}{w_c'}\,X_0 - \frac{1}{2}\,\frac{w_c''}{w_c'}\,\eta^2\,\frac{\mathrm{d}^2 X_0}{\mathrm{d}\eta^2}. \tag{1.2.9}$$

Four linearly independent solutions of equations (1.2.8) are $X_0^{(1)} = 1$, $X_0^{(2)} = \eta$, $X_0^{(3)} = \int_{+\infty}^{\eta} \mathrm{d}\eta \int_{+\infty}^{\eta} \eta^{\frac{1}{2}} H_{\frac{1}{3}}^{(1)} \left[\frac{2}{3} (\mathrm{i}\eta)^{\frac{3}{2}}\right] \mathrm{d}\eta$, $X_0^{(4)} =$

$$= \int_{-\infty}^{\eta} \mathrm{d}\eta \int_{-\infty}^{\eta} \eta^{\frac{1}{2}} H_{\frac{1}{3}}^{(2)} \left[\frac{2}{3} (\mathrm{i}\eta)^{\frac{2}{3}}\right] \mathrm{d}\eta,$$

where $H_{\frac{1}{3}}^{(1)}$ and $H_{\frac{1}{3}}^{(2)}$ are Hankel functions of the first and second kind respectively. Since from (1.2.1) and (1.2.2) we deduce that

$$\varphi_1 = \varepsilon\eta\,[1 + a_2\varepsilon\eta + \ldots],$$

$$\varphi_2 = 1 + \varepsilon\left[b_1\eta + \frac{w_c''}{w_c'}\eta(\ln\eta + \ln\varepsilon)\right] + \ldots, \tag{1.2.10}$$

it follows that the first term in an expansion with respect to ε of φ_2 and φ_1 will be $X_0^{(1)}$ and $X_0^{(2)}$ respectively. Therefore $X_0^{(1)}$ will represent the correction (of order ε) so that as $y \to y_c$ (i.e. as $\eta \to 0$), the corresponding solution of the Orr-Sommerfeld equation has no more singularities. $X_1^{(1)}$ satisfies the equation

$$\mathrm{i}\frac{\mathrm{d}^4 X_1^{(1)}}{\mathrm{d}\eta^4} + \eta\frac{\mathrm{d}^2 X_1^{(1)}}{\mathrm{d}\eta^2} = \frac{w_c''}{w_c'},$$

whence

$$X_1^{(1)\prime\prime} = \frac{w_c''}{w_c'}\frac{\pi}{6}\eta^{\frac{1}{2}}\left\{H_{\frac{1}{3}}^{(2)}\int^{\eta}\eta^{\frac{1}{2}}H_{\frac{1}{3}}^{(1)}\,\mathrm{d}\eta - H_{\frac{1}{3}}^{(1)}\int^{\eta}\eta^{\frac{1}{2}}H_{\frac{1}{3}}^{(2)}\,\mathrm{d}\eta\right\},$$

and by integrating this equation we find two solutions $X_1^{(1)}$ among which we shall retain only that one having the behaviour of φ_2 for large η, i.e. being proportional to $\eta\ln\eta$. The asymptotic behaviour of $X_1^{(1)}$ for large η is obtained using asymptotic expansions for the Hankel functions [82]. It is found that the critical point of the corresponding solution of the Orr-Sommerfeld equation will be a regular point. Tollmien's analysis exposed in the above has been carried out by Holstein [21].

Denote by $\Phi_1 = 1 + \varepsilon X_1^{(1)} + \ldots$, $\Phi_2 = \eta + \varepsilon X_1^{(2)} + \ldots$ the slow solutions of the Orr-Sommerfeld equation corresponding to φ_1 and φ_2 and let $\Phi_3 = X_0^{(3)} + \varepsilon X_1^{(3)} + \ldots$, $\Phi_4 = X_0^{(4)} + \varepsilon X_1^{(4)} + \ldots$, be fast solutions (for a viscous fluid). Taking into account the expressions of $X_0^{(3)}$ and $X_0^{(4)}$, we find that $\lim_{\eta\to\infty} X_0^{(4)} = \infty$, and $\lim_{\eta\to-\infty} X_0^{(3)} = \infty$. Note that this last behaviour holds in the exterior of the considered domain of motion ($0 < y < \infty$).

Therefore the general solution of the Orr-Sommerfeld equation is

$$\varphi = A_1\Phi_1 + A_2\Phi_2 + A_3\Phi_3 + A_4\Phi_4, \tag{1.2.11}$$

where $A_1, ..., A_4$ are unknown constants. Imposing now to φ to satisfy the four boundary conditions (1.1.26), from the condition that $A_1, ..., A_4$ should vanish we obtain a null determinant and, accordingly, a relation that relates α, Re and c and is *called the characteristic equation*. For assigned c, it provides the neutral curve's equation in the (α, Re) plane. Owing to the intricate form of the solution φ, Tollmien simplified the problem as follows. Inside the boundary layer he considered three zones, namely the one near the flat plate, that around the critical point and that near the frontier of the boundary layer. In this last zone, $y \to 1.015$ and consequently η is very large, so that, since $X_0^{(4)} \to \infty$, the solution Φ_4 is physically unacceptable. Hence $A_4 = 0$. On the other hand, since outside the boundary layer the fluid is inviscid (Re $= \infty$) one-dimensional, uniformly parallel to the Ox-axis and has the velocity U_{max}, it follows that near the frontier of the boundary layer φ is a solution of the equation

$$(U_{max} - c)(\varphi'' - \alpha^2\varphi) = 0$$

(deduced from the Orr-Sommerfeld equation for $\bar{u} = U_{max}$ and Re$\to\infty$). Therefore φ is proportional to $e^{-\alpha y}$ or $e^{\alpha y}$, this latter being again physically unacceptable (it becomes infinite as $y \to \infty$). Since $\lim_{\eta\to\infty} X_0^{(3)} = 0$ it follows that on the frontier of the boundary layer we have $\Phi_3 \to 0$. In this way, if the index m stands for the boundary frontier $y = 1.015$, we have $\frac{\varphi'_m}{\varphi} = -\alpha$ and consequently $\frac{A_1\Phi'_{1m} + A_2\Phi'_{2m}}{A_1\Phi_{1m} + A_2\Phi_{2m}} = -\alpha$. Equivalently

$$A_1\tilde{\Phi}_{1m} + A_2\tilde{\Phi}_{2m} = 0, \tag{1.2.12}$$

where

$$\tilde{\Phi}_{1m} = \Phi'_{1m} + \alpha\Phi_{1m}, \quad \tilde{\Phi}_{2m} = \Phi'_{2m} + \alpha\Phi_{2m}.$$

Note that in deducing these conditions, the viscosity did not appear (since it entered only in the solution Φ_3) and the Blasius velocity profile has been approximated by the second-order polynomial (1.1.27)'.

Near the flat plate, at $y = 0$, Tollmien takes $w \approx -c$ and $w'' \approx w''_p$, the index p indicating the value taken at the distance y_p from the wall. Then the Orr-Sommerfeld equation becomes

$$-i\alpha\text{Re}\{c(\varphi'' - \alpha^2\varphi) + w''_p\varphi\} = \varphi^{IV} - 2\alpha^2\varphi'' + \alpha^4\varphi,$$

whose solutions are of the form $\varphi = e^{ky}p$, where

$$k^2 = \frac{1}{2}\{-i\alpha\text{Re}\, c + 2\alpha^2 \pm \sqrt{-\alpha^2\,\text{Re}^2 c^2 - 4w''_p\, i\alpha\text{Re}}\}.$$

Denote by Φ_{1p}, Φ_{2p} and Φ_{3p} the solutions Φ_1, Φ_2, Φ_3 corresponding to y_p. Then, near the wall, we have the conditions

$$A_1\Phi_{1p} + A_2\Phi_{2p} + A_3\Phi_{3p} = 0,$$

$$A_1\Phi'_{1p} + A_2\Phi'_{2p} + A_3\Phi'_{3p} = 0,$$

or, equivalently,

$$A_1\Phi_{1p} + A_2\Phi_{2p} - \frac{\Phi_{3p}}{\Phi'_{3p}}(A_1\Phi'_{1p} + A_2\Phi'_{2p}) = 0. \tag{1.2.13}$$

Put $\dfrac{\Phi_{3p}}{\Phi'_{3p}} = -\,\varepsilon D(\eta_p)$, where η_p is the distance $\eta_p = -\dfrac{c}{\varepsilon w'_c}$ of the critical point from the wall; we see that this ratio does not depend on the velocity profile of the considered basic flow. D is called the Tietjens function (cf. [79] and [6]) and it is represented in fig. 1.2.1.

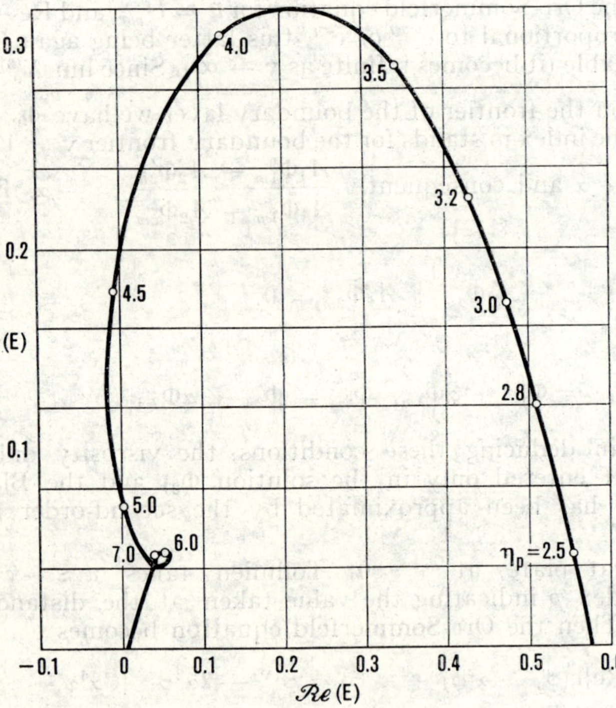

Fig. 1.2.1.

The boundary conditions of the eigenvalue problem for the Orr-Sommerfeld equation have been reduced to the forms (1.2.12) and (1.2.13).

Imposing the condition $(A_1, A_2) = (0, 0)$ we obtain the secular equation

$$-\varepsilon D = \frac{\widetilde{\Phi}_{1m}\Phi_{2p} - \widetilde{\Phi}_{2m}\Phi_{1p}}{\widetilde{\Phi}_{1m}\Phi'_{2p} - \widetilde{\Phi}_{2m}\Phi'_{1p}}$$

or

$$-\frac{D}{\eta_p} = -\frac{w'_c}{c}\,\frac{\widetilde{\Phi}_{1m}\Phi_{2p} - \widetilde{\Phi}_{2m}\Phi_{1p}}{\widetilde{\Phi}_{1m}\Phi'_{2p} - \widetilde{\Phi}_{2m}\Phi'_{1p}}$$

where the right-hand side in a known function $E(\alpha, c)$ and the left-hand side is a function of η_p. Consider now a system of coordinates where $\mathcal{Re}\left(-\frac{D}{\eta_p}\right)$ is the abscissa and $\mathcal{Im}\left(-\frac{D}{\eta_p}\right)$ is the ordinate. Choosing a value for c and, plotting E in terms of α we find Re. In this way, it follows the neutral curve Re = Re (α) (figs. 1.2.2 and 1.2.3).

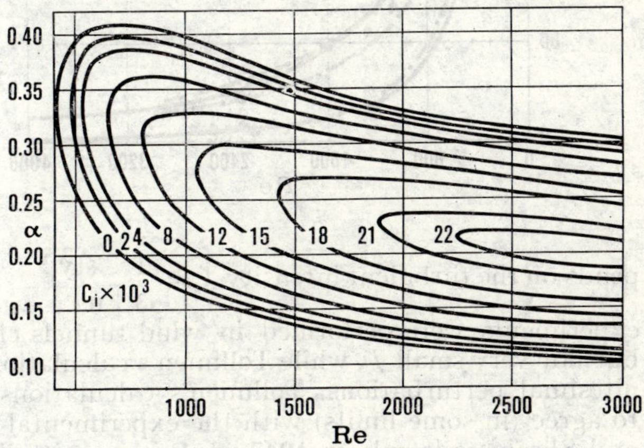

Fig. 1.2.2.

Tollmien obtained $\text{Re}_{cr} = \left(\frac{U_\infty \delta_1}{\nu}\right) = 420$, $(\alpha\delta_1)_{cr} = 0.37$, $c_r/U_\infty = 0.4$ where δ_1 is the displacement thickness. The smallest unstable wavelength λ_{min} corresponds to $(\alpha\delta_1)_{max} = 0.41$ and Re = 557 such that $\lambda_{min} = \frac{2\pi\delta_1}{0.41} = 15.3\ \delta_1 \approx 5.2\ \delta$. To compare Tollmien's results with experiments let us introduce the Reynolds number Re_x in terms of downstream distance x; we have $\delta_1 = 1.72\ \sqrt{\nu x U_\infty^{-1}}$ such that $\text{R}_{x\,cr} = \left(\frac{U_\infty x}{\nu}\right)_{cr} \approx 6.10^4$. Experiments showed that the transition "point" (in fact there is a whole zone of transition), where laminar flows turn into turbulent, is $\left(\frac{U_\infty x}{\nu}\right)_{tr} = 0\ (10^6)$ and it strongly de-

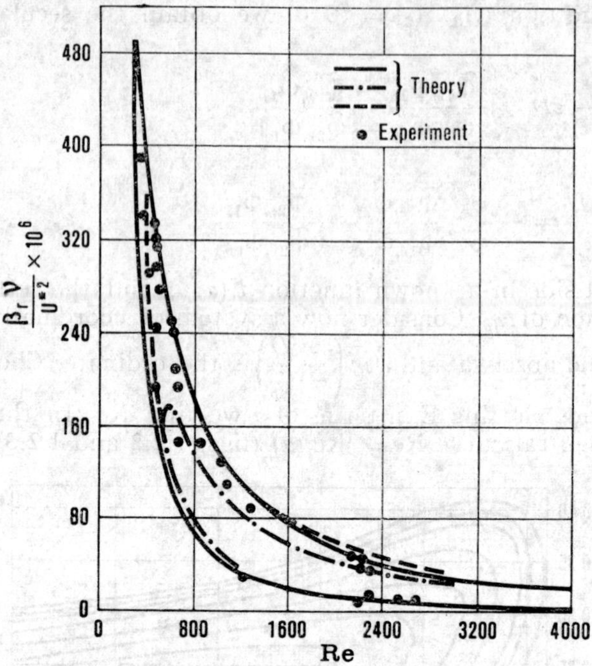

Fig. 1.2.3.

pends on the turbulent intensity $T = \frac{1}{3} \frac{\sqrt{|\mathbf{v}|^2}}{|\bar{\mathbf{u}}|_{|x| \to \infty}}$ in the basic flow. The experiments were performed in wind tunnels characterized by small but not very small T, while Tollmien's calculations were based on infinitesimal perturbations. Tollmien's calculations of 1929 were found to agree (in some limits) with the experimental results of Schubauer and Skramstadt only in 1947 when it was possible to built fine wind tunnels of very low turbulent intensity. Schubauer and Skrasmstadt found also that in the critical layer ($y \approx y_c$) the velocity exhibited the phase change predicted by Tollmien (i.e. for $y > y_c$ in (1.2.2) and, therefore, in (1.2.10) too, the correct branch of $\log(y - y_c)$ is $\log|y - y_c|$ while for $y < y_c$ it is $\log|y - y_c| - \pi i$). So, a very good qualitative agreement of Tollmien's results with experiments were found. Nevertheless, there was not such a good quantitative agreement. This motivates further investigations for the Orr-Sommerfeld equations (Appendix 6).

1.2.3. Relationship between Tollmien's and Heisenberg's solutions

The fundamental difference between the two types of solutions of the Rayleigh equation consists in the fact that the first uses expansions in powers of $y - y_c$ and the second employs expansions in powers of α^2; thus is it natural that the relationship between them should be derived from a double expansion in powers of α^2 and $y-y_c$. Let us write the solutions (1.2.1), (1.2.2) and (1.2.6), (1.2.7) in the form

$$(1.2.14)\quad \begin{cases} \varphi_{1T} = A_{ij}(y-y_c)^i\,\alpha^{2j}, \\ \varphi_{2T} = B_{ij}(y-y_c)^i\,\alpha^{2j} + 2x_2A_{ij}(y-y_c)^i\,\alpha^{2j}\log(y-y_c), \end{cases}$$

respectively,

$$(1.2.15)\quad \begin{cases} \varphi_{1H} = A'_{ij}(y-y_c)^i\,\alpha^{2j} + B'_{ij}(y-y_c)^i\,\alpha^{2j}\log(y-y_c), \\ \varphi_{2H} = A''_{ij}(y-y_c)^i\,\alpha^{2j} + B''_{ij}(y-y_c)^i\,\alpha^{2j}\log(y-y_c), \end{cases}$$

and let us look for a relationship of the form

$$(1.2.16)\quad \begin{cases} \varphi_{1H} = C_k\alpha^{2k}\varphi_{1T} + D_k\alpha^{2k}\varphi_{2T}, \\ \varphi_{2H} = E_k\alpha^{2k}\varphi_{1T} + F_k\alpha^{2k}\varphi_{2T}, \end{cases}$$

where the constant matrices C_k, D_k, E_k și F_k depend on y_c and y_1, and the sum runs from 0 to ∞ (we put $a_0 = 0$, $x_0 = 0$).

Introducing expressions (1.2.14) and (1.2.15) in (1.2.16) we obtain

$$D = \frac{1}{2\,x_2}A_i^{-1}B'_i,$$

$$C = A_i^{-1}[A'_i - DB_i],$$

$$E = A_i^{-1}[A''_i - FB_i],$$

$$F = (2x_2)^{-1}A_i^{-1}B''_i,$$

where i is an arbitrary index and

$$D = \begin{pmatrix} D_0 \\ D_1 \\ D_2 \\ \vdots \end{pmatrix},\quad A_i = \begin{pmatrix} A_{i0} & 0 & 0 & 0 & \dots \\ A_{i1} & A_{i0} & 0 & 0 & \dots \\ A_{i2} & A_{i1} & A_{i0} & 0 & \dots \\ \dots & \dots & \dots & \dots & \dots \\ \dots & \dots & \dots & \dots & \dots \end{pmatrix},\quad A'_i = \begin{pmatrix} A'_{i0} \\ A'_{i1} \\ A'_{i2} \\ \vdots \end{pmatrix}.$$

From (1.2.3)′ and (1.2.4)′ we deduce that A_1 is the identity matrix and B_1 the null matrix. Taking $i = 1$, we obtain

$$D = (2x_2)^{-1}B'_1,\ C = A'_1,\ F = (2x_2)^{-1}B''_1,\ E = A''_1.$$

Consequently in order to establich relation (1.2.16), it is sufficient to know the coefficient of $y-y_c$ in (1.2.15), whose determination can be carried out by elementary but fairly long calculations. The coefficients up to terms in α^2 can be found in [12]. For the case of the Blasius profile approximated by $w = -y^2 + 2y - c$, $a = \sqrt{1-c}$, $y_c = 1 - a$, $y_1 = -y_c$, we have

$$\begin{cases} \varphi_{1H} = 2a\left\{\varphi_{1T} + \alpha^2\left[\left(\alpha^2 + \frac{\alpha_1}{2a} + \frac{\alpha_1}{a}\ln a\right)\varphi_{1T} - \alpha_1\varphi_{2T}\right]\right\}, \\ \varphi_{2H} = (2a)^{-1}\left\{\left(\beta_1 + \frac{1}{2a} + \frac{1}{a}\ln a\right)\varphi_{1T} - \varphi_{2T} + \alpha^2\left[\left(\beta_3 + \frac{\beta_2}{2a} + \right.\right.\right. \\ \qquad \left.\left.\left. + \frac{\beta_2}{a}\ln a\right)\varphi_{1T} - \beta_2\varphi_{2T}\right]\right\}, \end{cases}$$

where

$$\alpha_1 = \frac{-8\,a^5 + 15\,a^4 - 10\,a^2 + 3}{60\,a^2}$$

$$\alpha_2 = -\frac{12a^5 - 15a^4 - 16a^3 + 10a^2 + 12a - 3}{120\,a^3} +$$

$$+\left(\frac{4a^2}{15} + \frac{\alpha_1}{a}\right)\ln\left(\frac{a+1}{2a}\right) - \frac{\alpha_1}{a}\ln(a-1),$$

$$\beta_1 = -\frac{a^2 + 4a - 1}{2a(1-a^2)} + \frac{1}{a}\ln\left(\frac{a+1}{2a(a-1)}\right), \quad \beta_2 = \beta_1\alpha_1 - \alpha_2,$$

$$\beta_3 = \frac{21a^3 + a^2 - 9a + 3}{60a(a+1)} - \frac{3a^3 - 11a^2 - 3a + 3}{30a(a-1)}\ln\frac{a+1}{2a} -$$

$$-\frac{a^3 - 7a^2 + 3a + 3}{30a(a+1)}\ln(a-1) + \frac{4a}{15}\ln(a-1)\ln\frac{a+1}{2a} +$$

$$+\frac{8a}{15}\left[\frac{a-1}{2a} + \frac{1}{2^2}\left(\frac{a-1}{2a}\right)^2 + \ldots\right].$$

§ 1.3. CRITERIA OF HYDRODYNAMIC STABILITY

In this paragraph we shall analyse various types of linear or nonlinear hydrodynamic stability criteria (stressing the main ideas which serve to justify them) and the best numerical results obtained. An important consequence of these criteria will be the proof of the fact for small Reynolds numbers, every motion is stable, whence the adequacy of asymptotic methods for the study of instability corresponding to high values of this number. We also give bounds for the eigenvalues of the problems attached to hydrodynamic stability theory.

1.3.1. **Serrin's universal criteria**

For practical purposes it is important to know the upper bound Re_G up to which the Reynolds number can vary such that a certain basic flow $\bar{\mathbf{u}}$ remains stable. The energy method provides a weaker bound of stability in the mean $\widetilde{\mathrm{Re}}$ ($\widetilde{\mathrm{Re}} < \mathrm{Re}_G$), reducing the stability problem to a difficult eigenvalue problem; a still weaker bound of asymptotic stability in the mean shall be given in the sequel. Its derivation has the advantage of being extremely simple.

Thus, let us write the energy equation

$$\frac{\mathrm{d}}{\mathrm{d}t}\int_\Omega \frac{|\mathbf{v}|^2}{2}\,\mathrm{d}x = -\int_\Omega (\mathbf{v}\,\mathrm{grad})\,\bar{\mathbf{u}}\cdot\mathbf{v}\,\mathrm{d}x - \nu\int_\Omega |\mathrm{grad}\,\mathbf{v}|^2\,\mathrm{d}x, \tag{1.1.15}$$

under the equivalent form

$$\frac{\mathrm{d}}{\mathrm{d}t}\int_\Omega \frac{|\mathbf{v}|^2}{2}\,\mathrm{d}x = -\int_\Omega \mathbf{v}\cdot\Phi\cdot\mathbf{v}\,\mathrm{d}x - \nu\int_\Omega |\mathrm{grad}\,\mathbf{v}|^2\,\mathrm{d}x, \tag{1.3.1}$$

where Φ is the rate of the deformation tensor of the basic flow $\bar{\mathbf{u}}\left(\Phi_{ij} = \frac{1}{2}\left(\frac{\partial \bar{u}_i}{\partial x_j} + \frac{\partial \bar{u}_j}{\partial x_i}\right)\right)$. Denoting by $-m$ the lower bound with respect to $t \in [0, \tau]$ of the characteristic values of the matrix (Φ_{ij}), we have

$$\mathbf{v}\cdot\Phi\cdot\mathbf{v} \geqslant -m|\mathbf{v}|^2. \tag{1.3.2}$$

Now let α be a number such that the inequality

$$\alpha l^{-2}\int_\Omega |\mathbf{v}|^2\,\mathrm{d}x \leqslant \int_\Omega |\,\mathrm{grad}\,\mathbf{v}|^2\,\mathrm{d}x, \tag{1.3.3}$$

holds in the class of vector functions $\mathbf{v}$ possessing continuous space derivatives and satisfying conditions (1.1.10) and (1.1.11), where l is the diameter of the sphere which contains the three-dimensional bounded domain Ω. Serrin [62] found that

$$\alpha = \frac{3+\sqrt{13}}{2}\,\pi^2 \approx 32.6,$$

a better estimation being given by Velte [83] under the assumption that Ω is contained in a cube of edge l,

$$6\pi^2 \leqslant \alpha \leqslant 6.33\pi^2.$$

The best constant α in (1.3.3), which we denote by $\tilde{\alpha}$, can be deduced from the variational problem

$$(1.3.4) \qquad -\int_\Omega |\mathbf{v}|^2\, \mathrm{d}x = \text{maxim}$$

in the class (1.1.20), (1.1.11), $\tilde{\alpha}^{-1}l^2$ being the extremum of the functional (1.3.4). By Theorem 1.1.1, $\tilde{\alpha}^{-1}l^2$ is the least eigenvalue of the equation

$$\alpha^{-1}l^2\, \Delta\mathbf{v} = -\mathbf{v} - \operatorname{grad} \lambda(\mathbf{x}, t)$$

in the class (1.1.20), (1.1.11). For spherical domains, Payne and Weinberger [46] found that $\tilde{\alpha} = 8.98^2$. For other domains of motion, bounds for the constant $\tilde{\alpha}$ can be found in [62] and [83].

Using (1.3.2) and (1.3.3), from (1.3.1) we deduce that the rate of change of the energy $K(t) = \int_\Omega \frac{|\mathbf{v}|^2}{2}\, \mathrm{d}x$ of the difference motion $\mathbf{v}$ admits the estimation

$$\frac{\mathrm{d}}{\mathrm{d}t}\int_\Omega \frac{|\mathbf{v}|^2}{2}\, \mathrm{d}x \leqslant (m - \alpha\nu l^{-2})\int_\Omega |\mathbf{v}|^2\, \mathrm{d}x,$$

or

$$\frac{\mathrm{d}K}{\mathrm{d}t} \leqslant 2(m - \alpha\nu l^{-2})\, K,$$

whence

$$\frac{\mathrm{d}}{\mathrm{d}t}\left(K\mathrm{e}^{-2(m-\alpha\nu l^{-2})\, t}\right) \leqslant 0,$$

which integrated with respect to t from 0 to τ, gives

$$(1.3.5) \qquad K(t) \leqslant K(0)\mathrm{e}^{2(m-\alpha\nu l^{-2})t}.$$

Serrin's first criterion [62]. *If the Reynolds number* $\mathrm{Re} = \frac{\mathrm{m}l^2}{\nu}$ *is smaller than* 32.6, *then* $\mathrm{K(t)} \to 0$ *as* $\mathrm{t} \to \infty$. *Hence the basic flow* $\bar{\mathbf{u}}$ *is asymptotically stable in the mean.*

This criterion provides a universal stability bound valid for every basic flow, every perturbation and every bounded domain of motion. Note that such a sufficient condition for stability can be given every time α can be determined; that is why universal criteria can be proved for motions in (unbounded) tubes or channels, plane flows in bounded domains etc. In short these criteria are proved for those domains of motions for which the embedding theorem 2.1.2 corresponding to the relation (1.3.3) (Section 2.6.1) holds. The best criterion for bounded three-dimensional domains is that for which Serrin's value $\alpha = 32.6$ is replaced by $\tilde{\alpha} = 8.98^2$ (Weinberger and Payne [46]).

For a certain domain of motion, consequently for a certain l, the motion of a fluid possessing an assigned viscosity depends then, through the Reynolds number, only on m, that is on the gradient of the basic flow. This is the reason why Serrin's first criterion is useful in those experiments in which one looks for a stability bound expressed by means of this gradient.

Let us now derive another criterion, useful for determining a stability bound expressed in terms of the modulus of the maximum velocity of the basic flow. To this purpose, we shall consider the equation

$$(1.1.15)' \quad \frac{d}{dt}\int_\Omega \frac{|\mathbf{v}|^2}{2}\,dx = \int_\Omega (\mathbf{v}\cdot\operatorname{grad})\,\mathbf{v}\cdot\bar{\mathbf{u}}\,dx - \nu\int_\Omega |\operatorname{grad}\mathbf{v}|^2\,dx.$$

From the identity

$$\mathbf{T}:\mathbf{T} - 2\mathbf{v}\cdot\mathbf{T}\cdot\bar{\mathbf{u}}| + |\mathbf{v}|^2|\bar{\mathbf{u}}|^2 \equiv (\mathbf{T}-\mathbf{v}\bar{\mathbf{u}})\cdot(\mathbf{T}-\mathbf{v}\bar{\mathbf{u}}) \geqslant 0,$$

where: stands for the tensor product and $\mathbf{v}\mathbf{u}$ is the tensor of components $v_i\bar{u}_j$; substituting $\mathbf{T} = \nu$ gard $\mathbf{v}$, we deduce that

$$(\mathbf{v}\cdot\operatorname{grad})\,\mathbf{v}\cdot\bar{\mathbf{u}} \leqslant 1/2\,(\nu|\operatorname{grad}\mathbf{v}|^2 + |\bar{\mathbf{u}}|^2|\mathbf{v}|^2/\nu).$$

Using inequality (1.3.3) and denoting $V = \max\limits_{t\in(0,\tau)} |\bar{\mathbf{u}}|$, from (1.1.15)' we deduce that

$$\frac{dK}{dt} \leqslant \frac{1}{\nu}(V^2 - \alpha\nu^2 l^{-2})\,K,$$

whence, as for the first criterion,

$$(1.3.6) \qquad K(t) \leqslant K(0)\,e^{(V^2-\alpha\nu^2 l^{-2})t\nu^{-1}},$$

which proves

Serrin's second criterion [62]. *If the Reynolds number* $\frac{\mathrm{Vl}}{\nu}$ *is smaller that* 5.71, *then* $\mathrm{K(t)} \to 0$ *as* $\mathrm{t} \to \infty$, *hence the basic flow is asymptotically stable in the mean.*

The best criterion is obtained by replacing Serrin's value $\sqrt{\alpha} = 5.71$ by $\sqrt{\bar{\alpha}} = 8.98$. In the literature, a large number of papers exists which extend the two Serrin's criteria to the case when apart from body forces, the fluid is subjected to the temperature action, to an electromagnetic field etc. Some generalized alternatives of Serrin's criteria, as well as other criterion shall be treated in § 2.6.

Before Serrin's work [62] it was known that for small $\max\limits_t |\bar{\mathbf{u}}|$ and $|\operatorname{grad}|\bar{\mathbf{u}}|$, every motion is stable [77], [22], but no estimation existed for the Reynolds number under which that motion is certainly stable. Serrin has given such an estimation and he has also indicated how to obtain the maximum value of Re by reducing the variational problem to an eigenvalue problem; this last idea has been used in the energy method (Section 1.1.2) when $\frac{1}{\widetilde{\mathrm{Re}}}$ was defined as the greatest eigenvalue of the problem (1.1.21), (1.1.20), (1.1.11).

The idea of using isoperimetric inequalities corresponding to (1.1.3) by replacing α with the best constant $\widetilde{\alpha}$ will (in particular) be extremely useful in deriving criteria of linear stability.

An important consequence of relations (1.3.5) and (1.3.6) is

Theorem 1.3.1. (*Serrin* [62]). *In the domain* Ω *consider two steady motions* $\bar{\mathbf{u}}$ *and* $\mathbf{u}$ *which satisfy the same condition on* $\partial\Omega$. *Then the two motions coincide if*

$$\frac{ml^2}{\nu} \leqslant \widetilde{\alpha} \text{ or } \frac{Vl}{\nu} \leqslant \sqrt{\widetilde{\alpha}}.$$

For domains Ω with rigid wall and for $\bar{\mathbf{u}} = 0$, from (1.3.5) and (1.3.6) it follows that

$$K(t) \leqslant K(0)\mathrm{e}^{-2\widetilde{\alpha}\nu l^{-2}t}, \tag{1.3.7}$$

which has as a consequence the global existence of a generalized solution of the Navier-Stokes equation. Estimations of the type (1.3.7) had been obtained before by Leray [62] for plane flows, Kampé de Fériet [33] and Berker [2] for three-dimensional flows, and for small values of α. Rayleigh [50] had proved that as $t \to \infty$ then $K(t) \to 0$, but without estimating the rapidity of this convergence.

If instead of taking an arbitrary Ω we take a certain particular domain of motion then a greater value for α is found and for the corresponding flow *particular criteria* of asymptotic stability in the mean can be deduced. Such a criterion was obtained by Serrin [62] for the Couette motion between rotating cylinders; in the following sections, another particular criteria of linear stability will be derived.

1.3.2. Synge's criterion for Couette flow between rotating cylinders

The linear stability of the Couette flow between rotating cylinders to infinitesimal perturbations with symmetry of rotation as normal modes containing the factor $\mathrm{e}^{\sigma\xi+i\lambda\zeta}$, is governed by the following eigenvalue problem, written in dimensionless coordinates (Section 1.1.3)

$$(L - \lambda^2 - \sigma)(L - \lambda^2) f = -2\lambda \operatorname{Re}\left(a + \frac{b}{\xi^2}\right) g, \tag{1.1.35}$$

$$(L - \lambda^2 - \sigma) g = -2\lambda \operatorname{Re} af, \tag{1.1.36}$$

$$f = \frac{\mathrm{d}f}{\mathrm{d}\xi} = g = 0 \text{ for } \xi = 1 \text{ and } \xi = \alpha. \tag{1.1.37}$$

$L = \frac{d}{d\xi^2} + \frac{1}{\xi}\frac{d}{d\xi} - \frac{1}{\xi^2}$ is the azimuthal velocity in the θ-direction, $\bar{v}_\theta = a\xi + \frac{b}{\xi}$, and $\mathrm{Re} = \frac{\omega_1 r_1^2}{\nu}$ represents the Reynolds number.

We shall deduce now sufficient conditions of linear stability, using the method of positively-defined integrals. To this purpose, we shall multiply (1.1.35) by $\xi\bar{f}$; integrating it by parts over $[1, \alpha]$ and taking into account (1.1.37), we get

$$\sigma(\mathcal{J}_1^2 + \mathcal{J}_0^2 + \lambda^2 K_0) + I_2^2 + 3(I_1^2 - I_0^2) +$$

(1.3.8) $$+ 2\lambda^2(\mathcal{J}_1^2 + \mathcal{J}_0^2) + \lambda^4 K_0^2 = -2\,\mathrm{Re}\,\lambda\int_1^\alpha (a + b\xi^{-2})\,\xi\bar{f}g\,d\xi,$$

where we employed the notations

$$I_2^2 = \int_1^\alpha \xi f''\bar{f}''d\xi,\; I_1^2 = \int_1^\alpha \xi^{-1}f'\bar{f}'\,d\xi,\; \mathcal{J}_1^2 = \int_1^\alpha \xi f'\bar{f}'\,d\xi,\; I_0^2 =$$

$$= \int_1^\alpha \xi^{-3}f\bar{f}\,d\xi,\; \mathcal{J}_0^2 = \int_1^\alpha \xi^{-1}f\bar{f}\,d\xi,\; K_0^2 = \int_1^\alpha \xi f\bar{f}\,d\xi.$$

By multiplying (1.1.36) by $(a + b\xi^{-2})\xi\bar{g}$ and integrating on $[1, \alpha]$ we find that

$$\sigma\int_1^\alpha \left(1 + \frac{b}{a\xi^2}\right)\xi g\bar{g}\,d\xi + \int_1^\alpha \left(1 + \frac{b}{a\xi^2}\right)(\xi g'\bar{g}' + \xi^{-1}g\bar{g} + \lambda^2\xi g\bar{g})\,d\xi -$$

(1.3.9) $$-\frac{2b}{a}\int_1^\alpha \xi^{-2}\bar{g}g'\,d\xi = 2\lambda\,\mathrm{Re}\int_1^\alpha \left(a + \frac{b}{\xi^2}\right)\xi f\bar{g}\,d\xi.$$

Adding (1.1.38) to its complex-conjugate, we get

$$\mathcal{R}e\{\sigma\}(\mathcal{J}_1^2 + \mathcal{J}_0^2 + \lambda^2 K_0^2) + I_2^2 + 3(I_1^2 - I_0^2) + 2\lambda^2(\mathcal{J}_1^2 +$$

(1.3.10) $$+ \mathcal{J}_0^2) + \lambda^4 K_0^2 = -\lambda\,\mathrm{Re}\int_1^\alpha (a + b\xi^{-2})\,\xi(f\bar{g} + \bar{f}g)\,d\xi.$$

Carrying out the same operations as we did for (1.3.9) and taking into account that

(1.3.11) $$\int_1^\alpha \xi^{-2}(\bar{g}g' + g\bar{g}')\,d\xi = 2\int_1^\alpha \xi^{-3}g\bar{g}\,d\xi,$$

we obtain

$$\mathcal{Re}\{\sigma\}\int_1^\alpha\left(1+\frac{b}{a\xi^2}\right)g\bar{g}\xi\,\mathrm{d}\xi+\int_1^\alpha\left(1+\frac{b}{a\xi^2}\right)(\xi g'\bar{g}'+\lambda^2 g\bar{g}+\xi^{-1}g\bar{g})\,\mathrm{d}\xi-$$

$$-2\,ba^{-1}\int_1^\alpha \xi^{-3}g\bar{g}\,\mathrm{d}\xi=\lambda\,\mathrm{Re}\int_1^\alpha(a+b\xi^{-2})\xi(f\bar{g}+\bar{f}g)\,\mathrm{d}\xi. \tag{1.3.12}$$

Finally, adding (1.3.10) with (1.3.12) and putting

$$P=\mathcal{J}_1^2+\mathcal{J}_0^2+\lambda^2K_0^2+\int_1^\alpha(1+ba^{-1}\xi^{-2})\xi g\bar{g}\,\mathrm{d}\xi,$$

$$L=I_2^2+3(I_1^2-I_0^2)+2\lambda^2(\mathcal{J}_1^2+\mathcal{J}_0^2)+\lambda^4K_0^2,$$

$$M=\int_1^\alpha(1+ba^{-1}\xi^{-2})(\xi g'\bar{g}'+\xi^{-1}g\bar{g})\,\mathrm{d}\xi-2\,ba^{-1}\int_1^\alpha\xi^{-3}g\bar{g}\,\mathrm{d}\xi,$$

$$N=\lambda^2\int_1^\alpha(1+ba^{-1}\xi^{-2})\,\xi g\bar{g}\,\mathrm{d}\xi,$$

we have

$$\mathcal{Re}\{\sigma\}\cdot P+L+M+N=0. \tag{1.3.13}$$

On the other hand, for every real constant X, the inequality

$$\int_1^\alpha\xi^{-1}(f'+Xf\xi^{-1})\,(\bar{f}'+X\bar{f}\xi^{-1})\,\mathrm{d}\xi\geqslant 0 \tag{1.3.14}$$

holds. Whence

$$I_1^2+X\int_1^\alpha\xi^{-2}(f\bar{f}'+\bar{f}f')\,\mathrm{d}\xi+X^2I_0^2\geqslant 0$$

or, taking into account (1.3.11),

$$I_1^2+2XI_0^2+X^2I_0^2\geqslant 0,$$

from which, putting $X=-1$, we obtain

$$I_1^2\geqslant I_0^2. \tag{1.3.15}$$

Consequently $L>0$.

Suppose now that $a>0$, which may be written as $\dfrac{\alpha^2\beta-1}{\alpha^2-1}>0$, or

$$\omega_2 r_2^2>\omega_1 r_1^2. \tag{1.3.16}$$

Then, since $b=-\dfrac{\alpha^2(\beta-1)}{\alpha^2-1}$, we have

$$\bar{v}_\theta\xi^{-1}a^{-1}=1+ba^{-1}\xi^{-2}=1+\frac{\alpha^2(\beta-1)}{\xi^{-2}(\alpha^2\beta-1)}=\frac{\alpha^2\beta(\xi^2-1)+\alpha^2-\xi^2}{\xi^2(\alpha^2\beta-1)}$$

and since $1 \leqslant \xi \leqslant \alpha$ taking into account (1.3.15) it follows that $\bar{v}_\theta \xi^{-1} a^{-1} > 0$. Therefore $P > 0$ and $N > 0$. As above, for every real constant X we have

$$(1.3.17) \qquad \int_1^\alpha \left(\xi + \frac{b}{a\xi}\right)\left(g' - \frac{Xg}{\xi}\right)\left(\bar{g}' - \frac{X\bar{g}}{\xi}\right) \mathrm{d}\xi \geqslant 0.$$

Taking into account (1.3.11), it follows that $M \geqslant 0$ for $X = 1$. From (1.3.13) we deduce that every eigenvalue has negative real part and consequently that the Couette motion is stable.

Synge's criterion [76] [5]. *If*

$$\omega_2 r_2^2 > \omega_1 r_1^2 > 0,$$

then the Couette flow between two cylinders rotating in the same sense is linearly asymptotically stable to perturbations with symmetry of rotation.

We note that the possibility of deriving criteria of hydrodynamic stability depends on that of establishing some inequalities between positively defined integrals containing functions f from a certain class, and positively defined integrals which contain the derivatives of the same functions. Synge was the first who deduced such inequalities starting with relations of the form

$$(1.3.18) \qquad \int_a^b \left[L_1(f) + \sum_{i=2}^n X_i L_i(f)\right]\left[\overline{L_1(f)} + \sum_{i=2}^n X_i \overline{L_i(f)}\right] \mathrm{d}x \geqslant 0,$$

where L_i $(i = 1, ..., n)$ are differential operators and X_i are arbitrary real constants. In the case of the Couette flow, Synge uses the relations (1.3.14) and (1.3.17). Additionally Serrin set the problem of using the best inequality of the specified type, therefore of finding the constant $\tilde{\alpha}$ from a corresponding isoperimetric inequality. The method according to which the stability criteria are obtained by means of positively defined integrals and of some inequalities of the form (1.3.18) is called *Synge's method*.

1.3.3. Synge's criteria for plane parallel flows

Consider the plane parallel steady basic flow $(U(y), 0)$ whose linear stability to Tollmien-Schlichting two-dimensional wave-like perturbations containing the factor $\mathrm{e}^{i\alpha(x-ct)}$, is governed by the eigenvalue problem for the Orr-Sommerfeld equation

$$(1.1.25) \qquad (U - c)(\varphi'' - \alpha^2\varphi) - U''\varphi = \frac{1}{\mathrm{i}\,\alpha\,\mathrm{Re}}(\varphi^{\mathrm{IV}} - 2\alpha^2\varphi'' + \alpha^4\varphi),$$

$$(1.1.26) \qquad \varphi(0) = \varphi'(0) = \varphi(1) = \varphi'(1) = 0.$$

[5] This solution has been obtained first by means of intuitive arguments concerning the kinetik energy of the perturbations by Rayleigh [51].

Put $I_j^2 = \int_0^1 |\varphi^{(j)}|^2 \, dy$, $(j = 0, 1, 2)$; multiplying (1.1.25) by $\overline{\varphi}$, integrating on [0, 1] and taking into account conditions (1.1.26), we obtain the relation

$$c(I_1^2 + \alpha^2 I_0^2) - \int_0^1 U |\varphi'|^2 \, dy - \int_0^1 U'\varphi'\overline{\varphi} \, dy - \alpha^2 \int_0^1 U |\varphi|^2 \, dy -$$

$$- \int_0^1 U'' |\varphi|^2 \, dy = \frac{1}{i \alpha \operatorname{Re}} (I_2^2 + 2\alpha^2 I_1^2 + \alpha^4 I_0^2), \tag{1.3.19}$$

which, added to its complex conjugate, gives the expression of the velocity of propagation of the perturbation

$$c_r = \left\{ \int_0^1 \left[U |\varphi'|^2 + \left(\alpha^2 U + \frac{U''}{2} \right) |\varphi|^2 \right] dy \right\} [I_1^2 + \alpha^2 I_0^2]^{-1}. \tag{1.3.20}$$

From this equality, we can easily derive bounds for c_r [74], [68], [45]; the best estimations will be given in Section 1.3.4.

Subtracting the complex-conjugate of (1.3.19) from the relation (1.3.19) we obtain the expression for the amplification factor

$$c_i = [I_1^2 + \alpha^2 I_0^2]^{-1} \left\{ \frac{i}{2} \int_0^1 U'\varphi\overline{\varphi}' dy - \frac{i}{2} \int_0^1 U'\overline{\varphi}\varphi' dy - \right.$$

$$\left. - \frac{1}{\alpha \operatorname{Re}} (I_2^2 + 2\alpha^2 I_1^2 + \alpha^4 I_0^2) \right\}. \tag{1.3.21}$$

Consequently every condition which determines a negative sign to the right-hand side of this relation is a linear stability criterion. The best sufficient condition is that for which the expression inside the brackets from (1.3.21) has a negative maximum. But since the corresponding Euler equation

$$\varphi^{IV} - 2\alpha^2\varphi'' + \alpha^4\varphi + i \alpha \operatorname{Re} \left(\varphi' U' + \frac{U''}{2} \varphi \right) = 0$$

is difficult to be solved, we shall look for a weaker stability condition. Thus, denoting $q = \max_{y \in [0,1]} |U'(y)|$ and applying Schwarz's inequality in (1.3.21), we obtain Synge's bound

$$c_i \leqslant \frac{q I_0 I_1 - (\alpha \operatorname{Re})^{-1} (I_2^2 + 2\alpha^2 I_1^2 + \alpha^4 I_0^2)}{I_1^2 + \alpha^2 I_0^2}. \tag{1.3.22}$$

Whence we deduce the sufficient condition of stability

$$\alpha \, \text{Re} < \max \frac{I_2^2 + 2\alpha^2 I_1^2 + \alpha^4 I_0^2}{I_1 I_0 q}, \tag{1.3.23}$$

which is valid for every basic flow having the maximum of the velocity gradient equal to q; note the analogy between the derivation of Serrin's criterion (1.1.19) from the energy equation and the derivation of Synge's criterion (1.3.23) from equation (1.3.19). The reduction of the variational problem (1.3.23) to an eigenvalue one is due to Joseph (Section 1.3.4); Synge [74] looked for bounds of the stability domain, as was done later on by Serrin for derivation of his first criterion. To this purpose, one looks for estimates of I_2 in terms of I_1 and I_0 by means of an inequality analogous to (1.3.18)

$$\int_0^1 (\varphi + x_1 y \varphi' + x_2 \varphi'')\,(\bar{\varphi} + x_1 y \bar{\varphi}' + x_2 \bar{\varphi}'')\,\mathrm{d}y \geqslant 0, \tag{1.3.18$'$}$$

whence it follows

$$x_2^2 I_2^2 \geqslant I_1^2(x_1 x_2 - x_1^2 + 2x_2) + I_0^2(x_1 - 1).$$

Therefore, instead of (1.3.22), we obtain a weaker inequality

$$\begin{aligned} c_i \alpha \, \text{Re} x_2^2 (I_1^2 + \alpha^2 I_0^2) \leqslant x_2^2 q \alpha \text{Re} I_0 I_1 - I_1^2(2\alpha^2 x_2^2 + x_1 x_2 - x_1^2 + \\ + 2x_2) - I_0^2(x_2^2 \alpha^4 + x_1 - 1). \end{aligned} \tag{1.3.22$'$}$$

This inequality holds for example if its right-hand side is a negatively-defined form with respect to I_0 and I_1, which occurs if

$$(x_2^2 q \alpha \text{Re})^2 < 4(2\alpha^2 x_2^2 + x_1 x_2 - x_1^2 + 2x_2)\,(\alpha^4 x_2^2 + x_1 - 1), \tag{1.3.24}$$

$$2\alpha^2 x_2^2 + x_1 x_2 - x_1^2 + 2x_2 > 0, \; \alpha^4 x_2^2 + x_1 - 1 > 0. \tag{1.3.25}$$

Thus, for every x_1 and x_2 satisfying (1.3.25), inequality (1.3.24) furnishes a stability criterion of Synge's type. Among the infinitely many possibilities concerning the choice of x_1 and x_2, Synge considered three, namely: 1) $x_1 = x_2 = 1$, for which (1.3.24) becomes $q \, \text{Re} < \sqrt{8\alpha^2(\alpha^2 + 1)} = \dfrac{a_1(\alpha)}{\alpha}$; 2) $x_1 = x_2 = 2$, when from (1.3.24) we obtain $q \, \text{Re} < \sqrt{(2\alpha^2 + 1)\,(4\alpha^4 + 1)} = \dfrac{a_2(\alpha)}{\alpha}$; and 3) $x_1 = x_2 = \dfrac{1}{\alpha}$, for which (1.3.24) becomes $q \, \text{Re} < \sqrt{8(1 - \alpha^2 + \alpha^3 + \alpha^4)} = \dfrac{a_3(\alpha)}{\alpha}$. In this way, we are led to the following stability condition.

Synge's criterion. *The basic flow* (U(y), 0) *is linearly stable to Tollmien-Schlichting perturbations in the domain of the plane* (α, Re) *delimited by the inequality*

$$\alpha \,\mathrm{Re} < \frac{g(\alpha)}{q} \stackrel{\mathrm{def}}{=} \frac{\max\{a_1, a_2, a_3\}}{q}, \tag{1.3.26}$$

where a_1, a_2, a_3 *are the curves introduced above.*

Synge's criteria are extremely important in hydrodynamic stability; they led, among others, to Serrin's theorem and criteria and to those of Joseph. Unfortunately, their practical value is reduced, since they provide very weak stability bounds; thus, from (1.3.26) it follows that the smallest Reynolds number up to which we have certain linear stability is 2.74 q^{-1}, while, for instance, for the boundary layer, instead of 2.74 we have 420. Improved numerical values can be obtained deriving from (1.3.18)′ the best inequalities satisfied by I_0, I_1 and I_2 [12]:

$$I_2^2 \geqslant \frac{9}{4} I_1^2,\ I_1^2 \geqslant \frac{1}{4} I_0^2,\ I_2^2 \geqslant 1.2121\ I_0^2. \tag{1.3.27}$$

1.3.4. Joseph's theorems

A method of deriving some criteria important in practice is that due to Joseph which combine Synge's and Serrin's methods. By this method, the stability criteria are deduced from Synge's expressions (1.3.22) and (1.3.23) by means of three *isoperimetric inequalities*

$$I_1^2 \geqslant \lambda_1^2 I_0^2, \quad I_2^2 \geqslant \lambda_2^2 I_1^2, \quad I_2^2 \geqslant \lambda_3^2 I_0^2. \tag{1.3.28}$$

Let H be the space of four times continuously differentiable functions on [0, 1] taking complex values and satisfying the conditions (1.1.26), and let $\overline{H}$ be the Hilbert space obtained by completing H in the norm I_2. Then in the Hilbert space of real-valued functions corresponding to H, inequalities (1.3.28) where [6)]

$$\lambda_1 = \pi, \quad \lambda_2 = 2\pi, \quad \lambda_3 = (4.73)^2$$

are valid.

[6)] By continuation of φ outside [0, 1] up to a periodic function with the period 1, it follows that φ can be expanded in a Fourier convergent series $\varphi(y) = \frac{a_0}{2} + \sum_{n=1}^{\infty} (a_n \cos 2n\pi y + b_n \sin 2n\pi y)$, hence $I_2^2 = 16\pi^4 \cdot \sum_{n=1}^{\infty} n^4 \frac{a_n^2 + b_n^2}{2}$, and $I_1^2 = 4\pi^2 \sum_{n=1}^{\infty} n^2 \frac{a_n^2 + b_n^2}{2}$; but for $n \geqslant 1$ it follows that $n^4 \geqslant n^2$, therefore, $I_2^2 \geqslant 4\pi^2 I_1^2$. An analogous calculation shows that $\lambda_1 = \pi$; λ_3^2 is the least eingenvalue of the problem $\varphi^{IV} = \mu\varphi$, $\varphi''(0) = \varphi''(1) = \varphi'''(0) = \varphi'''(1) = 0$ [52].

It is easy to see that the following relations also hold

$$\frac{I_0 \cdot I_1}{I_1^2 + \alpha^2 I_0^2} = \frac{1}{2\alpha} \cdot \frac{2\alpha\, I_0 I_1}{I_1^2 + \alpha^2 I_0^2} \leqslant \frac{1}{2\alpha} \cdot \frac{I_1^2 + \alpha^2 I_0^2}{I_1^2 + \alpha^2 I_0^2} = \frac{1}{2\alpha},$$

$$\frac{I_2^2 + 2\alpha^2 I_1^2 + \alpha^4 I_0^2}{I_1^2 + \alpha^2 I_0^2} = \frac{I_2^2 + \alpha^2 I_1^2 + \alpha^2(I_1^2 + \alpha^2 I_0^2)}{I_1^2 + \alpha^2 I_0^2} \geqslant \frac{\lambda_2 I_1^2 + \alpha^2 \lambda_3^2 I_0^2}{I_1^2 + \alpha^2 I_0^2} +$$

$$+ \alpha^2 \geqslant \min(\lambda_2^2, \lambda_3^2) + \alpha^2 = 4\pi^2 + \alpha^2.$$

Since $I_2^2 \geqslant \lambda_3 \lambda_2 I_0 I_1$, we have

$$\frac{I_2^2 + 2\alpha^2 I_1^2 + \alpha^4 I_0^2}{I_0 \cdot I_1} \geqslant \lambda_2\lambda_3 + \frac{2\alpha^2}{I_0 I_1}\left[\left(I_1 - \frac{\alpha}{\sqrt{2}} I_0\right)^2 + \frac{2\alpha}{\sqrt{2}} I_0 I_1\right] \geqslant$$

$$\geqslant \lambda_2\lambda_3 + \frac{2\alpha^2\, 2\alpha}{I_0 I_1 \sqrt{2}} I_0 I_1 = \lambda_2\lambda_3 + 2\sqrt{2}\alpha^3,$$

and, on the other hand,

$$\frac{I_2^2 + 2\alpha^2 I_1^2 + \alpha^4 I_0^2}{I_0 I_1} \geqslant \lambda_2\lambda_3 + 2\alpha^2 \frac{I_1}{I_0} +$$

$$+ \alpha^4 \frac{I_0}{I_1} \geqslant \lambda_2\lambda_3 + 2\alpha^2 \frac{I_1}{I_0} \geqslant \lambda_2\lambda_3 + 2\alpha^2\lambda_1.$$

Denoting $M_1(\alpha) = \lambda_2\lambda_3 + 2\sqrt{2}\alpha^3$, $M_2(\alpha) = \lambda_2\lambda_3 + 2\alpha^2\lambda_1$ and taking into account the values of λ_1, λ_2 and λ_3, we may state following theorem

Theorem 1.3.2 (*Joseph* [26]). *Let* c(α, Re) *be an eigenvalue of problem* (1.1.25), (1.1.26). *Then the following inequality holds*

(1.3.29) $$c_i < \frac{q}{2\alpha} - \frac{4\pi^2 + \alpha^2}{\alpha \mathrm{Re}}.$$

In addition, no amplified perturbation ($c_i > 0$) *of* (1.1.25), (1.1.26) *can exist if*

(1.3.30) $$\alpha \mathrm{Re} q < f(\alpha) \stackrel{\mathrm{def}}{=} \max[M_1, M_2].$$

In [26] Joseph uses the value $\lambda_2 = \pi$; subsequently [27] he uses a better value $\lambda_2 = 2\pi$. The domain of stability in the plane (qRe, α) is located to the left of the curve $\frac{f(\alpha)}{\alpha}$, which draws near the exact linear stability curve. The minimum value of qRe below which every infinitesimal perturbation is certainly damped out, is 45.52, for $\lambda_2 = \pi$ and 72.26 (compared with Synge's value 2.74) for $\lambda_2 = 2\pi$ (fig. 1.3.1).

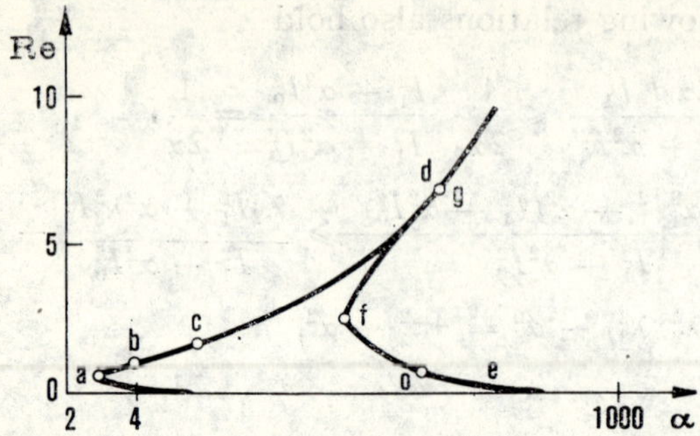

Fig. 1.3.1.

Besides the stability criterion (1.3.30), Theorem 1.3.2 provides a bound for the imaginary part of the eigenvalue c; the following theorem gives a bound for the real part of c too.

Theorem 1.3.3 (*Joseph* [26]). *Let* c(α, Re) *be an eigenvalue of* (1.1.25), (1.1.26). *Then the following inequalities hold*

(a) $$U''_{\min} \geqslant 0,\; U_{\min} < c_r < U_{\max} + \frac{U''_{\max}}{2(\pi^2 + \alpha^2)},$$

(b) $$U''_{\min} \leqslant 0 \leqslant U''_{\max},\; U_{\min} + \frac{U''_{\min}}{2(\pi^2 + \alpha^2)} < c_r < U_{\max} + \frac{U''_{\max}}{2(\pi^2 + \alpha^2)},$$

(c) $$U''_{\max} \leqslant 0,\; U_{\min} + \frac{U''_{\min}}{2(\pi^2 + \alpha^2)} < c_r < U_{\max},$$

where $U_{\max}$, $U''_{\max}$, $U_{\min}$ *and* $U''_{\min}$ *are the maximal, respectively the minimal values of* $U(y)$ *and* $U''(y)$ *for* $y \in [0, 1]$.

Proof. Applying the mean theorem in (1.3.20), we have

(1.3.20)′ $$c_r = U(y_1) + \frac{\frac{1}{2}U''(y_2)}{I_1^2 I_0^{-2} + \alpha^2},$$

where $y_1, y_2 \in (0,1)$. Using the inequalities $\pi^2 \leqslant I_1^2$, $I_0^{-2} \leqslant \infty$ the hypotheses imply that

(a) $$0 < \frac{U''(y_2)}{I_1^2 I_0^{-2} + \alpha^2} < \frac{U''_{\max}}{\pi^2 + \alpha^2},$$

(b) $$\frac{U''_{\min}}{\pi^2 + \alpha^2} < \frac{U''(y_2)}{I_1^2 I_0^{-2} + \alpha^2} < \frac{U''_{\max}}{\pi^2 + \alpha^2},$$

(c) $$\frac{U''_{\min}}{\pi^2 + \alpha^2} < \frac{U''(y_2)}{I_1^2 I_0^{-2} + \alpha^2} < 0,$$

which, together with (1.3.20)′, prove the theorem.

In the case of the boundary layer on the half-bounded flat plate, the linear stability problem in dimensionless coordinates can be expressed by the Orr-Sommerfeld equation (1.1.25) together with the conditions on the plate

$$\varphi(0) = \varphi'(0) = 0 \tag{1.3.31}$$

and the conditions at infinity

$$\varphi(\infty) = \varphi'(\infty) = 0. \tag{1.3.31$'$}$$

The studies which use asymptotic methods, approximate this problem in unbounded domains by another problem in the bounded range [0, 1]: it is assumed that outside the boundary layer ($y \geqslant 1$), $U(y) = 1$, $U' \equiv 0$ and $\varphi = \tilde{\varphi} = A\mathrm{e}^{-\alpha y} + B\mathrm{e}^{-\beta y}$, where $\beta = = \mathrm{i}\,\alpha\,\mathrm{Re}(1 - c) + \alpha^2$.

Under the hypothesis that the part of $\tilde{\varphi}$ which depends on viscosity decreases faster than that which is not influenced by viscosity, $\tilde{\varphi}$ can be approximated (for $y \geqslant 1$) by $\mathrm{e}^{-\alpha y}$. Then, from the condition that the velocity is continuous across the frontier of the boundary layer, we obtain

$$\varphi'(1) + \alpha\varphi(1) = 0\,; \tag{1.3.32}$$

and from the condition that the motion is inviscid on this frontier, it follows that

$$\varphi''(1) - \alpha^2\varphi(1) = 0. \tag{1.3.33}$$

The second condition reflects the continuity of the vorticity across the frontier of the boundary layer. Consequently the eigenvalue problem associated with the Orr-Sommerfeld equation in a bounded domain is (1.1.25), (1.3.31)—(1.3.33).

A priori inequalities analogous to the isoperimetric inequalities (1.3.28) are difficult to establish for this problem, so Joseph [26] considered, instead of (1.3.33) a condition weaker than the continuity of the vortex, namely

$$\varphi'''(1) + \alpha\varphi''(1) = 0. \tag{1.3.34}$$

For the eigenvalue problem (1.1.25), (1.3.31), (1.3.32) and (1.3.34) by Joseph method the bounds for eigenvalues c and stability criteria can be obtained as above. But this time, we shall use isoperimetric inequalities in which λ_i depend on the wave number α,

$$I_1^2 \geqslant \lambda_1^2(\alpha) I_0^2,\ I_2^2 \geqslant \lambda_2^2(\alpha) I_1^2,\quad I_2^2 \geqslant \lambda_3^2(\alpha) I_0^2, \tag{1.3.35}$$

where

$$\lambda_1^2(\alpha) = \min_{\varphi} \frac{\int_0^1 (\varphi')^2\,dy + \alpha^2\varphi^2(1)}{\int_0^1 \varphi^2\,dy}, \quad \lambda_2^2(\alpha) = \min_{\varphi} \frac{\int_0^1 (\varphi'')^2\,dy}{\int_0^1 (\varphi')^2\,dy + \alpha\varphi^2(1)},$$

$$\lambda_3^2(\alpha) = \min_{\varphi} \frac{\int_0^1 (\varphi'')^2\,dy}{\int_0 \varphi^2\,dy}.$$

Writing the corresponding Euler equations, the solutions may be found explicitly and therefore, taking into account the boundary conditions, the characteristic equations may be written; from these equations, λ_1, λ_2 and λ_3 can then be calculated in terms of α (their representations are given in [27]). Using the isoperimetric inequalities (1.3.35), one then finds that the minimal value of $q\dfrac{\mathrm{Re}}{\alpha}$ below which the stability is certain is 25.

Now consider the basic flow $(0, 0, U(r))$ in tubes, subjected to three-dimensional perturbations in the form of normal modes $\mathbf{v}(r, \theta, z, t, \mathrm{Re}) = \mathbf{u}(r, \alpha, N, c, \mathrm{Re})e^{i(\alpha z + N\theta - c\alpha t)}$, where $\mathbf{u} = (w, v, u)$. As is known, the single steady parallel flow in a tube is the Hagen-Poiseuille flow, which has a parabolic velocity profile $U(r)$; nevertheless, as in the case of the Orr-Sommerfeld equation, it is assumed that the eigenvalue problem, governing the linear stability of the parallel and almost parallel flows in a tube, is expressed in cylindrical coordinates r, θ, z by the relations

$$(1.3.36)\qquad \begin{cases} i\,\alpha\,\mathrm{Re}(U - c)\,w = -Dp + \mathfrak{L}_N w - 2\,i\,N v r^{-2}, \\ i\,\alpha\,\mathrm{Re}(U - c)\,v = -\,i\,N r^{-1} p + \mathfrak{L}_N v + 2\,i\,N w r^{-2}, \\ i\,\alpha\,\mathrm{Re}(U - c)\,u + \mathrm{Re}\,w\,DU = -i\,\alpha\,p + \mathfrak{L}_N u, \\ \dfrac{D(rw)}{r} + \dfrac{iN}{rv}\,i\,\alpha\,u = 0, \end{cases}$$

$$(1.3.37)\qquad u(1) = v(1) = w(1) = 0.$$

To the conditions (1.3.37), we add the condition that u, v, w and p have bounded Dirichlet norm [7]. In (1.3.36) we used the notation

$$D = \frac{d}{dr};\quad \mathfrak{L}_N = \frac{1}{r}D(rD) - \frac{N^2 + 1}{r^2} - \alpha^2,\quad L_N = \mathfrak{L}_N + \frac{1}{r^2}.$$

---

7) By definition, the Dirichlet norm of the function a is $\sqrt{\int_\Omega |\,\mathrm{grad}\, a\,|^2\,dx}$.

Eliminating the pressure from the system (1.3.36), we obtain the relations

$$\text{(1.3.38)} \quad \mathrm{i}\,\mathrm{Re}(NG - \alpha rB) = \mathrm{i}\,NL_N u - \mathrm{i}\,\alpha\, r\mathfrak{L}_N v + 2\alpha\, Nwr^{-1},$$

$$\text{(1.3.39)} \quad \mathrm{Re}[-\mathrm{i}\,NA + \mathrm{D}(rB)] = -\mathrm{i}\,N\mathfrak{L}_N w - 2\,N^2 vr^{-2} + \\ + \mathrm{D}(r\mathfrak{L}_N v) + 2\,\mathrm{i}\,N\mathrm{D}(wr^{-1}),$$

where

$$A = \mathrm{i}\,\alpha w(U - c), \quad B = \mathrm{i}\,\alpha v(U - c), \quad H = \mathrm{i}\,\alpha u(U - c) + w\mathrm{D}U.$$

These relations play the role of the Orr-Sommerfeld equation. The derivations of the expressions for c_i and c_r may be carried out in the same manner as the derivation given at the beginning of Section 1.3.3. So, multiplying the relation $\overline{(1.3.38)}$ (the bar stands for the complex-conjugation) by ur, the relation $\overline{(1.3.39)}$ by wr, integrating on [0, 1], substracting the results and taking into account the equalities

$$-\mathrm{i}\,\alpha\,\langle ru\bar{B}\rangle + \langle w\mathrm{D}(r\bar{B})\,\rangle = \mathrm{i}\,N\,\langle v\bar{B}\rangle,$$

$$-\mathrm{i}\,\alpha\,\langle ur\mathfrak{L}_N\bar{v}\rangle + \langle w\mathrm{D}(r\mathfrak{L}_N\bar{v})\rangle = \mathrm{i}\,N\langle v\mathfrak{L}_N\bar{v}\rangle,$$

$$\alpha\,\langle u\bar{w}r^{-1}\rangle + \mathrm{i}\,\langle \bar{w}\mathrm{D}(\bar{w}r^{-1})\rangle = -N\,\langle \bar{w}r^{-2}v\rangle,$$

we obtain

$$\text{(1.3.40)} \quad -\mathrm{Re}(\langle w\bar{A}\rangle + \langle v\bar{B}\rangle + \langle u\bar{G}\rangle) = \mathrm{D}_{\alpha N}(\mathbf{u}),$$

where

$$\langle a\rangle = \int_0^1 ra\,\mathrm{d}r, \text{ and } D_{\alpha N}\;(\mathbf{u}) = -\langle w\mathfrak{L}_N\bar{w}\rangle - \langle v\mathfrak{L}_N\bar{v}\rangle - \langle u\mathfrak{L}_N\bar{u}\rangle -$$

$$-2\,\mathrm{i}\,N\left(\left\langle\frac{\bar{v}w}{r^2}\right\rangle - \left\langle\frac{\bar{w}}{r}\frac{v}{r}\right\rangle\right) = \langle |\,\mathrm{D}w\,|^2 + |\,\mathrm{D}v\,|^2 +$$

$$+ |\,\mathrm{D}u\,|^2 + \alpha^2(|\,w\,|^2 + |\,v\,|^2 + |\,u\,|^2)\rangle +$$

$$+ N^2\left\langle\left|\frac{u}{r}\right|^2\right\rangle + \left\langle\frac{|Nw + \mathrm{i}v|^2}{r^2} + \frac{|Nv - \mathrm{i}w|^2}{r^2}\right\rangle.$$

Taking the real and imaginary parts of (1.3.40) we obtain the following expressions for c_i and c_r:

$$c_i = -\{\langle \mathrm{D}U(u\bar{w} + w\bar{u})\rangle + 2\,\mathrm{Re}^{-1}\,D_{\alpha N}\}\,/2\,\alpha\,\langle |\,\mathbf{u}\,|^2\rangle,$$

$$c_r = \{2\,\alpha\,\langle U\,|\,\mathbf{u}\,|^2\rangle + \mathrm{i}\,\langle \mathrm{D}U(u\bar{w} - w\bar{u})\rangle\}\,/2\,\alpha\,\langle |\,\mathbf{u}\,|^2\rangle.$$

Using the isoperimetric inequality

$$\langle \mathrm{D}\varphi^2\rangle \geqslant \eta_0^2\langle\varphi^2\rangle,$$

(valid for $\varphi(1) = 0$ where $\eta_0 = 2.405$ is the first zero of the Bessel function $J(\eta)$), an argument similar to that which led to Theorems 1.3.2 and 1.3.3 allows us now to state the following theorem

Theorem 1.3.4 [27]. *Let* c(α, N, Re) *be an eigenvalue of the problem* (1.3.36), (1.3.37). *Then*

$$c_i \leqslant \frac{q}{2\alpha} - \frac{\eta_0^2 + \alpha^2 + (N-1)^2}{\alpha \mathrm{Re}},$$

$$U_{min} - \frac{q}{2\alpha} < c_r < U_{max} + \frac{q}{2\alpha},$$

where $q = \max_{r \in [0,1]} |DU|$.

Much better estimations for c_i and c_r and corresponding stability criteria may be deduced in the case of the axi-symmetric perturbations ($v = 0$, $N = 0$), when we find

(1.3.41) $$c_i \leqslant \frac{q}{2\alpha} - (S_2^2 + 2\alpha^2 S_1^2 + \alpha^4 S_0^2)\,[\alpha \mathrm{Re}(S_1^2 + \alpha^2 S_0^2)]^{-1},$$

$$c_r = \left\{ \int_0^1 \left[U\,(|\,\mathrm{D}w\,|^2 + |\,wr^{-1}\,|^2 + \alpha^2\,|\,w\,|^2) + \right.\right.$$

$$\left.\left. + \frac{1}{2}\Lambda\,|\,w\,|^2 \right] r\,\mathrm{d}r \right\} \{S_1^2 + \alpha^2 S_0^2\}^{-1},$$

where

$$\Lambda = r^3 \mathrm{D}\{r^{-1}\mathrm{D}U\},\ S_0^2 = \langle |\,w\,|^2\rangle,\ S_1^2 = \langle\,|\,\mathrm{D}w\,|^2 +$$

$$+ |\,wr^{-1}\,|^2\rangle = -\langle \bar{w}, \widetilde{\mathfrak{L}}_0\,w\rangle,$$

$$S_2^2 = \langle |\,\mathrm{D}^2 w\,|^2\rangle + 3(\langle |\,\mathrm{D}w/r\,|^2\rangle - \langle\,|\,wr^{-2}\,|^2\rangle) =$$

$$= \langle |\,\widetilde{\mathfrak{L}}_0 w\,|^2\rangle,\ \widetilde{\mathfrak{L}}_0 = \mathfrak{L}_0 + \alpha^2,\ \mathfrak{L}_0 = \mathfrak{L}_N$$

for $N = 0$.

Let $\widehat{H}$ (respectively H) be the Hilbert space whose elements w are complex (respectively real) valued functions for which the norm $\langle |\,\widetilde{\mathfrak{L}}_0 w\,|^2\rangle$ is finite and satisfying the boundary conditions $w(1) = \mathrm{D}w(1) = 0$, $w(0)$ bounded. Then the following isoperimetric inequalities hold,

$$S_1^2 \geqslant \eta_1^2 S_0^2, \quad S_2^2 \geqslant \eta_2^2 S_1^2, \quad S_2^2 \geqslant \eta_3^2 S_0^2,$$

where $\eta_1^2 = 3.83^2$, $\eta_2^2 = 5.13^2$, $\eta_3^2 = 4.61^4$; η_1^2, η_2^2, η_3^2 are the minimal values of some ratios of functionals for which the corresponding Euler equations admit as solutions some linear combinations of Bessel functions [27]. Consequently, owing to the analogy between (1.3.41) and (1.3.22), the enuntiations of Theorem 1.3.2 (and hence of the stability criterion implied by it) and 1.3.3 may be written out word by word if λ_1, λ_2 and λ_3 are replaced by η_1, η_2 and respectively η_3 and U'' by Λ. Joseph's method has been applied to many other configurations for various fluid flows. As could be expected, the numerical calculations showed that it is more adequate to the study of motions whose stability depends on their global characteristics [12], [14] (for instance, in the case of the Couette flow [13], the linear stability bound is 1708 and by Joseph's method one obtains 1585).

1.3.5. The envelope method

In the previous section, the derivation of bounds for c_i and c_r has been reduced to the search of some majorations of the ratios $\dfrac{I_2^2 + 2\alpha^2 I_1^2 + \alpha^4 I_0^2}{I_1^2 + \alpha^2 I_0^2}$ and $\dfrac{I_2^2 + 2\alpha^2 I_1^2 + \alpha^4 I_0^2}{I_1 I_0}$ by means of three isoperimetric inequalities which related the functionals I_0, I_1 and I_2. Hence arises the problem (analysed in detail in [12] and [11]) of finding the best estimates for these ratios. Thus, since

$$I_1^2 \leqslant \lambda_2^{-2} I_2^2 \text{ and } I_0^2 \leqslant \lambda_3^{-2} I_2^2,$$

we have

$$\frac{I_2^2}{I_1^2 + \alpha^2 I_0^2} \geqslant \frac{I_2^2}{(\lambda_2^{-2} + \alpha^2 \lambda_3^{-2})\, I_2^2} = \frac{\lambda_2^2 \lambda_3^2}{\lambda_3^2 + \alpha^2 \lambda_2^2}, \tag{1.3.42}$$

and, from $I_1^2 = I_1^2$ and $I_0^2 \leqslant \lambda_1^{-2} I_1^2$, it follows that

$$\frac{\alpha^2 I_1^2}{I_1^2 + \alpha^2 I_0^2} \geqslant \frac{\alpha^2 I_1^2}{(1 + \alpha^2 \lambda_1^{-2})\, I_1^2} = \frac{\alpha^2 \lambda_1^2}{\lambda_1^2 + \alpha^2}. \tag{1.3.43}$$

Finally, taking into account $\dfrac{\alpha^2 I_1^2 + \alpha^4 I_0^2}{I_1^2 + \alpha^2 I_0^2} = \alpha^2$ and relations (1.3.42), (1.3.43) we obtain

$$\frac{I_2^2 + 2\alpha^2 I_1^2 + \alpha^4 I_0^2}{I_1^2 + \alpha^2 I_0^2} \geqslant \frac{\lambda_2^2 \lambda_3^2}{\lambda_3^2 + \alpha^2 \lambda_2^2} + \frac{\alpha^2 \lambda_1^2}{\lambda_1^2 + \alpha^2} + \alpha^2,$$

so that instead of the estimation (1.3.29) we have

$$c_i < \frac{q}{2\alpha} - \left(\frac{\lambda_2^2 \lambda_3^2}{\lambda_3^2 + \alpha^2 \lambda_2^2} + \frac{\alpha^2 \lambda_1^2}{\lambda_1^2 + \alpha^2} + \alpha^2 \right) (\alpha \operatorname{Re})^{-1}.$$

This estimation becomes even better than (1.3.29) as α draws closer to 4.73/2 and gives an improvement less good on the measure of its approaching zero.

To obtain an improvement of the above mentioned second ratio, we introduce a real parameter a, $0 \leqslant a \leqslant 2$, such that

$$\frac{2\alpha^2 I_1^2 + \alpha^4 I_0^2}{I_0 I_1} = \frac{a\alpha^2 I_1^2 + \alpha^2(2 - a) I_1^2 + \alpha^4 I_0^2}{I_0 I_1};$$

then, taking into account the isoperimetric inequalities (1.3.28) and the inequality $\alpha^2(2-a) I_1^2 + \alpha^4 I_0^2 \geqslant 2\alpha^3 \sqrt{2-a}\, I_0 I^1$, the estimate

$$\frac{I_2^2 + 2\alpha^2 I_1^2 + \alpha^4 I_0^2}{I_0 I_1} \geqslant \lambda_2\lambda_3 + a\alpha^2\lambda_1 + 2\alpha^3\sqrt{2-a} = \widetilde{M}(a, \alpha) \tag{1.3.44}$$

follows, where $\widetilde{M}(a, \alpha)$ is a family of curves depending on α; for $a = 0$ and $a = 2$ we obtain Joseph's curves $M_1(\alpha) = \widetilde{M}(0, \alpha)$ and respectively $M_2(\alpha) = \widetilde{M}(2, \alpha)$. It is easy to see that all curves of the family $\widetilde{M}(a, \alpha)$ are located above the curve $M_1(\alpha)$ for $0 \leqslant \alpha \leqslant \lambda_1\sqrt{2}$ and cut the curve $M_2(\lambda)$ at a point $(\alpha, M_2(\alpha))$ with $\alpha \leqslant \frac{\lambda_1}{\sqrt{2}}$ and the curve $M_1(\alpha)$ at a point $(\alpha, M(\alpha))$ with $\frac{\lambda_1}{\sqrt{2}} \leqslant \alpha \leqslant \lambda_1\sqrt{2}$. For $\alpha \geqslant \lambda_1\sqrt{2}$ the upper curve of this family is $M_1(\alpha)$. Among all curves of the family $\widetilde{M}(a, \alpha)$ we are interested in that for which the inequality (1.3.44) is best; hence that curve must be located above all the other. Clearly, for $0 \leqslant \alpha \leqslant \lambda_1\sqrt{2}$ this curve is the envelope $\widetilde{M}(\alpha)$ of the family $\widetilde{M}(a, \alpha)$, where $\widetilde{M}(\alpha) = \lambda_2\lambda_3 + 2\alpha^2\lambda_1 + \alpha^4\lambda_1^{-1}$; therefore instead of Joseph's stability criterion (1.3.30) we have

$$\alpha \operatorname{Re} q < h(\alpha) = \max\{\widetilde{M}(\alpha), M_1(\alpha)\}, \tag{1.3.30$'$}$$

which ensures the linear stability of plane parallel flows for $q\mathrm{Re} < 74.28$ (fig. 1.3.2).

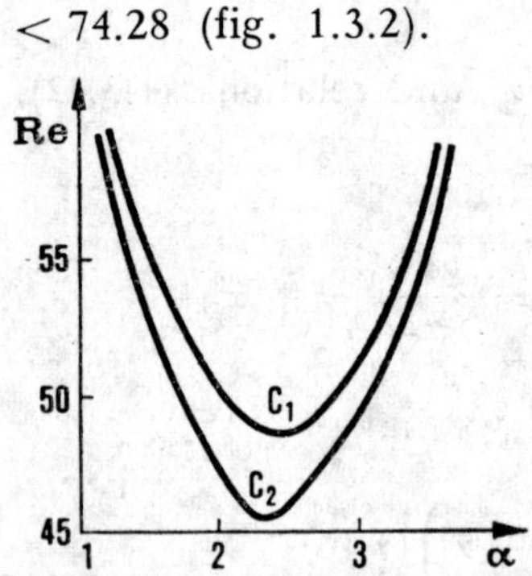

Fig. 1.3.2.

Note that the envelope method leads to better bounds for qRe below which stability is ensured, as well as to larger domains of stability in the (qRe, α) plane.

The best stability criteria may be obtained using the parameters $a, b, c, d, e,$ instead of a single parameter, a, and writing the ratio $K = \dfrac{I_2^2 + 2\alpha^2 I_1^2 + \alpha^4 I_0^2}{I_0 I_1}$ in the form

$$K = [aI_2^2 + bI_2^2 + (1 - a - b)\, I_2^2 + c\alpha^2 I_1^2 + d\alpha^2 I_1^2 +$$

$$+ (2 - c - d)\, \alpha^2 I_1^2 + e\alpha^4 I_0^2 + (1 - e)\, \alpha^4 I_0^2]\, (I_0 I_1)^{-1}$$

in the conditions

$$a, b, c, d, e,\ 1 - a - b,\ 1 - e,\ 2 - c - d \geqslant 0.$$

Then, grouping together in a suitable manner the terms in K and applying the inequality $x^2 + y^2 \geqslant 2xy$, instead of (1.3.44), we obtain

$$\frac{I_2^2 + 2\alpha^2 I_1^2 + \alpha^4 I_0^2}{I_0 I_1} \geqslant 2\alpha \sqrt{ac}\lambda_3 + 2\alpha^2 \sqrt{be}\lambda_2 + (1 - a - b)\, \lambda_2\lambda_3 +$$

$$+ 2\alpha^3 \sqrt{d(1 - e)} + (2 - c - d)\, \alpha^2 \lambda_1. \tag{1.3.44$'$}$$

If $\lambda_3 > \lambda_1\lambda_2$, $\lambda_3 < \lambda_2^2$ and $\lambda_1 < \lambda_2$ this leads to the stability criterion

$$\alpha \operatorname{Re} q < k(\alpha) = \{\max P(\alpha), S(\alpha)\} =$$

$$= \begin{cases} P(\alpha) \text{ for } 0 \leqslant \alpha^2 \leqslant \lambda_3^2\lambda_2^{-2} + \sqrt{\lambda_3^4\lambda_2^{-4} + \lambda_3^2} \\ S(\alpha) \text{ for } \alpha^2 \geqslant \lambda_3^2\lambda_2^{-2} + \sqrt{\lambda_3^4\lambda_2^{-4} + \lambda_3^2} \end{cases} \tag{1.3.30$''$}$$

where $P(\alpha) = \lambda_2\lambda_3 + 2\alpha^2\lambda_2\lambda_3^{-1} + \alpha^4\lambda_2\lambda_3^{-2}$ and $S(\alpha) = 2\alpha^2\sqrt{2\alpha^2 + \lambda_2^2}$.

REFERENCES

[1] Belyakova, V. K., *On the problem of viscous fluid flow stability in a circular straight pipe*, P.M.M, **14**, 105–110. 1950. (In Russian).

[2] Berker, R., *Inégalité verifiée par l'énergie cinétique d'un fluide visqueux incompressible occupant un domaine spatial borné*, "Bull. Tech. Univ. Istanbul", 2, 41–51 (1949).

[3] Betchov, R., W. O. Criminale Jr., *Stability of parallel flows*, Academic Press, New York, London, 1967.

[4] Birkhoff, G., *Helmholtz and Taylor Instability*, in "Proc. Symposia in appl. math.", vol. XIII, *Hydrodynamic instability* (ed. G. Birkhoff, R. Bellman, 1962).

[5] Chandrasekhar, S., *Hydrodynamic and hydromagnetic stability*, Clarendon Press, Oxford, 1961.

[6] Chen, T. S., Joseph, D. D., Sparrow, E. M., *Evolution of Tietjens function in stability calculations*, "Phys. Fluids", **9**, nr. 12, 2519–2521 (1966).

[7] Davies, S. J., White, C. M., *An experimental study of the flow of water in pipes of rectangular section*, "Proc. Roy. Soc.", **A 119**, nr. 781, 92–107 (1928).

[8] Di Prima, R. C., *Some variational principles for problems in hydrodynamic and hydromagnetic stability*, "Quart Appl. Math.", **18**, nr. 4, 375—385 (1961).

[9] Dryden, H. L., Murnaghan, F. D., Bateman, H., *Hydrodynamics*, Dover, New York, 1956.

[10] Eckhaus, W., *Studies in nonlinear stability theory*, Springer Verlag, New York, 1965.

[11] Georgescu, A., *Note on Joseph inequalities in stability theory*, "ZAMP", **21**, nr. 2, 258—260 (1970).

[12] Georgescu, A., *Contributions to the study of the linear stability of fluid flows*, "St. Cerc. Mat.", **22**, nr. 9, 1247—1333 (1970).

[13] Georgescu, A., *On the neutral stability of the Couette flow between two rotating cylinders*, "Rev. Roum. Math. Pures et Appl.", **16**, nr. 4, 499—503 (1971).

[14] Georgescu, A., *Linear Couette flow stability for arbitrary gap between two rotating cylinders*, "Rev. Roum. Math. Pures et Appl.", **17**, nr. 4, 507—518 (1972).

[15] Georgescu, A., *Stability of the Couette flow of a viscoelastic fluid*, "Rev. Roum. Math. Pures et Appl.", **18**, nr. 9, 1371—1374 (1973).

[16] Gershuni, G. Z., Jukovitski, E. M., *Convective stability of incompressible fluids*, Nauka, Moscow, 1972. (In Russian)

[17] Goldstein, S., *Modern developments in fluid dynamics*, Oxford University Press, Oxford, 1938.

[18] Greenspan, H. P., *The theory of rotating fluids*, Cambridge University Press, Cambridge, 1968.

[19] Heisenberg, W., *Über Stabilität und Turbulenz von Flüssigkeitsströmen*, "Ann. d. Phys.", Leipzig, (4), **74**, 577—627 (1924).

[20] Helmholtz, H., *Über discontinuierliche Flüssigkeitsbewegung*, "Monatsber. Kön. Akad.", Berlin, **23**, 215—228 (1869).

[21] Holstein, H., *Über die äussere und inner Reibungsschicht bei Störungen laminarer Strömungen*, "ZAMM", **30**, 25—49 (1950).

[22] Hopf, E., *On nonlinear partial differential equations*, Lecture series of the Symposium on partial differential equations, University of California, pp. 7—11, 1955.

[23] Jacob, C., *Introduction mathématique à la mécanique des fluides*, Gauthier-Villars, Bucarest, Paris, 1959.

[24] Joseph, D. D., *On the stability of the Boussinesq equations*, "Arch Rational Mech. Anal.", **20**, nr. 1, 59—71 (1965).

[25] Joseph, D. D., *Nonlinear stability of the Boussinesq equations by the method of energy*, "Arch. Rational Mech. Anal.", **22**, nr. 3, 163—184 (1966).

[26] Joseph, D. D., *Eigenvalue bounds for the Orr-Sommerfeld equation*, "J. Fluid. Mech.", **33**, nr. 3, 617—621 (1968).

[27] Joseph, D. D., *Eingenvalue bounds for the Orr-Sommerfeld equation*, part II, "J. Fluid Mech.", **36**, nr. 4, 721—734 (1969).

[28] Joseph, D. D., Carmi, S., *Stability of Poiseuille flow in pipes, annuli and channels*, "Quart. Appl. Math.", **24**, nr. 4, 575—599 (1969).

[29] Joseph, D. D., Munson, B. R., *Global stability of spiral flow*, "J. Fluid Mech.", **43**, nr. 3, 545—575 (1970).

[30] Joseph, D. D., *Global stability of fluid motions*, Dept. Aerospace Eng. and Mech., Univ. of Minnesota, Minneapolis, 1971.

[31] Jeffreys, H., *Some cases of instability in fluid motion*, "Proc. Roy. Soc.", **A 118**, 145—208 (1926).

[32] Jungclaus, G., "Appl. Mech. Rev.", **11**, 630 (1958).

[33] Kampé de Fériet, J., *Sur la décroissance de l'énergie cinétique d'un fluide visqueux incompressible occupant un domaine borné ayant pour frontière des parois solides fixes*, "Ann. Soc. Sci. Bruxelles", **63**, 35—46 (1949).

[34] Kármán, Th. von, *Über die Stabilität der Laminarströmung und die Theorie der Turbulenz*, Proc. 1st Int. Congr. Appl. Mech., Delft, 1924, pp. 97—112.

[35] Langer, R. E., *On the stability of the laminar flow of a viscous fluid*, "Bull. Amer. Math. Soc.", **46**, 257—263 (1940).

[36] Langer, R. E., *On the asymptotic solutions of a class of ordinary differential equations of the fourth order with special reference to an equation of hydrodynamics*, "Trans. Amer. Math. Soc.", **84**, 144 (1957).

[37] Lin, C. C., *On the stability of two-dimensional parallel flows*, "Proc. Nat. Acad. Sci. Washington", **30**. 316—323 (1944).

[38] Lin C. C., *The theory of hydrodynamic stability*, Cambridge University Press, Cambridge, 1955.

[39] Meksyn, D., *Stability of viscous flow between rotating cylinders*, I, II, III, "Proc. Roy. Soc.", **A 187**, 115—128, 480—491, 492—504 (1946).

[40] Meksyn, D., *New methods in boundary layer theory*, cap. 17—26, Pergamon Press, 1961.

[41] Monin, A. S., Yaglom, A. M., *Hydrodynamic instability and the appearance of turbulence*, Prikl. Meh. Teh. Fiz., **5**, 3—38 (1962). (In Russian)

[42] Monin, A. S., Yaglom, A. M., *Statistical hydromechanics*, Moscow, 1965. (In Russian)

[43] Nield, D. A., *Geophysical and astrophysical applications of thermal instability theory*, "New Zealand Science Review", **23**, nr. 6, 86 (1965).

[44] Orr, W. Mc F., *The stability or instability of the steady motion of a liquid*, Part II, *A viscous liquid*, "Proc. Roy. Irish Acad.", **A 27**, 69—138 (1907).

[45] Pai, S. I., *On a generalization of Synge's criteria for suficient stability of plane parallel flows*, "Quart. Appl. Math.", **12**, 203—206 (1954).

[46] Payne, L. E., Weinberger, H. F., *An exact stability bound for Navier-Stokes flow in a sphere. Nonlinear Problems*, ed. R. E. Langer, University of Wisconsin Press, Madison, 1963.

[47] Potter, M. C., *Stability of plane Couette-Poiseuille flow*, "J. Fluid Mech.", **24**, nr. 3, 609—619 (1966).

[48] Lord Rayleigh, *On the instability of jets*, *"Scientific Papers"*, Cambridge Univ. Press, **1**, 361—371 (1878).

[49] Lord Rayleigh, *On the stability or instability of certain fluid motions*, "Scientific Papers", Cambridge Univ. Press, **1**, 474—487 (1880).

[50] Lord Rayleigh, *On the motion of a viscous fluid*, "Phil. Mag.", (**6**), **26**, 776—786 (1913), "Papers", **6**, 187—197 (1913).

[51] Lord Rayleigh, *On the dynamics of revolving fluids*, "Scientific Papers", **6**, 447—453 (1916).

[52] Lord Rayleigh, *Theory of sound*, Dover Publications Inc., New York, 1945.

[53] Reid, W. H., *The stability of parallel flows*, in *Basic developments in fluid dynamics*, **1**, ed. M. Holt, Academic Press, New York, 1965, pp. 249—307.

[54] Reid, W. H., *Asymptotic approximations in hydrodynamic stability*, in *Non Equilibrium Thermodynamics, Variational Techniques and Stability*, University of Chicago Press, Chicago, 1966.

[55] Reynolds, O., *An experimental investigation of the circumstances which determine whether the motion of water shall be direct or sinuous and the law of resistance in parallel channels*, "Phil. Trans. Roy. Soc. London", **174**, 935—982 (1883).

[56] Reynolds, O., *On the dynamical theory of incompressible viscous fluids and the determination of the criterion*, "Phil. Trans. Roy. Soc. London (A)", **186**, 123—164 (1895), "Scientific Papers", **2**, 535—577 (1895).

[57] Rotta, J. C., *Experimenteller Beitrag zur Entstehung turbulenter Strömung im Rohr*, "Ing. Arch.", **24**, nr. 4, 258—281 (1956).

[58] Schiller, L., *Das Turbulenzproblem und verwandte Fragen*, "Physikalische Zeitschrift", **26**, 566—595 (1925).

[59] Schlichting, H., *Boundary-layer theory*, IVth ed. McGraw-Hill, New York, 1960.

[60] Schubauer, G. B., Skramstad, H. K., *Laminar boundary-larger oscillations and transition on a flat plate*, "J. Aero. Sci.", **14**, nr. 2, 69—78 (1947).

[61] Segel, L., *Nonlinear hydrodynamic stability theory and its applications to themal convection and curved flow*, in *Non-equilibrium, Thermodynamics, Variational Techniques and Stability*, University of Chicago Press, Chicago, 1966.

[62] Serrin J., *On the stability of viscous fluid motions*, "Arch. Rational Mech. Anal.", **3**, nr. 1, 1—13 (1959).

[63] Serrin, J., *Mathematical principles of classical fluid mechanics*, Springer-Verlag Berlin (West) — Göttingen — Heidelberg, 1959.

[64] Sexl, T., *Über dreidimensionale Störungen der Poiseuilleschen Strömung*, "Ann. Phys.", Lpz., **84**, nr. 4, 807—822 (1927).

[65] Shen, S. F., *Calculated amplified oscillations in plane Poiseuille flow and Blasius flows*, "J. Aero. Sci.", **21**, nr. 1, 62–64 (1954).
[66] Shen, S. F., *Stability of laminar flows*, in *Theory of laminar flows*, High Speed Aerodynamics and Jet Propulsion, Princeton, 1964.
[67] Sommerfeld, A., *Ein Beitrag zur hydrodynamischen Erklärung turbulenter Flussigkeitsbewegung*, "Proc. 4th Int. Congr. Math. Rome", **2**, 116–124 (1908).
[68] Solberg, H., *Zum Turbulenzproblem*, in *Proc. First Int. Congress Appl. Mech.*, Delft, 1924, pp. 387–394.
[69] Squire, H. B., *On the stability of the three-dimensional disturbances of viscous flow between parallel walls*, "Proc. Roy. Soc.", **A 142**, 621–628 (1933).
[70] Stuart, J. T., *On the non-linear mechanics of hydrodynamic stability*, "J. Fluid Mech.", **4**, nr. 1, 1–21 (1958).
[71] Stuart, J. T., *On the non-linear mechanics of disturbances in stable and unstable parallel flows* (part I, *The basic behaviour in plane Poiseuille flow*), "J. Fluid Mech.", **9**, 353 (1960).
[72] Stuart, J. T., *Hydrodynamic stability*, in *Laminar boundary layers*, ed. L. Rosenhead, Oxford, London, 1963, pp. 629–670.
[73] Stuart, J. T., *Nonlinear stability theory*, "Annual Review of Fluid Mechanics", **3**, (1971).
[74] Synge, J. L., *Hydrodynamic stability*, "Semicentennial Publications of the American Mathematical Society", **2**, 227–269 (1938).
[75] Synge, J. L., *On the stability of a viscous liquid between rotating coaxial cylinders*, "Proc. Roy. Soc.", **A 167**, 250–256 (1938).
[76] Taylor, G. I., *Stability of a viscous fluid contained between two rotating cylinders*, "Phil, Trans.", **A 223**, 289–343 (1923).
[77] Thomas, T. Y., *On the uniform convergence of the solutions of the Navier-Stokes equation*, "Proc. Nat. Acad. Sci. U.S.A.", **29**, 243–246 (1943).
[78] Thomas, L. H., *The stability of plane Poiseuille flow*, "Phys. Rev.", **91**, nr. 2, 780–783 (1953).
[79] Tietjens, O., *Beiträge zur Entstehung der Turbulenz*, "Nachr. Ges. Wiss. Göttingen, Math.-phys. Klasse", 21–44 (1922).
[80] Tollmien, W., *Über die Entstehung der Turbulenz*, "Nachr. Ges. Wiss. Göttingen, Math.-phys. Klasse", **1**, 21–44 (1929).
[81] Tollmien, W., *Ein allgemeines Kriterium der Instabilität laminarer Geschwindigkeitsverteilungen*, "Nachr. Ges. Wiss. Göttingen, Math.-phys. Klasse", **50**, 79–114 (1935).
[82] Wasow, W., *Asymptotic expansions of solutions of ordinary differential equations*, Mir, Moscow, 1968. (In Russian).
[83] Velte, V., *Über ein Stabilitätskriterium der Hydrodynamik*, "Arch. Rational Mech. Anal.", **9**, 9–20 (1962).
[84] Watson, J., *Three-dimensional disturbances in flow between parallel planes*, "Proc. Roy. Soc.", **A 254**, nr. 1279, 562–569 (1960).
[85] Watson, J., *On the nonlinear mechanics of wave disturbances in stable and unstable parallel flows*, (Part II, *The development of a solution for plane Poiseuille flow and plane Couette flow*, "J. Fluid Mech.", **9**, 371–389 (1960).
[86] Yih, C. S., *Dynamics of nonhomogeneous fluids*, Macmillan Series in Advanced Mathematics and Theoretical Physics, The Macmillan Company, New York, Collier-Macmillan Ltd. London, 1965.

Chapter 2

GENERALIZED SOLUTIONS IN HYDRODYNAMIC STABILITY

This chapter surveys the results obtained in the generalized frame of hydrodynamic stability theory. After an introductory paragraph (2.1), which deals with the mathematical facts required in the sequel, the generalized solutions which appear in the study of hydrodynamic stability are discussed in § 2.2. Then two problems related to linear stability, which justify from the mathematical point of view the investigations of classical linear stability theory, are discussed: the completeness of the normal modes (§ 2.3) and the linearization principle (§ 2.4). In § 2.5 the principle of exchange of stabilities connecting hydrodynamic stability with the branching of solutions of the Navier-Stokes equations is analysed, and § 2.6 is devoted to nonlinear stability theory, namely, to universal criteria of hydrodynamic stability.

§ 2.1. FUNCTION SPACES

The spaces considered in this chapter (apart from section 2.2.4, in which we have to deal also with complex normed spaces) shall be real normed spaces. Assume that two normed spaces X and Y are given. Then a linear map $f: X \to Y$ is called *bounded* if there exists M, $M \in R^+$ such that

$$\| f(u) \|_Y \leqslant M \| u \|_X, \quad (u \in X);$$

this property is equivalent to the continuity of f in the topologies defined by norms. If, further, f is also injective, we say that it is an *embedding* of X into Y; the embeddings which preserve the norm are called *isometries.*

The set of bounded linear maps from X to Y determines a normed space denoted by $(X \to Y)$. In particular, the space $(X \to R)$ denoted by X^* is called the *conjugate space* of X and its elements are referred to as bounded linear functionals on X.

Apart from the topology induced by the norm (called the strong topology), a weak topology is also defined on X according to which $u_n \xrightarrow{\text{weakly}} u$ iff $f(u_n) \to f(u)$ for every $f \in X^*$.

We recall that a Banach space is a complete normed space, that is a space Y where every fundamental sequence has a limit in Y. A linear subspace Y' of a complete space Y is complete iff it is closed. The closure $\overline{Y}'$ of Y' into Y is a linear subspace of Y, hence this closure is the smallest complete space which contains Y'.

Given an isometry $f: X \to Y$ of a normed space X into a Banach space Y, we shall identify X with its image $Y' = f(X)$. By this convention, we shall say that the *completion* of X (denoted by $\overline{X}^{\|\cdot\|}$) is the closure of $f(X)$ in Y. This definition is, up to an isomorphism, independent of the given isometry.

Two norms $\|\cdot\|_1$ and $\|\cdot\|_2$ on the same linear space X being given, we say that $\|\cdot\|_1$ is stronger than $\|\cdot\|_2$ (and write $\|\cdot\|_1 \geqslant \|\cdot\|_2$) if there exists $M \in R^+$ such that $\|u\|_2 \leqslant M\|u\|_1$, $(u \in X)$. Then every sequence converging in the norm $\|\cdot\|_1$ converges in the norm $\|\cdot\|_2$ too and $\overline{X}^{\|\cdot\|_1}$ is embedded into $\overline{X}^{\|\cdot\|_2}$. In particular if $\|\cdot\|_1$ and $\|\cdot\|_2$ are equivalent (that is if $\|\cdot\|_1 \geqslant \|\cdot\|_2$ and $\|\cdot\|_2 \geqslant \|\cdot\|_1$), the spaces $\overline{X}^{\|\cdot\|_1}$ and $\overline{X}^{\|\cdot\|_2}$ may be identified as linear spaces, the corresponding norms remaining still equivalent.

2.1.1. Spaces of continuous functions

Let Ω be a domain of R^n and let k be an integer $(0 \leqslant k \leqslant \infty)$. By definition, $C^k(\Omega)$ is the linear space whose elements are real functions having continuous partial derivates up to the order k in Ω. $C_0(\Omega)$ is the linear space of continuous functions with compact support[1)] in Ω, and C_0^k is the space $C_0(\Omega) \cap C^k(\Omega)$.

On $C_0^k(\Omega)$ we can introduce the norm

$$\|u\|_p = \left[\int_\Omega |u(x)|^p \, \mathrm{d}x\right]^{\frac{1}{p}}, \quad (p \text{ real} \geqslant 1) \tag{2.1.1}$$

or, more general, the norm

$$\|u\|_{l,p} = \left[\int_\Omega \sum_{0 \leqslant |\alpha| \leqslant l} |\mathrm{D}^\alpha u|^p \, \mathrm{d}x\right]^{\frac{1}{p}}, \quad (l \text{ integer} \leqslant k), \tag{2.1.2}$$

where $\alpha = (\alpha_1, \ldots, \alpha_n)$ is a system of non-negative integers, $|\alpha| = \alpha_1 + \ldots \alpha_n$ and $\mathrm{D}^\alpha = \dfrac{\partial^{|\alpha|}}{\partial x_1^{\alpha_1} \ldots \partial x_n^{\alpha_n}}$. For $p = \infty$ we define

$$\|u\|_\infty = \sup_{x \in \Omega} |u(x)|$$

---

1) The support of a function $u : G \to R$ is the closure of the set $\{x \in G \mid u(x) \neq 0\}$ hence the functions with compact support are those functions vanishing on the complementary set of a compact set in G. If G is open, then u is nonvanishing only on a bounded subdomain of G having a strictly positive distance up to the boundary of G. If G is compact, then every function defined on G has compact support.

and, similarly,

$$\|u\|_{l,\infty} = \sup_{x\in\Omega}\Big(\sum_{0\leqslant|\alpha|\leqslant l} |D^\alpha u|\Big).$$

Now let $\hat{C}^{l,p}(\Omega)$ be the normed space whose elements are functions $u \in C^l(\Omega)$ for which $\|u\|_{l,p} < \infty$. Evidently, the space $C_0^l(\Omega)$ with the norm $\|\cdot\|_{l,p}$ is a subspace of $\hat{C}^{l,p}(\Omega)$; if Ω is bounded, a space lying between $C_0^l(\Omega)$ and $\hat{C}^{l,p}(\Omega)$ is the space $C^l(\bar{\Omega})$ of the functions $u \in C^l(\Omega)$ which may be extended by continuity, together with their derivatives up to order l, to $\bar{\Omega}$.

Remark 2.1.1. The norm (2.1.2) is equivalent to the norm

$$\|u\|'_{l,p} = \max_{0\leqslant|\alpha|\leqslant l} \|D^\alpha u\|_p.$$

Hence convergence in the norm $\|\cdot\|_{l,p}$ of a sequence $\{u_n\}$ of functions in $\hat{C}^{l,p}$ means convergence in the norm $\|\cdot\|_p$ of the sequence $\{u_n\}$ and of their derivatives up to the order l.

Among the normed spaces introduced in the above, the only ones which are complete are the spaces $C^l(\bar{\Omega})$ with the norm $\|\cdot\|_{l,\infty}$. In the following, we construct completions of the spaces $C_0^l(\Omega)$ and $\hat{C}^{l,p}(\Omega)$ in the norms $\|\cdot\|_{l,p}$ with $p < \infty$, starting with the case of the norm $\|\cdot\|_{o,p} = \|\cdot\|_p$.

2.1.2. The L^p spaces

By analogy with relations (2.1.1) and (2.1.1)′, to every measurable function $u : \Omega \to R$, we can associate the expressions

$$(2.1.3)\quad \|u\|_p = \begin{cases} \left[\int_\Omega |u(x)|^p \,dx\right]^{\frac{1}{p}}, \text{ if } 1 \leqslant p < \infty, \\ \operatorname{ad}\max_{x\in\omega} |u(x)| = \inf \ \{M \in R \mid |u(x)| \leqslant M \text{ almost} \\ \text{everywhere in } \Omega\}, \text{ if } p = \infty, \end{cases}$$

where $\|u\|_p$ is a non-negative real number, or $\|u\|_p = \infty$, and $\int_\Omega |u(x)|^p\,dx$ is the Lebesgue integral of the function $|u(x)|^p$.

For $p < \infty$ let $\mathfrak{L}^p(\Omega)$ be the linear space of functions which are p-summable on Ω (i.e. measurable functions whose p^{th} power is Lebesgue integrable on Ω), and for $p = \infty$ let $\mathfrak{L}^\infty(\Omega)$ be the space of the measurable functions which are essentially bounded on Ω, ($\operatorname{ad}\max_{x\in\omega} |u(x)| < \infty$). Denote by $\mathfrak{M}(\Omega)$ the space of all measurable functions on Ω. From the definitions, it follows that

$$\mathfrak{L}^p(\Omega) = \{u \in M(\Omega) \mid \|u\|_p < \infty\}, \ (1 \leqslant p \leqslant \infty).$$

Two numbers p and q are called conjugate if $\frac{1}{p}+\frac{1}{q}=1$ (by definition, the conjugate of 1 is ∞). For every pair of conjugate numbers $1 \leqslant p,\ q \leqslant \infty$ the functions u and v, measurable on Ω, satisfy Hölder's inequality

$$(2.1.4) \qquad \int_\Omega |u(x)v(x)|\, \mathrm{d}x \leqslant \| u \|_p \cdot \| v \|_q,$$

which in particular shows that if $u \in \mathfrak{L}^p(\Omega)$ and $v \in \mathfrak{L}^q(\Omega)$, then $uv \in \mathfrak{L}^1(\Omega)$. If $p_2 \geqslant p_1$, and Ω has Lebesgue measure $\int_\Omega \mathrm{d}x = c < \infty$, from the same inequality it is also implied that $\mathfrak{L}^{p_2}(\Omega) \subset \mathfrak{L}^{p_1}(\Omega)$, and for $u \in \mathfrak{L}^{p_2}(\Omega)$, the relation

$$(2.1.5) \qquad \| u \|_{p_1} \leqslant \| u \|_{p_2} \cdot c^{\frac{p_2-p_1}{p_1 p_2}}$$

holds.

Ω being arbitrary, it follows that every p_2-summable function is also p_1 locally summable (i.e. p_1-summable on every compact in Ω) when $p_1 \leqslant p_2$.

For $u \in \mathfrak{L}^p(\Omega)$, the relation $\| u \|_p = 0$ is satisfied iff $u(x) = 0$ almost everywhere on Ω, hence $\| \cdot \|_p$ does not define a norm on $\mathfrak{L}^p(\Omega)$. To obtain a normed space, the equivalence relation $u \sim v$ if $u(x) = v(x)$ almost everywhere on Ω is introduced in the algebra of functions measurable on Ω. Let M (Ω) be the algebra of equivalence classes with respect to this relation, and let $L^p(\Omega)$ $(1 \leqslant p \leqslant \infty)$ be the linear subspaces of $M(\Omega)$ corresponding to the subspace $\mathfrak{L}^p(\Omega)$ of $\mathfrak{M}(\Omega)$. Then, for every $u \in \mathfrak{L}^p(\Omega)$, the number $\| u \|_p$ depends only on the class of u in $L^p(\Omega)$, and the function $\| \cdot \| : L^p(\Omega) \to R$ defines a norm on the space $L^p(\Omega)$.

We obtain an isometry of the space $C_0^0(\Omega)$ normed by $\| \cdot \|_p$ $(1 \leqslant p < \infty)$ in the space $L^p(\Omega)$ defined as above, by composing the inclusion of $C_0^0(\Omega)$ in $\mathfrak{L}^p(\Omega)$ with the passage to the classes from $\mathfrak{L}^p(\Omega)$ to $L^p(\Omega)$. From the Lebesgue integral theory it is known that $\mathfrak{L}^p(\Omega)$ is a complete space and the subspace $C_0^\infty(\Omega)$ of $C_0^0(\Omega)$ is dense in $L^p(\Omega)$; this last space is the completion of all spaces $C_0^l(\Omega)$ in the norm $\| \cdot \|_p$

$$L^p(\Omega) = \overline{C_0^l(\Omega)}^{\| \cdot \| p}, \quad (1 \leqslant p < \infty,\ 0 \leqslant l \leqslant \infty).$$

This is the reason why elements of $L^p(\Omega)$ are still referred to as functions and the symbols used to denote them are the letters u, v, ... employed also for the elements of $C_0^0(\Omega)$.

For $u \in L^p(\Omega)$, the expression $\int_\Omega [u(x)]^p \, dx$ makes sense in an evident manner and it satisfies the inequality

$$\left| \int_\Omega [u(x)]^p \, dx \right| \leqslant \int_\Omega |u(x)|^p \, dx = (\|u\|_p)^p.$$

In particular for $p = 1$, the map $\psi : L^1(\Omega) \to R$, $\psi(u) = \int_\Omega u(x) \, dx$ is a bounded linear functional on $L^1(\Omega)$.

As above, for $u \in L^p(\Omega)$ and $v \in L^q(\Omega)$, the product uv (performed in $M(\Omega)$) belongs to $L^1(\Omega)$ and satisfies the Hölder inequality

$$\text{(2.1.4)}' \qquad \|uv\|_1 \leqslant \|u\|_p \cdot \|v\|_q, \quad \frac{1}{p} + \frac{1}{q} = 1,$$

and when $p_1 \leqslant p_2$ we have $L^{p_2}(\Omega) \subset L^{p_1}_{\text{loc}}(\Omega)$ (the space of equivalence classes of measurable functions which are locally p_1-summable on every compact in Ω functions); if Ω is bounded, then inequality (2.1.5) still holds.

The map $\varphi : L^p(\Omega) \times L^q(\Omega) \to L^1(\Omega)$, $\varphi(u, v) = u \cdot v$, is clearly bilinear, hence for every fixed $v \in L^p(\Omega)$, it defines a linear map $\varphi_v : L^p(\Omega) \to L^1(\Omega)$, $\varphi_v(u) = uv$, which is bounded too as it follows from (2.1.4)′. Composing φ_v with the functional $\psi : L^1(\Omega) \to R$, we get a functional $\Phi_v \in (L^p(\Omega))^*$, such that

$$\text{(2.1.6)} \qquad \Phi_v(u) = \int_\Omega u(x) v(x) \, dx, \quad (u \in L^p(\Omega)).$$

Since the strong convergence implies the weak one, if $u_n \xrightarrow{L^p} u$, then, for every $v \in L^q(\Omega)$, we have $\Phi_v(u_n) = \int_\Omega u_n(x) \, v(x) \, dx \to \int_\Omega u(x) \, v(x) \, dx = \Phi_v(u)$. In other words the passage to the limit under the integral sign is allowed.

The map $\Phi : L^q(\Omega) \to (L^q(\Omega))^*$, $\Phi(v) = \Phi_v$, $(v \in L^q(\Omega))$, is linear and, for $1 \leqslant p < \infty$, it can be proved that it is bijective (Riesz-Fréchet theorem). Hence (2.1.6) represents the general form of the linear continuous functionals on $L^p(\Omega)$.

It follows that for $1 < p < \infty$, the space $L^p(\Omega)$ is reflexive. The generalized solutions of the hydrodynamic problem are constructed by means of the space $L^2(\Omega)$, which is a Hilbert space with the norm defined by the scalar product

$$(u, v) = \int_\Omega u(x) \cdot v(x) \, dx,$$

correponding to the right-hand side of relation (2.1.6), for $p = q = 2$[2]. In this case inequality (2.1.4)′ coincides with the Schwarz inequality

$$(2.1.4)'' \qquad |(u, v)| \leqslant \|u\|_2 \cdot \|v\|_2.$$

An important property of the L^p spaces is that they are *separable Banach spaces,* namely they contain a countable everywhere dense set; this follows easily from the Stone-Weierstrass theorem of approximation by polynomials of continuous functions defined on a compact set.

Every element $u \in L^p(\Omega)$ may be approximated by a family $u_\varepsilon(\varepsilon > 0)$ of continous functions, called mean functions of u, defined by the relation

$$(2.1.7) \qquad u_\varepsilon(x) = \int_\Omega \omega_\varepsilon(x - y)\, u(y)\, \mathrm{d}y = \omega_\varepsilon * u,$$

where ω_ε are non-negative indefinitely differentiable functions defined on R^n, with $\omega_\varepsilon(x) = 0$ for $|x| \geqslant \varepsilon$, (hence $\omega_\varepsilon \in C_0^\infty(R^n)$) and $\int_{R^n} \omega_\varepsilon(z)\, \mathrm{d}z = c\varepsilon^n$, where c is a constant[3].

The mean functions have the following properties:

1° $u_\varepsilon(x) \in C^\infty(R^n)$; its partial derivatives are obtained differentiating the function $\omega_\varepsilon(x - y)$ with respect to the variable x under the integral sign in (2.1.7).

2° for $1 \leqslant p < \infty$, we have $\|u_\varepsilon\|_p \leqslant \|u\|_p$, (hence $u_\varepsilon \in \hat{C}^{0,p}(\Omega)$), and as $\varepsilon \to 0$, the functions $u_\varepsilon(x)$ converge towards $u(x)$ in the norm $\|\cdot\|_p$

$$\int_\Omega |u_\varepsilon(x) - u(x)|^p\, \mathrm{d}x \to 0 \text{ as } \varepsilon \to 0.$$

2.1.3. Generalized derivatives

For every $u \in C^{|\alpha|}(\Omega)$, $\varphi \in C_0^{|\alpha|}(\Omega)$ and every system of non-negative integers $\alpha = (\alpha_1, \ldots, \alpha_n)$ with $|\alpha| \leqslant l$, we have the formula

$$\int_\Omega u(x)\, (\mathrm{D}^\alpha \varphi)\, (x)\, \mathrm{d}x = (-1)^{|\alpha|} \int_\Omega (\mathrm{D}^\alpha u)\, (x)\, \varphi(x)\, \mathrm{d}x.$$

---

[2] When complex valued functions and accordingly complex Banach spaces, are considered, the scalar product in $L^2(\Omega)$ is defined by the equality $(u, v) = \int_\Omega u(x)\, \bar{v}(x)\, \mathrm{d}x$.

[3] We can take $\omega_\varepsilon = \begin{cases} e^{\frac{|x|^2}{|x|^2 - \varepsilon^2}} & \text{for } |x| < \varepsilon \\ 0 & \text{for } |x| \geqslant \varepsilon \end{cases}$; c is then $\int_{|x| \leqslant 1} e^{\frac{|x|^2}{|x|^2 - 1}}\, \mathrm{d}x$.

This suggests the following definition for the generalized derivative of a function u, locally integrable on Ω.

Definition 2.1.1. *Given two functions* u, v $\in L^1_{loc}(\Omega)$, *we say that* v *is the generalized derivative of type* α *of* u, *if for every* $\varphi \in C_0^{|\alpha|}(\Omega)$, *the following relation holds*

$$\int_\Omega u D^\alpha \varphi \, dx = (-1)^{|\alpha|} \int_\Omega v\varphi \, dx. \tag{2.1.8}$$

From this definition it follows that if a function of $L^1_{loc}(\Omega)$ admits a generalized derivative, this is unique in $L^1_{loc}(\Omega)$. In particular it makes sense to speak of the generalized derivative of a p-summable or locally p-summable function on Ω.

From Definition 2.1.1, the following properties of the generalized derivatives are immediate:

1° If u_1 and u_2 have the derivatives $D^\alpha u_1$ and $D^\alpha u_2$, then, for every scalars a_1 and a_2, $a_1u_1 + a_2u_2$ has the derivative

$$D^\alpha(a_1u_1 + a_2u_2) = a_1D^\alpha u_1 + a_2D^\alpha u_2.$$

If u_1 and $Du_1 \in L^p(\Omega)$, u_2 and $Du_2 \in L^q(\Omega)$ and p and q are conjugate integers, then u_1u_2 has the generalized derivative $D(u_1u_2) = u_1Du_2 + u_2Du_1$. When $\alpha = (\alpha_1,\dots,\alpha_n)$ is arbitrary, a formula analogous to that of Leibnitz from the classical case can be obtained for $D^\alpha(u_1u_2)$.

2° If u has the derivative $D^\alpha u$, and $D^\alpha u$ has the derivative $D^\beta(D^\alpha u)$, then u has the derivative $D^{\beta+\alpha}u = D^\beta(D^\alpha u)$[4].

3° If $D^\alpha u$ is the derivative of u in Ω, then it is the derivative of u in every subdomain of Ω. The converse of this proposition also holds. (Therefore the generalized derivative has the same local character as the classical derivative.)

4° If $\{u_n\}$ is a sequence of elements of $L^p(\Omega)$, having generalized derivatives $D^\alpha u_n = v_n \in L^p(\Omega)$ and if $u_n \xrightarrow{L^p} u$, $v_n \xrightarrow{L^p} v$, then the generalized derivative $D^\alpha u$ is equal to v. (This follows passing to the limit under the integrals in the equality $\int_\Omega u_n D^\alpha\varphi \, dx = (-1)^{|\alpha|}\int_\Omega v_n\varphi \, dx$.)

5° The mean function v_ε of the generalized derivative $v = D^\alpha u$ coincides with the derivative $D^\alpha u_\varepsilon$ of the mean function u_ε, at all points in Ω located at a distance greater then ε from the boundary of Ω. (This may be proved using property 1° of mean functions).

6° A sufficient condition for the generalized derivatives of type α of a function u to exist is that the set of derivatives of type α of the corresponding mean functions be bounded.

---

[4] Generally the existence of the generalized derivative $D^\alpha u$ does not imply the existence of the derivatives $D^\beta u$ with $|\beta| < |\alpha|$.

Proposition 2.1.1. *In order that the function* $v \in L^p_{loc}(\Omega)$ *be the generalized derivative of type* α *of the function* $u \in L^p_{loc}(\Omega)$, *it is necessary and sufficient that, for every compact* $K \subset \Omega$ *a sequence* $u_n \in C^{|\alpha|}(\Omega)$ *exists, such that* $\int_K |u_n - u|^p\, dx \to 0$ *and* $\int_K |D^\alpha u_n - v|^p\, dx \to 0$ (*i.e.* $u_n|_K \xrightarrow{L^p(K)} u$, $D^\alpha u_n|_K \xrightarrow{L^p(K)} v$)[5]).

Proof. Using the local character of the generalized derivative, the sufficiency of the condition follows from the above property 4°. To prove the necessity, let $\delta > 0$ be the distance from K to the boundary of Ω, $\{\varepsilon_n\}$ a sequence such that $\varepsilon_n \to 0$ and $\varepsilon_n < \delta$, and u_n and v_n respectively be the mean functions u_ε and v_ε of u and respectively v for $\varepsilon = \varepsilon_n$. From the property 5° it follows that $D^\alpha u_n(x) = v_n(x)$ for $x \in K$. Hence, $D^\alpha u_n|_K \xrightarrow{L^p(K)} v|_K$ as $n \to \infty$ (property 2° of the mean functions). Therefore the sequence $\{u_n\}$ satisfies the condition of the theorem.

Remark 2.1.2. If the domain Ω is bounded, it can be shown [52] that the sequence $\{u_n\}$ may be chosen in such a way that $D^\alpha u_n$ converges to v in $L^p(\Omega)$; assuming further that Ω is starlike, the functions u_n may be chosen from $C^p(\overline{\Omega})$ [87].

2.1.4. Sobolev spaces

For any non-negative integer l and any real $p \geqslant 1$, let $W^{l,p}(\Omega) = W^{l,p}$ be the linear space of the functions $u \in L^p = L^p(\Omega)$ admitting generalized derivatives $D^\alpha u \in L^p (0 \leqslant |\alpha| \leqslant l)$; clearly, $W^{0,p} = L^p$[6]).

On $W^{l,p}$ we may consider the norm $\|\cdot\|_{l,p}$, introduced in Section 2.1.1 for $\hat{C}^{l,p}$, $W^{l,p}$ is complete with respect to this norm. This follows from property 4° of generalized derivatives using the completeness of L^p in the norm $\|\cdot\|_p$ and Remark 2.1.1.

As in the case of the space $C^0_0(\Omega)$, we can isometrically embed the linear space $\hat{C}^{l,p} = \hat{C}^{l,p}(\Omega)$ into L^p, composing the inclusion of $\hat{C}^{l,p}$ into $\mathfrak{L}^p$ with the passage to the classes $\mathfrak{L}^p \to L^p$. Clearly, $\hat{C}^{l,p}$ so embedded into L^p is contained in $W^{l,p}$.

Consider

$$H^{l,p} = \overline{\hat{C}^{l,p}}^{\|\cdot\|_{l,p}}.$$

---

5) Sometimes the derivative in Definition 2.1.1 is called the weak derivative or the derivative in the sense of distributions, and v-having the properties of Proposition 2.1.1 is called the strong derivative.

6) For $l > 1$, $W^{l,p}(\Omega)$ is a proper subspace of $L^p(\Omega)$ and it is dense in this last one (since it contains the subspace $C^l_0(\Omega)$ possessing this property); consequently, it is not closed in $L^p(\Omega)$, and hence is not complete in the norm $\|\cdot\|_p$.

By construction, $H^{l,p} \subset W^{l,p}$ and for bounded domains it follows, using the property quoted in Remark 2.1.2, that we even have the equality

$$H^{l,p} = W^{l,p}.$$

$\mathring{W}^{l,p}$ is defined as the Banach subspace of $H^{l,p}$ corresponding to the inclusion of C_0^∞ into $\hat{C}^{l,p}$

$$\mathring{W}^{l,p} = \overline{C_0^\infty}^{\|\cdot\|_{l,p}}.$$

Hence, $\mathring{W}^{l,p}$ is the completion in the norm $\|\cdot\|_{l,p}$ of each of the spaces $C_0^k (l \leqslant k \leqslant \infty)$, and the integration by parts formula (2.1.8) holds for every $u \in W^{l,p}(\Omega)$ and $\varphi \in \mathring{W}^{l,p}(\Omega)$.

For bounded domains Ω, we also introduce the subspace

$$\hat{H}^{l,p}(\Omega) = \overline{C^l(\overline{\Omega})}^{\|\cdot\|_{l,p}}$$

of $H^{l,p}$, which can be shown to coincide with $H^{l,p}$ if Ω is of class C^l [19], [65], or if Ω is starlike (Remark 2.1.2).

The Sobolev spaces defined above are separable Banach spaces; if $p = 2$, then they become Hilbert spaces.

Proposition 2.1.2. *Let* $u \in \hat{H}^{l,p}(\Omega)$, $(l \geqslant 1)$, $\{u_n\}$ *be a sequence of elements from* $C^l(\overline{\Omega})$ *whose limit is* u, *and let* S *be the intersection of* $\overline{\Omega}$ *with a linear manifold (or, more generally, a smooth surface from* R^n). *Then* [87] *the sequence* $\{u_n|_S\}$ *tends to an element* $v \in L^p(S)$ *which can be called the "restriction" of* u *to* S.

Remark 2.1.3. If Ω is of class C^l and $u \in \hat{H}^{l,p}(\Omega)$ $(l \geqslant 1)$, then the element $u|_{\partial\Omega}$ is in particular well-defined and $\mathring{W}^{l,p}(\Omega)$ coincides with the set of vectors from $\hat{H}^{l,p}(\Omega)$ which vanish on the boundary of Ω.

2.1.5. Embedding theorems

The embedding theorems of Sobolev give sufficient conditions for the space $W^{l,p}(\Omega)$ to be embedded into $C(\overline{\Omega})$, respectively into $L^{q^*}(\Omega)$, $(q^* > p)$. If $W^{l,p}(\Omega)$, $C(\overline{\Omega})$ and $L^{q^*}(\Omega)$ are thought of as subspaces of $M(\Omega)$ (the space of classes of measurable functions on Ω), then the corresponding embeddings become inclusions. We state these theorems for starlike bounded domains $\Omega \subset R^n$ (such that $W^{l,p}(\Omega) = \hat{H}^{l,p}(\Omega)$; hence we may apply Proposition 2.1.2). We mention that these theorems have been extended to more general domains [1], [68], [97].

Theorem 2.1.1. *If* $lp > n$, $p > 1$ *and* $u \in W^{l,p}(\Omega)$, *then* $u \in C(\overline{\Omega})$ *and we have*

(2.1.9)
$$\sup_{x \in \overline{\Omega}} |u(x)| \leqslant c \|u\|_{l,p},$$

where the constant c *depends only on* n, l, p *and* Ω.

Theorem 2.1.2. *Let* S *be the intersection of* Ω *with an* m-*dimensional linear manifold from* R^n. *If* $lp \leqslant n$, $p > 1$, $m > n - lp$ *and* $q^* < \frac{mp}{n - lp}$, *and* $u \in W^{l,p}(\Omega)$, *then* $u|_S \in L^{q^*}(S)$ *and we have*

$$\| u \|_{q^*, S} < c \| u \|_{l, p, \Omega}, \tag{2.1.10}$$

where the constant c *depends only on* m, n, l, p. q^*, Ω and S[7].

Inequalities (2.1.9) and (2.1.10) show the boundedness of the inclusion maps $W^{l,p}(\Omega) \to C(\overline{\Omega})$, respectively $W^{l,p}(\Omega) \to L^{q^*}(S)$.

Theorem 2.1.2 holds also for $q^* = mp\,(n - lp)^{-1}$ but in this case, the inclusion $W^{l,p}(\Omega) \to L^{q^*}(S)$ is no longer completely continuous (cf. Theorem 2.1.4). In particular, for $m = n$, the theorem 2.1.2 gives an embedding of $L^p(\Omega)$ into $L^{q^*}(\Omega)$.

These theorems have the following

Corollaries. *Let* $u \in W^{l,p}(\Omega)$ *and* $|\alpha| = j$.

1) *If* $lp > n$ *and* $0 \leqslant j < l - \frac{n}{p}$, *then every derivative* $D^\alpha u$ *is a function of* $C(\overline{\Omega})$ *and we have*

$$\sup_{x \in \overline{\Omega}} |D^\alpha u| < c \| u \|_{l, p}. \tag{2.1.11}$$

2) *If* $j \geqslant 0$ *and* $j \geqslant l - \frac{n}{p}$, *then every derivative* $D^\alpha u|_S$ *is a function of* $L_{q_j}(S)$, *where* $q_j < \frac{mp}{n - (l - j)\,p}$, $m > n - (l - j)\,p$ *and we have*

$$\| D^\alpha u \|_{q_j, S} \leqslant c \| u \|_{l, p, \Omega}. \tag{2.1.12}$$

From Theorem 2.1.1 and Corollary 1 it also follows the

Lemma *(Sobolev). If* $u \in W^{\left[\frac{n}{2}\right] + 1 + r, 2}(\Omega)$, *then* $u \in C^r(\overline{\Omega})$[8].

By means of this lemma, the regularity of the generalized solution of an rth order equation may be proved; if this solution belongs to $W^{l,2}(\Omega)$ with $l \geqslant r + 1 + \left[\frac{n}{2}\right]$, then it is also a classical one.

The corollaries imply the inclusion of $W^{l,p}(\Omega)$ into $W^{[t],q}(S)$ in the following form.

---

7) We have denoted $\| u|_S \|_{L^{q^*}(S)} = \| u \|_{q^*, S}$, $\| u \|_{W^{l,p}(\Omega)} = \| u \|_{l, p, \Omega}$.

8) Sobolev's lemma holds for Erling's domains [64]; [ ] represents the integral part.

Theorem 2.1.3. *If* $u \in W^{l,p}(\Omega)$ *and* $0 < t = l - \frac{n-m}{p} - \left(\frac{1}{p} - \frac{1}{q}\right) m$, $1 < p < q < \infty$, *then* $u|_S \in W^{[t],q}(S)$ *and we have*

$$\| u \|_{[t], q, S} \leqslant c \| u \|_{l, p, \Omega}, \tag{2.1.13}$$

where the constant c *depends on* m, n, l, p, q, t, S *and* Ω.

In the mathematical problem of hydrodynamics the dimension of Ω is $n = 2$ or 3; then, from $u \in W^{2,2}(\Omega)$, Theorem 2.1.1 shows that $u \in C(\bar{\Omega})$ and Inequality (2.1.9) holds, while if $u \in W^{1,2}(\Omega)$, we can no longer assert that u is continous. Note also that the embeddings of the Sobolev spaces depend on n and in some cases they may hold for $n = 2$ but not for $n = 3$. Consequently, more results are available in the two-dimensional case than in the three-dimensional one.

Using integral representations for the functions from $C_0^\infty(R^n)$, we may prove certain inequalities in which, taking into account that $C_0^\infty(\Omega) \subset C_0^\infty(R^n)$ and $C_0^\infty(\Omega)$ is dense in $\mathring{W}^{1,2}$, remain valid for functions from $\mathring{W}^{1,2}$. We give below some of these inequalities [52], which are often used in the study of Navier-Stokes equations.

If the domain Ω belongs to R^2 or R^3, then $\mathring{W}^{1,2}(\Omega) \subset L^4(\Omega)$ and we have the inequalities

$$\| u \|_4^4 \leqslant 2 \| u \|_2^2 \cdot \| Du \|_2^2 \text{ [9]}, \text{ for } \Omega \subset R^2, \tag{2.1.14}$$

respectively,

$$\| u \|_4^4 \leqslant \left(\frac{4}{3}\right)^{\frac{3}{2}} \| u \|_2 \cdot \| Du \|_2^3, \tag{2.1.15}$$

$$\| u \|_q \leqslant (48)^{\frac{\alpha}{6}} \| u \|_2^{1-\alpha} \cdot \| Du \|_2^\alpha, \text{ for } \Omega \subset R^3, \tag{2.1.16}$$

where $q \in [2, 6]$, and $\alpha = \frac{3}{2} - \frac{3}{q}$ (hence $\alpha \in [0, 1]$). For the unbounded domains $\Omega \subset R^n$ and $u \in \mathring{W}^{1,2}(\Omega)$, we have Leray's inequalities [53]

$$\int_\Omega \frac{u^2(x)}{|x-y|^2} \, dx \leqslant \left(\frac{2}{n-2}\right)^2 \int_\Omega |Du(x)|^2 \, dx, \; n \neq 2, \tag{2.1.17}$$

and for $n = 2$ and $\Omega = \{1 \leqslant x < \infty\}$,

$$\int_{|x| \geqslant 1} \frac{u^2(x)}{|x|^2 \ln^2 |x|} \, dx \leqslant 4 \int_{|x| \geqslant 1} |Du(x)|^2 \, dx. \tag{2.1.18}$$

---

[9] We have used the notation $\| Du \|_2 = \sqrt{\sum_{i=1}^{2} \left\| \frac{\partial u}{\partial x_i} \right\|^2}$.

If Ω is bounded, then the elements of the space $\mathring{W}^{1,2}$ satisfy an inequality of the form [9])

$$\| u \|_{1,2} \leqslant c \| \mathrm{D}u \|_2, \quad (u \in \mathring{W}^{1,2}); \tag{2.1.19}$$

and when, in addition, Ω is of class C^2 and $\mathring{W}^{1,2}$ is replaced by $\mathring{W}^{2,2}$, the inequality

$$\| u \|_{2,2} \leqslant c \| \mathrm{D}^2 u \|_2, \quad (u \in \mathring{W}^{2,2}) \tag{2.1.20}$$

holds.

Remark 2.1.4 From Inequalities (2.1.19) and (2.1.20) it follows that in $\mathring{W}^{1,2}$ the norm $\| u \|_{1,2} = \sqrt{\| u \|_2^2 + \| \mathrm{D}u \|_2^2}$ is equivalent to the norm $\| u \| = \| \mathrm{D}u \|_2$, defined by the scalar product

$$((u, v)) = (\mathrm{D}u, \mathrm{D}v)_2,$$

and in $\mathring{W}^{2,2}$ the norm $\| u \|_{2,2}$ is equivalent to the norm $\| \mathrm{D}^2 u \|_2$, defined by the scalar product

$$[u, v] = (\mathrm{D}^2 u, \mathrm{D}^2 v)_2.$$

2.1.6. Compactness in the L^p spaces

A set A of elements belonging to a separable topological space X is called *compact* (respectively *relatively compact*) if from every open covering of A (respectively $\bar{A}$) a finite covering can be extracted. If X is a normed space, then this condition is equivalent to the following: every sequence of elements from A contains a subsequence convergent to a limit from A (respectively from X). The set A is called *weakly compact* (respectively *relatively weakly compact*) if every sequence of elements from A contains a subsequence weakly converging to an element from A (respectively from X).

An important class of linear continuous maps $f : X \to Y$ (where X and Y are two Banach spaces) is that of *completely continuous* maps, characterized by the property that they transform bounded sets in relatively compact sets. If f is a completely continuous map and $\{x_n\}$ is a sequence weakely convergent to an element x, then [10] the sequence $\{f(x_n)\}$ strongly converges towards $f(x)$ (Appendix 1).

Theorem 2.1.4 *(Kondrashev). The embedding maps of the spaces* $\mathrm{W}^{1,p}$ *into* C *and* L^{q^*} *(Theorems* 2.1.1 *and* 2.1.2*) are completely continuous.*

Corollary. *If* $\Omega \subset \mathrm{R}^n$, *then every bounded set of elements from* $\mathrm{W}^{2,2}(\Omega)$ *(respectively* $\mathrm{W}^{1,2}(\Omega)$*) is relatively compact in* $\mathrm{C}(\bar{\Omega})$ *(respectively in* $\mathrm{L}^{q^*}(\Omega)$*), when* $q^* < \infty$ *(for* $n = 2$*) or when* $q^* < 4$ *(for* $n = 3$*).*

Theorem 2.1.5 *(Kolmogorov). Let* Ω *be a bounded domain from* R^n. *A set* $A \subset L^p(\Omega)$ $(1 \leqslant p < \infty)$ *is relatively compact iff the following conditions are satisfied:*

1° A *is bounded;*

2° A *is equicontinuous of order* p *in the mean, i.e. for every* $\varepsilon > 0$, *there exists* $\eta(\varepsilon) > 0$ *such that for every* $u \in A$ *we have* $\int_\Omega |u(x+y) - u(x)|^p \, dx < \varepsilon$ *for* $|y| < \eta(\varepsilon)$.

Theorem 2.1.6. *A set* $A \subset L^p(\Omega)$, $(1 < p < \infty)$, *is relatively weakly compact iff it is bounded* (this follows from the Eberlein-Smuljan theorem [104] which asserts that a Banach space X is reflexive iff in X the weak compactness is equivalent with boundedness).

2.1.7. Spaces of vector functions

If X is a normed linear space, we denote by $C^l(\Omega, X)$ the linear space whose elements are functions $f : \Omega \to X$ possessing continuous partial derivatives up to order l and by $C_0^l(\Omega, X)$ — the space of the functions from $C^l(\Omega, X)$ with compact support in Ω. For $p \geqslant 1$, the expression

$$(2.1.1)' \qquad \| u \|_p = \left[\int_\Omega \| u(x) \|_X^p \, dx \right]^{\frac{1}{p}}$$

defines a norm on the space $C_0^0(\Omega, X)$. By this notion

$$(2.1.2)' \qquad \| u \|_{l,p} = \left[\sum_{|\alpha| \leqslant l} (\| D^\alpha u \|_p)^p \right]^{\frac{1}{p}}$$

is a norm on $C_0^l(\Omega, X)$.

By definition, $\mathfrak{L}^p(\Omega, X)$ is the linear space of functions $f : \Omega \to X$ such that f is p-summable in the norm on Ω

$$\mathfrak{L}^p(\Omega, X) = \{ u : \Omega \to X \mid \| u(\cdot) \|_X \in L^p(\Omega) \},$$

and the space of the equivalence classes of functions from $\mathfrak{L}^p(\Omega, X)$ (with respect to the relation of almost everywhere coincidence in Ω), denoted by $L^p(\Omega, X)$, is the completion of $C_0^l(\Omega, X)$ in the norm $\| \cdot \|_p$.

Replacing u and v in (2.1.4)′ by $\| u(\cdot) \|_X$ and $\| v(\cdot) \|_X$ respectively, we obtain Hölder's inequality in the form

$$(2.1.4)''' \qquad \int_\Omega \| u(x) \|_X \| v(x) \|_X \, dx \leqslant \| u \|_p \| v \|_q.$$

If X is a normate algebra such that the product function uv is well-defined and we have $\| u(x) \, v(x) \|_X \leqslant \| u(x) \|_X \| v(x) \|_X$ then Hölder's inequality shows that $uv \in L^1(\Omega, X)$ when $u \in L^p(\Omega, X)$ and $v \in L^q(\Omega, X)$. The inequality

$$\| uv \|_1 \leqslant \| u \|_p \| v \|_q \tag{2.1.4)'}$$

is still satisfied.

In the study of hydrodynamic stability, the following two particular cases occur:

1° $\Omega = (0, T) \subset R^1$, $X = L^2(\Omega', R)$ [10] or X is a subspace of $L^2(\Omega', R)$, with $\Omega' \subset R^n$;

2° $\Omega \subset R^n$, $X = R^m$ (with the Euclidean norm). In this case, a function defined on Ω with values in X is of the form $\mathbf{u} = (u_1, \ldots, u_m)$. Putting $L^p(\Omega, R) = L^p(\Omega)$, the condition $\mathbf{u} \in L^p(\Omega, R^m)$ is equivalent with $u_i \in L^p(\Omega)$ for $1 \leqslant i \leqslant m$. Whence $L^p(\Omega, R^m) = [L^p(\Omega)]^m$ and

$$\| \mathbf{u} \|_p = \Big(\sum_i \| u_i |_p^p \Big)^{\frac{1}{p}}.$$

By definition, $\mathbf{v} = (v_1, \ldots, v_m)$ is the generalized derivative D^α of $\mathbf{u}$ if for $1 \leqslant i \leqslant m$, we have $D^\alpha u_i = v_i$; $W^{l,p}(\Omega, R^m) = [W^{l,p}(\Omega)]^m$ is then the subspace of $L^p(\Omega, R^m)$ consisting of the vectors admitting generalized derivatives up to order l belonging to $L^p(\Omega, R^m)$. Analogously, we define $\mathring{W}^{l,p}(\Omega, R^m) = [\mathring{W}^{l,p}(\Omega)]^m$.

The embedding and compactness theorems for the scalar case can be easily extended to the finite-dimensional vectorial case.

When no possibility of confusion exists, we denote the spaces of vector functions $L^p(\Omega, R^m)$ and $W^{l,p}(\Omega, R^m)$ by the same symbols $L^p(\Omega)$ and $\mathring{W}^{l,p}(\Omega)$ used for the scalar case.

2.1.8. Solenoidal vectors

The space $\mathring{W}^{l,2}(\Omega)$ is a Hilbert space obtained by taking the clousure of $C_0^\infty(\Omega)$ in the norm $\| \cdot \|_{l,2}$ given by the scalar product

$$(\mathbf{u}, \mathbf{v})_{l,p} = \int_\Omega \sum_{|\alpha| \leqslant l} D^\alpha \mathbf{u} \cdot D^\alpha \mathbf{v} \, dx \left(\text{where } D^\alpha \mathbf{u} \cdot D^\alpha \mathbf{v} = \sum_{i=1}^m D^\alpha u_i \cdot D^\alpha v_i \right).$$

In what follows we assume that $n = m$ $\left(\text{in order to define the divergence of a vector } \mathbf{u} \text{ from } C^1(\Omega), \operatorname{div} \mathbf{u} = \sum_{i=1}^n \frac{\partial u_i}{\partial x_i}\right)$ and that Ω is a bounded domain of class C^2 from R^n.

---

[10] I.e., we shall consider spaces of the form $C((0, T), L^2(\Omega))$, $L^2((0, T), L^2(\Omega))$.

We recall that a vector $\mathbf{u} \in C^1(\Omega)$ is called *solenoidal* if $\operatorname{div} \mathbf{u} = 0$ and it is a *potential* if $\mathbf{u} = \operatorname{grad}\varphi = \left(\frac{\partial \varphi}{\partial x_1}, \ldots, \frac{\partial \varphi}{\partial x_n}\right)$, $(\varphi \in C^1(\Omega))$. Since the incompressibility of viscous fluids can be expressed by the solenoidality of velocities, the study of classical solutions of the Navier-Stokes equations for incompressible fluids requires the introduction of the spaces

$$\mathcal{C}^n_*(\Omega) = \{\mathbf{u} \in C^n(\overline{\Omega}) \mid \operatorname{div} \mathbf{u} = 0,\ \mathbf{u}|_{\partial\Omega} = 0\},$$

$$\mathfrak{N}(\Omega) = \{\mathbf{u} \in C_0^\infty(\Omega) \mid \operatorname{div} \mathbf{u} = 0\}$$

and, analogously, for generalized solutions, the subspaces $N(\Omega)$ of $L^2(\Omega)$, obtained by the completion of $\mathfrak{N}(\Omega)$,

$$N(\Omega) = \overline{\mathfrak{N}(\Omega)}^{\|\cdot\|_2}$$

whose elements are called solenoidal vectors in the generalized sense. The vectors in $L^2(\Omega)$, which are orthogonal to $N(\Omega)$, determine a closed subspace, denoted by $G(\Omega)$, such that

$$L^2(\Omega) = N(\Omega) \oplus G(\Omega).$$

Taking into account the formula

$$\int_\Omega (\operatorname{div} \mathbf{u})\, \varphi \,\mathrm{d}x = -\int_\Omega \mathbf{u} \cdot \operatorname{grad} \varphi \,\mathrm{d}x,$$

valid for every $\varphi \in L^1_{loc}(\Omega)$ and admitting first-order generalized derivatives, for every $\mathbf{u} \in C_0^1(\Omega)$, it can be seen that every potential vector in the generalized sense, that is every element $\mathbf{v} \in L^2(\Omega)$ of the form grad φ such that $\varphi \in L^1_{loc}(\Omega)$, belongs to $G(\Omega)$. The following theorem shows that the converse statement is also true.

Theorem 2.1.7 *(Weyl [102]). The orthogonal complement of the subspace* N(Ω) *in* L²(Ω) *is the subspace of the potential vectors (in the generalized sense).*

In this way, every vector from $L^2(\Omega)$ can be expressed in an unique way as a sum of a solenoidal vector and a potential vector, both of them in the generalized sense.

For every integer $l \geqslant 1$, we put

$$N^l = \overline{\mathfrak{N}}^{\|\cdot\|_{l,2}}.$$

From the definition,

$$N^l = N \cap \mathring{W}^{l,2} = N^1 \cap W^{l,2}. \tag{2.1.21}$$

In particular, for $l = 1$, Remark 2.1.4 shows that for bounded domains the norm of N^1 is equivalent to the norm given by the scalar product $((\mathbf{u}, \mathbf{v})) \overset{\text{def}}{=} (\text{grad } \mathbf{u}, \text{ grad } \mathbf{v})$, and the norm of N^2 is equivalent to that defined by the scalar product $[\mathbf{u}, \mathbf{v}] = (\mathrm{D}^2\mathbf{u}, \mathrm{D}^2\mathbf{v})$. In the following paragraphs we shall denote by $\| \mathbf{u} \|$ and $((\mathbf{u}, \mathbf{v}))$ the norm and scalar product in N^1 defined above, and by $|\mathbf{u}|$ and $(\mathbf{u}, \mathbf{v})$ the norm and scalar product from N (which are those from L^2).

2.1.9. **Functions of time**

The functions which appear in the nonstationary problems of hydrodynamics have different integrability and differentiability properties with respect to the arguments t and x. So it is necessary to introduce some special spaces in oder to stress these differences.

Let $\Omega_T = \Omega \times (0, T)$, where Ω is a domain in R^n. By definition,

$$L^{q,r}(\Omega_T), \ (q, r \geqslant 1),$$

is a Banach space consisting of the functions $u(x, t)$ whose norm

$$\| u \|_{q,r} = \left(\int_0^T \left(\int_\Omega |u|^2 \,\mathrm{d}x \right)^{r/q} \mathrm{d}t \right)^{1/r}$$

is finite. It is easy to see that this norm can be also written in the form

$$\| u \|_{q,r} = \left(\int_0^T \| u(\cdot, t) \|^r_{L_q(\Omega)} \,\mathrm{d}t \right)^{1/r} .$$

The set $W_p^{l_1, l_2}(\Omega_T)$, ($p \geqslant 1$; l_1 and l_2 are non-negative integers) is the Banach space consisting of functions $u(x, t)$ which belong to $L_p(\Omega_T)$ together with their derivatives with respect to x up to order l_1 and with respect to t up to the order l_2. The norm in this space is

$$\| u \|_p^{l_1, l_2} = \left(\int_{\Omega_T} \left(\sum_{|\alpha|=0}^{l_1} |\mathrm{D}_x^\alpha u|^p + \sum_{k=0}^{l_2} |\mathrm{D}_t^k u|^p \right) \mathrm{d}x \,\mathrm{d}t \right)^{1/p} .$$

Finally, let $C^{l_1, l_2}(\Omega \times I)$, where I may be one of the intervals $[0, T]$, $(0, T)$, $[0, T)$, $(0, T]$, be the linear space of functions u whose partial derivatives are continuous up to the order l_1 with respect to x and up to the order l_2 with respect to t and let also

$$C_*^{l_1, l_2}(\Omega) = \{ u \in C^{l_1, l_2} | \, u(\cdot, t) \in C_*^{l_1}(\Omega) \text{ for every } t \in I \}.$$

The spaces C^{l_1, l_2} and $C_*^{l_1, l_2}$ are useful for the definition of classical solutions of the mathematical problem of hydrodynamics.

§ 2.2. TYPES OF SOLUTIONS IN HYDRODYNAMIC STABILITY THEORY

2.2.1. Classical solutions

Let Ω be a domain of the space R^n ($n = 2$ or 3) and consider the (mathematical problem

$$\frac{\partial \mathbf{u}}{\partial t} + (\mathbf{u} \text{ grad})\, \mathbf{u} = \mathbf{f} - \frac{1}{\rho} \text{grad}\, p + \nu \Delta \mathbf{u}, \tag{1.1.1}$$

$$\text{div}\, \mathbf{u} = 0, \tag{1.1.2}$$

$$\mathbf{u}\,|_{t=0} = \mathbf{u}_0, \tag{1.1.3}$$

$$\mathbf{u}|_{\partial\Omega} = \mathbf{w}, \tag{1.1.4}$$

$(\mathbf{w} \in C^0(\partial\Omega), \int_{\partial\Omega} \mathbf{w} \cdot \mathbf{n}\, dx = 0)$, associated to the motion of a viscous incompressible fluid. In the sequel, we treat separately the case when the boundary condition is homogeneous,

$$\mathbf{u}|_{\partial\Omega} = 0. \tag{1.1.4)*}$$

Equations (1.1.1) and (1.1.2) are satisfied in $\Omega \times (0, T)$, Condition (1.1.3) is valid for $\mathbf{x} \in \Omega$, and the condition (1.1.4) for $t \in (0, T)$. The classical solution of the problem (1.1.1) — (1.1.4) is a system of $n + 1$ scalar functions $\mathbf{u}(\mathbf{x}, t)$ and $p(\mathbf{x}, t)$. The function $p(\mathbf{x}, t)$ is determined up to an additive constant. In solving the above problem, this function is eliminated by application of the curl map (see Chap. 1). Hence the only vectorial unknown remains the velocity $\mathbf{u}$. Consequently, a classical solution $\mathbf{u}(\mathbf{x}, t)$ of the problem (1.1.1)—(1.1.4) satisfies the following conditions

(2.2.1) $\mathbf{u}(\cdot, t) \in C^2(\Omega) \cap C^0(\overline{\Omega})$ for t assigned in $(0, T)$;

(2.2.2) $\mathbf{u}(\mathbf{x}, \cdot) \in C^1(0, T)$, for $\mathbf{x}$ assigned in Ω;

(2.2.3) $\mathbf{u}(\cdot, 0) \in C^0(\Omega)$, for $\mathbf{x} \in \Omega$;

(2.2.4) $\mathbf{u}(\cdot, t)|_{\partial\Omega} \in C^0(\partial\Omega)$, for $t \in (0, T)$.

In the case of problem (1.1.1)—(1.1.4)*, the solution $\mathbf{u}(\mathbf{x}, t)$ is an element of the linear space

$$C_*^{2,1}(\Omega \times (0, T)) \cap C^{0,0}(\Omega \times (0, T)) \text{ [11]}. \tag{2.2.5}$$

---

[11] Note that in this class solenoidality condition (1.1.2) and the boundary condition are satisfied. Hence the mixed problem (1.1.1) — (1.1.4) in the class (2.2.1) — (2.2.4) becomes an initial-value problem (1.1.1), (1.1.3) in the class (2.2.5).

The classical solution of the problem (1.1.1)—(1.1.4) corresponds to a mathematical model in which it is assumed from the beginning that the functions occurring in it are of class C^2. Hence these functions and their gradients are continuous and bounded. This assumption corresponds to the laminar flow of viscous fluids which takes place at small t and large ν (hence for small Reynolds numbers). Most treatises of hydrodynamics present the above problem as if it were the only mathematical model deriving from the laws of mechanics. But, even at the beginning of this century, Oseen [69] has shown that these laws lead to integro-differential equations, where the second-order derivative of velocity does not appear. If in these equations the above disscussed smoothness properties are valid, then we obtain the Navier-Stokes equations. Therefore, for large values of t and Re, the mathematical problem itself (in the general form given by Oseen) shows that its solutions are not necessarily classical, since they are not bounded in Ω. On the other hand, it was found experimentally [83] that there exist big jumps of the velocity and its gradient which may be assimilated with some discontinuities. Therefore, the class of solutions imposed by experiment does not correspond to smooth functions. That is why the definition of generalized solutions, introduced and studied by Leray [53]—[55], appear natural.

Leray's remark that the solution of the mathematical problem of hydrodynamics in the form given by Oseen is not necessarily classical has led to the relatively easy proof of the existence theorem of the generalized solution for every t.

2.2.2. Generalized solutions of the linear problem

Let $A : C_0^\infty(\Omega) \longrightarrow L^2(\Omega)$ be a linear differential operator. The *weak solution* of the problem

$$\text{(2.2.6)} \qquad \begin{cases} A\mathbf{u} = -\dfrac{1}{\rho}\,\mathrm{grad}\, p + \mathbf{f}, \\ \mathrm{div}\, \mathbf{u} = 0, \end{cases}$$

with the homogeneous boundary condition

$$\text{(2.2.6 a)} \qquad \mathbf{u}|_{\partial\Omega} = 0,$$

is an element $\mathbf{u} \in N^1(\Omega)$ which satisfies the equation

$$\text{(2.2.7)} \qquad (\mathbf{u}, A^+ \mathbf{v}) = (\mathbf{f}, \mathbf{v})$$

for every $\mathbf{v} \in \mathcal{N}(\Omega)$, where A^+ is the operator formally conjugate to A (Appendix 1) and $(\cdot,\cdot)$ is the scalar product in $L^2(\Omega)$.

Let l be the order of the differential operator A; the operators A and A^+ can be defined on $\mathring{W}^{l,2}$ and they satisfy the relation

$$(2.2.8) \qquad (A\mathbf{u}, \mathbf{v}) = (\mathbf{u}, A^+\mathbf{v}), \quad (\mathbf{u}, \mathbf{v} \in \mathring{W}^{l,2}).$$

A *strong solution* of problem (2.2.6)—(2.2.6 a) is a pair $(\mathbf{u}, p)$, with $\mathbf{u} \in N^l(\Omega)$ and $\operatorname{grad} p \in G(\Omega)$, which satisfies the equation

$$(2.2.6)_1 \qquad A\mathbf{u} = -\frac{1}{\rho}\operatorname{grad} p + \mathbf{f}.$$

If $\mathbf{u}$ is a weak solution which belongs to the space $N^l(\Omega)$, then from (2.2.7) and (2.2.8) it follows that

$$(A\mathbf{u}, \mathbf{v}) = (\mathbf{f}, \mathbf{v}), \quad (\mathbf{v} \in \mathfrak{N}).$$

Consequently, $A\mathbf{u} - \mathbf{f}$ is orthogonal to the subspace $\mathfrak{N}$ of L^2, hence also to its closure N. According to the definition of the orthogonal complement, there exists an element $\operatorname{grad} p \in G(\Omega)$, such that $A\mathbf{u} - \mathbf{f} = -\frac{1}{\rho}\operatorname{grad} p$, which shows that the pair $(\mathbf{u}, p)$ is a strong solution. Analogously it can be seen that if $(\mathbf{u}, p)$ is a strong solution (or in particular, a classical solution) of the problem (2.2.6)—(2.2.6 a), then $\mathbf{u}$ is also a weak solution of this problem.

Consider problem (2.2.6) with the nonhomogeneous boundary condition

$$(2.2.6\text{ b}) \qquad \mathbf{u}|_{\partial\Omega} = \mathbf{w},$$

where Ω is of class C^2 and $\mathbf{w} \in L^2(\partial\Omega)$ satisfies the condition $\int_{\partial\Omega} \mathbf{wn}\, d\sigma = 0$.

Suppose that $\mathbf{w}$ may be extended to the interior of Ω to an element $\mathbf{u}^* \in N(\Omega) \cap W^{1,2}(\Omega)$, in the case of the weak solution, and to $\mathbf{u}^* \in N(\Omega) \cap W^{l,2}(\Omega)$ in the case of a strong solution. Condition (2.2.6 b) then becomes

$$(2.2.6\text{ c}) \qquad \mathbf{u}|_{\partial\Omega} = \mathbf{u}^*|_{\partial\Omega}\text{ [12]}.$$

A weak solution of the problem (2.2.6), (2.2.6 b) is then an element $\mathbf{u} \in N \cap W^{1,2}$ which satisfies relation (2.2.7) and the condition $\mathbf{u} - \mathbf{u}^* \in N^1$ [13].

---

[12] A sufficient condition for the existence of the element $\mathbf{u}^* \in N(\Omega) \cap W^{l,2}(\Omega)$ such that $\mathbf{u}^*|_{\partial\Omega} = \mathbf{w}$, is [8] that Ω is of class C^s ($s = \max(l, 2)$) and $\mathbf{w}$ belong to the space $W^{l-1/2,2}(\partial\Omega)$.

[13] Since we want that on the boundary $\mathbf{u}$ and $\mathbf{u}^*$ take the same values, by Remark 2.1.3, we must have $\mathbf{u} - \mathbf{u}^* \in \mathring{W}^{1,2}$; and since $\mathbf{u} - \mathbf{u}^* \in N$, from (2.1.21) it follows that $\mathbf{u} - \mathbf{u}^* \in N^1$.

We may define in an analogous way the strong solutions of the problem (2.2.6)—(2.2.6 b) as pairs of the form $(\mathbf{u}, p)$ which satisfy Equation $(2.2.6)_1$, where $\mathbf{u} \in N \cap W^{l,2}$ is an element with the property that $\mathbf{u} - \mathbf{u}^* \in N^l$.

From the linearity of the operator A it follows that every solution of the problem (2.2.6)—(2.2.6 c) is of the form $\mathbf{v} + \mathbf{u}^*$, where $\mathbf{v}$ is a solution of the problem

$$\begin{cases} A\mathbf{v} = -\dfrac{1}{\rho} \operatorname{grad} p + (\mathbf{f} - A\mathbf{u}^*), \\ \operatorname{div} \mathbf{v} = 0, \\ \mathbf{v}|_{\partial\Omega} = \mathbf{0}. \end{cases}$$

In this way, the solution of the problem (2.2.6)—(2.2.6 b) is reduced to a solution of a problem with a homogeneous boundary condition.

For $A = \nu\Delta$ $(\nu > 0)$, (2.2.6)—(2.2.6 b) becomes the Stokes linear problem

$$\text{(2.2.8 a)} \quad \begin{cases} \nu\Delta\mathbf{u} = -\dfrac{1}{\rho} \operatorname{grad} p + \mathbf{f}, \\ \operatorname{div} \mathbf{u} = 0, \end{cases}$$

$$\text{(2.2.8 b)} \quad \mathbf{u}|_{\partial\Omega} = \mathbf{w}.$$

Since in this case, A is formally self-adjoint, Condition (2.2.7) is written

$$\text{(2.2.9)} \quad (\mathbf{u}, \nu\Delta\mathbf{v}) = (\mathbf{f}, \mathbf{v})$$

or

$$\text{(2.2.9)}' \quad -\nu((\mathbf{u}, \mathbf{v})) = (\mathbf{f}, \mathbf{v}), \quad (\mathbf{v} \in \mathfrak{N}(\Omega)),$$

where $((\mathbf{u}, \mathbf{v})) = (\operatorname{grad} \mathbf{u}, \operatorname{grad} \mathbf{v})$ is the scalar product defined in Section 2.1.8.

Theorem 2.2.1. *If the domain Ω is bounded, then the Stokes problem* (2.2.8 a)—(2.2.8 b) *admits just one weak solution in* $\mathrm{N}^1(\Omega)$.

Proof. Without loss of generality, we may assume that $\mathbf{f} \in N(\Omega)$, $\nu = \rho = 1$ and $\mathbf{w} = 0$. Then relation (2.2.9)′ becomes

$$\text{(2.2.9)}'' \quad -((\mathbf{u}, \mathbf{v})) = (\mathbf{f}, \mathbf{v}).$$

In the above hypothesis on Ω, the scalar product $((\cdot,\cdot))$ defines in $N^1(\Omega)$ a norm equivalent with the norm of N^1 (Remark 2.1.4). Chosing in $\mathfrak{N}$ a sequence $\{\mathbf{v}_n\}$ converging towards an arbitrary element $\mathbf{v}$ from $N^1(\Omega)$ and passing to the limit, we see that equality (2.2.9)′′ still holds for $\mathbf{v} \in N^1(\Omega)$. If $\mathbf{u}_1$ and $\mathbf{u}_2$ are two weak solutions, each of them satisfy this equality for $\mathbf{v} = \mathbf{u}_1 - \mathbf{u}_2$. Subtracting the two resulting relations, it follows that $((\mathbf{u}_1 - \mathbf{u}_2, \mathbf{u}_1 - \mathbf{u}_2)) = 0$,) which proves the uniqueness of the weak solution.

To prove existence, consider the linear functional $l_f(\mathbf{v}) = -(\mathbf{f}, \mathbf{v})$. From inequality (2.1.19), if follows that

$$|(\mathbf{f}, \mathbf{v})| \leqslant |\mathbf{f}|\,|\mathbf{v}| \leqslant c\,|\mathbf{f}|\,\|\mathbf{v}\|.$$

Hence l_f is bounded in N^1. Applying the Riesz representation theorem, the existence of an element $\mathbf{u}_f \in N^1(\Omega)$, such that

$$((\mathbf{u}_f, \mathbf{v})) = -(\mathbf{f}, \mathbf{v}), \; (\mathbf{v} \in N^1)$$

follows.
Hence, $\mathbf{u}_f$ is a weak solution of the problem (2.2.8 a)—(2.2.8 b).

Clearly, a linear map $B : N \to N$, such that $R(B) \subset N^1$ is obtained in this way. Then for every $\mathbf{f} \in N$ and $\mathbf{v} \in N^1$ we have the equality

$$((B\mathbf{f}, \mathbf{v})) = -(\mathbf{f}, \mathbf{v}). \tag{2.2.10}$$

B is invertible, since it is injective: if to an $\mathbf{f} \in N$ there corresponds the solution $\mathbf{u} = B(\mathbf{f}) = 0$, then from (2.2.9)″ it follows that $\mathbf{f}$ is orthogonal to the subspace $\mathfrak{N}$ whose closure coincides with N; therefore $\mathbf{f} = 0$. In particular, from (2.2.10) we get the equality

$$(B\mathbf{f}, \mathbf{g}) = -((B\mathbf{f}, B\mathbf{g})) = (\mathbf{f}, B\mathbf{g}), \tag{2.2.11}$$

when $\mathbf{f}, \mathbf{g} \in \mathfrak{N}$, and since $\mathfrak{N}$ is dense in N, $(B\mathbf{f}, \mathbf{g}) = (\mathbf{f}, B\mathbf{g})$ for every $\mathbf{f}, \mathbf{g} \in N$, B is a symmetric operator defined on the whole Hilbert space N. By the first von Neumann's theorem (Appendix 1) it follows that B is self-adjoint. Denoting by $\tilde{\Delta}$ the inverse of B, $\tilde{\Delta}$ is a selfadjoint operator with $D(\tilde{\Delta}) = R(B) \subset N^1$ and $R(\tilde{\Delta}) = N$; it satisfies the relation

$$((\mathbf{u}, \mathbf{v})) = -(\tilde{\Delta}\mathbf{u}, \mathbf{v}), \; (\mathbf{u} \in D(\tilde{\Delta}), \; \mathbf{v} \in N^1). \tag{2.2.12}$$

In particular, if follows that $\tilde{\Delta}$ is negative definite in N^1.

On the other hand, for every $\mathbf{u} \in N^2$ and $\mathbf{v} \in N^1$, we have the relation

$$-((\mathbf{u}, \mathbf{v})) = (\Delta\mathbf{u}, \mathbf{v})$$

(Δ being the Laplacian in the generalized sense) which is easily deduced from the equality $N^2 = N \cap \mathring{W}^{1,2}$ and from the analogous relation for $\mathbf{u} \in \mathfrak{N}$. Denoting by P the orthogonal projection of L^2 onto N, we have

$$(\Delta\mathbf{u}, \mathbf{v}) = (P\Delta\mathbf{u}, \mathbf{v}).$$

Hence $P\Delta\mathbf{u}$ is an element of N having the property that

$$-((\mathbf{u}, \mathbf{v})) = (P\Delta\mathbf{u}, \mathbf{v}).$$

In this way $\mathbf{u} \in D(\tilde{\Delta})$, $P\Delta\mathbf{u} = \tilde{\Delta}\mathbf{u}$; and since $\mathbf{u}$ is arbitrary in N^2 it follows that $N^2 \subset D(\tilde{\Delta})$ and $\tilde{\Delta}|_{N_2} = P\Delta$ [14].

---

[14] It is easy to see that $-\tilde{\Delta}$ coincides with the Friederichs extension of the positive definite map $-P\Delta$.

Finally, note that taking $\mathbf{v} = B\mathbf{f}$ in (2.2.10) and using again inequality (2.1.19), we obtain

$$\| B\mathbf{f} \|^2 = -(\mathbf{f}, B\mathbf{f}) \leqslant | \mathbf{f} | \, | B\mathbf{f} | \leqslant c | \mathbf{f} | \, \| B\mathbf{f} \|.$$

Therefore

$$\| B\mathbf{f} \| \leqslant c | \mathbf{f} |, \tag{2.2.13}$$

and the map B transforms bounded sets from N into bounded sets from N^1. By the Corollary to Theorem 2.1.4, it is completely continuous. In this way we have proved

Theorem 2.2.2. *The map* $P\Delta : N^2 \to N$ *admits a negative definite selfadjoint extension* $\widetilde{\Delta}$ *such that* $\widetilde{\Delta}^{-1} : N \to N^1$ *is a completely continuous map.*

It can be shown [8], [98] that if the domain Ω is of class C^2, then problem (2.2.8 a)—(2.2.8 b) admits in $N^2(\Omega)$ a strong solution which satisfies the inequality

$$\| \mathbf{u} \|_{N^2} \leqslant k \, | \mathbf{f} | \tag{2.2.14}$$

for a certain constant k depending only on Ω. Owing to the uniqueness of the weak solution, it follows that this $\mathbf{u}$ will coincide with the solution $B\mathbf{f}$ given in Theorem 2.2.1. Therefore, $D(\widetilde{\Delta}) = R(B) \subset N^2$. Since the reverse inclusion is always true, we can state the following for the map $A = -P\Delta : N^2 \to N$

Theorem 2.2.3. *If the domain* Ω *is bounded and of class* C^2, *then the operator* $A : N^2 \to N$ *defined by the equality*

$$(A\mathbf{u}, \mathbf{v}) = ((\mathbf{u}, \mathbf{v})), \tag{2.2.15}$$

is self-adjoint and satisfies the inequality

$$\| \mathbf{u} \|_{N^2} \leqslant k \, | A\mathbf{u} |. \tag{2.2.14$'$}$$

Its inverse, $A^{-1} : N \to N^2$, *is completely continuous and everywhere defined.*

In Section 2.2.4 we use the following existence theorem for the solutions of the problem (2.2.6)—(2.2.6 a).

Theorem 2.2.4. *If the map* A^+ *satisfies for every* $\mathbf{u} \in N^1$ *and* $\mathbf{v} \in \mathfrak{N}$ *the conditions*

$$1^\circ \quad | (\mathbf{u}, A^+ \mathbf{v}) | \leqslant C_1 \| \mathbf{u} \| \, \| \mathbf{v} \|,$$

$$2^\circ \quad (\mathbf{v}, A^+ \mathbf{v}) \geqslant C_2 \| \mathbf{v} \|^2,$$

and the domain Ω *is bounded, then for every* $\mathbf{f} \in N$, *problem* (2.2.6)—(2.2.6 a) *admits at least one weak solution in* N^1.

Proof. Denoting $(\mathbf{u}, A^+\mathbf{v}) = G_{\mathbf{u}}(\mathbf{v})$, $(\mathbf{u} \in N^1)$, Condition 1° shows that $G_{\mathbf{u}}$ is a linear functional bounded on the dense subspace $\mathfrak{N}$ of N^1.

Hence, using a continuation Hahn-Banach's extension theorem for linear functionals, there exists a bilinear form $G(\cdot,\cdot)$ defined on N^1 such that $G(\mathbf{u}, \mathbf{v}) = (\mathbf{u}, A^+\mathbf{v})$ if $\mathbf{v} \in \mathfrak{N}$, and G satisfies the hypotheses of the Lax-Milgram lemma (Appendix 1). In N^1 the linear continuous functional $l_f(\mathbf{v}) = -((B\mathbf{f}, \mathbf{v}))$ is represented by means of an element $\mathbf{u}_f \in N^1$ such that

$$l_f(\mathbf{v}) = G(\mathbf{u}_f, \mathbf{v}), \ (\mathbf{v} \in N^1).$$

From (2.2.10), it follows that for every $\mathbf{v} \in \mathfrak{N}$, we have

$$G(\mathbf{u}_f, \mathbf{v}) = (\mathbf{u}_f, A^+\mathbf{v}) = (\mathbf{f}, \mathbf{v}).$$

Hence $\mathbf{u}_f$ is a weak solution of problem (2.2.6)—(2.2.6 a).

2.2.3. Generalized solutions of the nonlinear problem

Four types of generalized solutions are met in the study of Navier-Stokes equations:

a) Solutions satisfying an integral relation which does not contain the partial derivative $\dot{\mathbf{u}}$. These solutions will be called turbulent solutions, the name under which they were introduced by Leray [53].

b) Weak solutions, satisfying an integro-differential equation which contains the derivative $\dot{\mathbf{u}}$. This type of solutions were first analyzed by Kiselev and Ladyzhenskaya [44].

c) Intermediate solutions, introduced by Lions [56], which admit a franctional derivative $\dot{\mathbf{u}}$.

d) Strong solutions, introduced by Krein [45], which admit sufficient generalized derivatives in order to satisfy equation (2.2.1).

Turbulent solutions. If $\mathbf{u}$ is a classical solution of the problem (1.1.1)—(1.1.4)*, corresponding to the body forces deriving from a potential, then, integrating on Ω this equation multiplied by an arbitrary $\boldsymbol{\varphi} \in C_0^1((0, T), \mathfrak{N}(\Omega))$ and using Green's formula, we get the relation

$$\int_\Omega [\mathbf{u} \cdot \dot{\boldsymbol{\varphi}} + (\mathbf{u} \cdot \operatorname{grad})\, \boldsymbol{\varphi} \cdot \mathbf{u} + \mathbf{u} \cdot \Delta\boldsymbol{\varphi})]\, \mathrm{d}x = \frac{\mathrm{d}}{\mathrm{d}t} \int_\Omega \mathbf{u} \cdot \boldsymbol{\varphi}\, \mathrm{d}x,$$

which integrated with respect to t, becomes

$$(2.2.15)' \quad \int_0^T \{(\mathbf{u}, \dot{\boldsymbol{\varphi}}) + ((\mathbf{u} \cdot \operatorname{grad})\, \boldsymbol{\varphi}, \mathbf{u}) + \nu(\mathbf{u}, \Delta\boldsymbol{\varphi})\}\, \mathrm{d}t = -(\mathbf{u}_0, \boldsymbol{\varphi}(0)).$$

An element $\mathbf{u} \in L^2((0, T), N^1(\Omega))$ which for every $\boldsymbol{\varphi} \in C_0^1([0, T), \mathfrak{N}(\Omega))$ satisfies the integral relation (2.2.15)′ is by definition a *turbulent solution* of the problem (1.1.1)—(1.1.4)*.

All the methods proving the existence of turbulent solutions are characterized by the fact that the solution obtained satisfies the energy inequality [15])

$$\frac{1}{2}|\mathbf{u}|^2 + \nu\int_0^T |\mathbf{Du}|^2 \, dt \leqslant \frac{1}{2}|\mathbf{u}_0|^2. \tag{2.2.16}$$

Most authors (beginning with Leray) include the energy inequality into the definition of the turbulent solution, which is then sought in the space

$$L^2((0, T), N^1(\Omega)) \cap L^\infty((0, T), N(\Omega)).$$

Using this additional condition, various uniqueness theorems may be proved [86].

We mention Leray's theorem on the structure of turbulent solutions: every turbulent solution is a class of functions almost everywhere equal with a function $\mathbf{u} : (0, T) \longrightarrow N^1(\Omega)$, having the following properties:

a) The corresponding function p and the derivatives

$$\frac{\partial u_i}{\partial x_k}, \quad \frac{\partial^2 u_i}{\partial x_k \partial x_l}, \quad \frac{\partial u_i}{\partial t} \text{ and } \frac{\partial p}{\partial x_i}$$

are continuous in all their arguments up to a set M of measure zero in the interval $(0, T)$.

b) The energy $W(t) = \frac{1}{2}|\mathbf{u}(t)|^2 = \frac{1}{2}\int_\Omega \sum u_i^2(x, t) \, dx$ is a decreasing function on the set consisting of M and the initial moment.

c) The function $\mathbf{u}$ is continuous on $[0, t)$ with respect to the weak topology of $N(\Omega)$.

d) For $t \to t_0 \in M$, the squared mean of the velocity gradient and the velocity maximum on Ω tend to ∞.

In this way, in certain points, the union of which has measure zero, the only continuity condition imposed on the velocity $\mathbf{u}(t)$ is c) and the velocity and its gradient have at these points jumps experimentally observed. This confirms Oseen's hypothesis (Section 2.1.1).

For a given t, $\mathbf{u}(t)$ is an element from $L^2(\Omega)$ having vanishing generalized divergence, which corresponds to $L^2(\Omega)$ condition (1.1.2) imposed on the classical solutions. Since $\mathbf{u}(t) \in N^1(\Omega)$, $\mathbf{u}(t)$ belongs (in particular) to the space $\mathring{W}^{1,2}$ and Remark 2.1.3 shows that $\mathbf{u}|_{\partial\Omega} = 0$, i.e. the analogy of Condition (1.1.4). In this way the single condition which must be imposed on $\mathbf{u}$ is

$$\lim_{t \to 0+0} \mathbf{u}(t) = \mathbf{u}_0. \tag{2.2.17}$$

---

[15]) The first term on the left-hand side of this inequality represents the square of the second-order mean (with respect to $\mathbf{x}$) of the energy, and the second one – the square of the second-order mean (with respect to t and $\mathbf{x}$) of the velocity gradient.

This corresponds to relation (1.1.3) where $\mathbf{u}_0$ is an arbitrary element of $N(\Omega)$.

In the case of body forces that do not derive from a potential, the turbulent solution is [75] an element of the set

$$L^2((0, T), N^1(\Omega)) \cap L^\infty((0, T), N(\Omega)),$$

which satisfies the integro-differential relation [16)]

$$(2.2.18) \quad \int_0^T \{(\mathbf{u}, \dot{\boldsymbol{\varphi}}) - \nu((\mathbf{u}, \boldsymbol{\varphi})) - ((\mathbf{u} \cdot \text{grad})\, \mathbf{u}, \boldsymbol{\varphi})\}\, dt =$$

$$= -(\mathbf{u}_0, \boldsymbol{\varphi}(0)) - \int_0^T (\mathbf{f}, \boldsymbol{\varphi})\, dt$$

for every $\boldsymbol{\varphi} \in C((0, T), N^1(\Omega))$ having the properties

$$\dot{\boldsymbol{\varphi}} \in L^2((0, T), N(\Omega)), \ \boldsymbol{\varphi}(T) = 0$$

(the data of the problem being

$$\mathbf{u}_0 \in N(\Omega), \ \mathbf{f}(t) \in L^2((0, T), N(\Omega))).$$

For these relations Leray's structure theorem is still valid.

Weak solutions. The weak solutions of the problem (1.1.1)—(1.1.4)* are the elements $\mathbf{u} \in W^{1,2}(\Omega_T)$ where $\Omega_T = \Omega \times [0, T]$, with

$$\text{div}\, \mathbf{u} = 0, \ \mathbf{u}|_{\partial\Omega_T} = 0, \ \mathbf{u}|_{t=0} = \mathbf{u}_0,$$

such that the equation

$$(2.2.19) \quad \int_0^T \int_\Omega [\dot{\mathbf{u}}\boldsymbol{\varphi} + \nu \, \text{grad}\, \mathbf{u} \cdot \text{grad}\, \boldsymbol{\varphi} - (\mathbf{u} \cdot \text{grad})\, \boldsymbol{\varphi}\mathbf{u} - \mathbf{f}\boldsymbol{\varphi}] dx\, dt = 0$$

is satisfied for every vector $\boldsymbol{\varphi} \in L^2(\Omega_T)$ for which

$$D_x\boldsymbol{\varphi} \in L^2(\Omega_T), \text{div}\, \boldsymbol{\varphi} = 0, \ \boldsymbol{\varphi}|_{\partial\Omega_T} = 0.$$

Intermediate solutions. For $\mathbf{u} \in L^2((0, \infty), L^2(\Omega))$, denote also by $\mathbf{u}$ the extension by zero for $t < 0$ and by $\hat{\mathbf{u}}(z)$ the Fourier transform with respect to t of $\mathbf{u}$, $\hat{\mathbf{u}} \in L^2((-\infty, +\infty), L^2(\Omega))$. We say that $\mathbf{u}$ admits in $L^2((0, \infty), L^2(\Omega))$ a derivative of order γ (γ real > 0) with respect to t, if

$$|\tau|^\gamma \, \hat{\mathbf{u}} \in L^2((-\infty, +\infty), L^2(\Omega)).$$

16) Taking into account the definition of the scalar product $((\cdot,\cdot))$, we have $((\mathbf{u}, \boldsymbol{\varphi})) = -(\mathbf{u}, \Delta\boldsymbol{\varphi})$ and from $\text{div}\ \mathbf{u} = 0$, it follows that $((\mathbf{u} \cdot \text{grad})\, \boldsymbol{\varphi}, \mathbf{u}) = -((\mathbf{u}\ \text{grad})\, \mathbf{u}, \boldsymbol{\varphi})$. Hence, in the case of the body forces which derive from a potential, (2.2.18) is equivalent to (2.2.15).

In other words, the fractional derivative $D_t^\gamma \mathbf{u} \in L^2((0, \infty), L^2(\Omega))$ is defined [86] by the equality

$$(D_t^\gamma \mathbf{u})(t) = \frac{1}{\Gamma(1-\gamma)} \frac{d}{dt} \int_0^t \frac{\mathbf{u}(\tau)}{(t-\tau)^\gamma} d\tau,$$

where Γ is Euler's function. Since $D_t^0 \mathbf{u} = \mathbf{u}$ and $D_t^1 \mathbf{u} = D\mathbf{u}$, where D is the weak derivative, it follows that if $D_t^\alpha \mathbf{u}$ exists for $0 < \alpha < 1$, then $\mathbf{u}$ has properties stronger than those resulting from the fact that it belongs to $L^2((0, \infty), L^2(\Omega))$, and weaker properties than those which result from the fact that $D\mathbf{u}$ belongs to the same class.

Problem T_1. *For* $\mathbf{f} \in L^2((0, \infty), N(\Omega))$, $\mathbf{u}_0 \in N(\Omega)$, *find* $\mathbf{u} \in L^2((0, \infty), N(\Omega))$ *which satisfies the relation*

$$\int_0^\infty [-(\mathbf{u}, \dot{\boldsymbol{\varphi}}) + \nu((\mathbf{u}, \boldsymbol{\varphi})) - (\mathbf{u} \cdot \operatorname{grad}) \boldsymbol{\varphi}, \mathbf{u})] \, dt = \int_0^\infty (\mathbf{f}, \boldsymbol{\varphi}) dt + (\mathbf{u}_0, \boldsymbol{\varphi}(0)),$$

for every $\boldsymbol{\varphi} \in C^\circ((0, \infty), L^2(\Omega))$, *with* $\boldsymbol{\varphi}(t) = 0$ *for* t *large enough and* $\dot{\boldsymbol{\varphi}} \in L^2((0, \infty), L^2(\Omega))$.

Problem S_1. *If* $\mathbf{f} \in L^2((0, \infty), N(\Omega))$ *and* $\mathbf{u}_0 \in N(\Omega)$, *find* $\mathbf{u} \in L^2((0, \infty), N(\Omega))$ with $\dot{\mathbf{u}} \in L^2((0, \infty), L^2(\Omega))$, *which satisfies the relation*

$$\int_0^\infty [(\dot{\mathbf{u}}, \boldsymbol{\varphi}) + \nu((\mathbf{u}, \boldsymbol{\varphi})) - ((\mathbf{u} \cdot \operatorname{grad}) \boldsymbol{\varphi}, \mathbf{u})] \, dt = \int_0^\infty (\mathbf{f}, \boldsymbol{\varphi}) \, dt,$$

for every $\boldsymbol{\varphi} \in C^\circ((0, \infty), L^2(\Omega))$ *with* $\boldsymbol{\varphi}(t) = 0$ *for* t *large enough.*

Every element $\mathbf{u}$ which belongs to an intermediate space between the spaces of the solution of the problems T_1 and S_1 is called an intermediate solution of the problem (1.1.1)–(1.1.4)*. We give below a global existence theorem for the intermediate solutions; for some other existence, uniqueness and stability theorems, see refs. [59] and [58].

Theorem *(Lions)* [57]. *Let* Ω *be a domain from* R^n *with* $n \leqslant 4$. *Then there exists a solution of the problem* T_1 *which satisfies the additional condition* that $\mathbf{u} \in L^\infty((0, T), L^2(\Omega))$, *for every* T *finite, and for every* $\gamma < \frac{1}{2}$ *we have*

$$D_t^\gamma \mathbf{u} \in L^2((0, \infty), L^2(\Omega)).$$

Strong solutions. A pair $(\mathbf{u}, p)$ is called a strong solution of problem (1.1.1)–(1.1.4)* corresponding to the data

$$\mathbf{f} \in L^2((0, T), L^2(\Omega)), \ \mathbf{u}_0 \in N^1(\Omega),$$

if p admits a generalized gradient grad $p \in L^2((0, T), L^2(\Omega))$, and $\mathbf{u} \in C((0, T), N^1(\Omega))$ is such that

a) $\mathbf{u}$ admits the generalized derivative $\dot{\mathbf{u}} \in L^2((0, T), N(\Omega))$;
b) $\mathbf{u} \in L^2((0, T), N^2(\Omega))$;
and $(\mathbf{u}, p)$ satisfies Equation (1.1.1) and the initial condition (2.2.17).

It can be shown that if $\mathbf{f}$ derives from a potential, then the strong solutions are infinitely differentiable with respect to x and t. Hence, they are classical solutions [86], [39].

As in the case of the linearized problem, if a generalized solution (of one of the above defined types) satisfies Conditions a) and b), then it is a strong solution. For instance, if $\mathbf{u}$ is a weak solution, then Equation (2.2.19) may be written in the form

$$\int_0^T \int_\Omega [\dot{\mathbf{u}} - \nu\Delta\mathbf{u} + (\mathbf{u} \cdot \operatorname{grad})\,\mathbf{u} - \mathbf{f}]\,\boldsymbol{\varphi}\,\mathrm{d}x\,\mathrm{d}t = 0,$$

and from Theorem 2.1.7 the existence of a potential function in the generalized sense, $\operatorname{grad} p \in G(\Omega_T)$, such that

$$\dot{\mathbf{u}} + (\mathbf{u} \cdot \operatorname{grad})\,\mathbf{u} - \nu\Delta\mathbf{u} - \mathbf{f} = \operatorname{grad} p$$

follows. Hence $(\mathbf{u}, p)$ satisfies Equation (1.1.1).

Consider now problem (1.1.1)—(1.1.3) with the nonhomogeneous boundary condition (1.1.4), where $\mathbf{w} \in L^2(\partial\Omega)$ satisfies the condition $\int_{\partial\Omega} \mathbf{w} \cdot \mathbf{n}\,\mathrm{d}\sigma = 0$. As in the previous section, we assume that $\mathbf{w} = \mathbf{u}^*|_{\partial\Omega}$, where $\mathbf{u}^* \in N(\Omega) \cap W^{2,2}(\Omega)$; and the initial condition $\mathbf{u}_0$ is chosen such that $\mathbf{u}_0 - \mathbf{u}^* \in N^1(\Omega)$.

We say that a pair $(\mathbf{u}, p)$, where $p \in L^2((0, T), L^2(\Omega))$, $\mathbf{u}$ satisfies Condition a) and

b′) $\quad \mathbf{u} \in L^2((0, T),\ N(\Omega) \cap W^{2,2}(\Omega))$

b″) $\quad \mathbf{u} - \mathbf{u}^* \in L^2((0, T), N^2(\Omega))$ [17)]

is a strong solution of the problem (1.1.1)—(1.1.4) if $(\mathbf{u}, p)$ satisfies Equation (1.1.1) and the initial condition (2.2.17).

If in (1.1.1) and (2.2.17) we put $\mathbf{u} = \mathbf{u}^* + \mathbf{v}$ and $\mathbf{u}_0 = \mathbf{u}^* + \mathbf{v}_0$, we get the equation

$$\frac{\partial\mathbf{v}}{\partial t} + (\mathbf{u}^* \cdot \operatorname{grad})\,\mathbf{v} + (\mathbf{v} \cdot \operatorname{grad})\,\mathbf{u}^* + (\mathbf{v} \cdot \operatorname{grad})\,\mathbf{v} - \nu\Delta\mathbf{v} = -\frac{1}{\rho}\operatorname{grad} p + \mathbf{f} + \nu\Delta\mathbf{u}^* - (\mathbf{u}^* \cdot \operatorname{grad})\,\mathbf{u}^* \tag{2.2.20}$$

with the initial condition

$$\lim_{t \to 0+0} \mathbf{v}(t) = \mathbf{v}_0 \tag{2.2.21}$$

17) Condition b″) shows that on the boundary $\mathbf{u}$ takes (at every time) the same values $\mathbf{u}^*$ (Equality (2.1.23)).

the solution of which are pairs $(\mathbf{v}, p)$ with $\mathbf{v} \in C((0, T), N^1(\Omega))$ which satisfy Conditions a) and b). Following Krein [46], [45], in the Equations (1.1.1) and (2.2.20) the pressure is eliminated by applying on both sides the projection $P : L^2(\Omega) \longrightarrow N(\Omega)$. Assuming that $\mathbf{f} \in L^2((0, T), N(\Omega))$ and $\nu = 1$, we get the evolution differential equations in the Hilbert space $L^2((0, T), N(\Omega))$

$$\frac{d\mathbf{u}}{dt} + A\mathbf{u} = R(\mathbf{u}) + \mathbf{f}, \tag{2.2.22}$$

$$\frac{d\mathbf{v}}{dt} + A\mathbf{v} + M(\mathbf{u}^*, \mathbf{v}) = R(\mathbf{v}) + \mathbf{g}(t), \tag{2.2.23}$$

where $A = - P\Delta$, $R(\cdot) = - P((\cdot \operatorname{grad})\cdot)$, $M(\mathbf{u}^*, \mathbf{v}) = P[(\mathbf{u}^* \cdot \operatorname{grad})\, \mathbf{v} + (\mathbf{v} \cdot \operatorname{grad})\, \mathbf{u}^*]$, iar $\mathbf{g}(t) = P[\nu\Delta\mathbf{u}^* - (\mathbf{u}^* \cdot \operatorname{grad})\, \mathbf{u}^* + \mathbf{f}(t)]$.

It can be shown that if $\mathbf{u}$ (respectively $\mathbf{v}$) $\in C((0, T), N(\Omega))$ satisfying Conditions a) and b) is a solution of Equation (2.2.22) (respectively (2.2.23)), then there exists a function p such that $(\mathbf{u}, p)$, (respectively $(\mathbf{v} + \mathbf{u}, p)$ is a strong solution of the problem (1.1.1)—(1.1.4)* (respectively (1.1.1)—(1.1.4)).

The study of the stability of strong solutions of the Navier-Stokes equations by the method of fractional powers [23] of A, in connection with the analysis of existence and uniqueness, was the object of Sobolevskii's papers [91]—[94]. For the stability problem of differential equations in Banach spaces, see ref. [11].

2.2.4. Existence of generalized solutions

Leray's method. For the case $\Omega = R^3$, Leray [54] constructed the turbulent solutions of the problem (1.1.1)—(1.1.4)* with the data $\mathbf{f} = 0$, $\mathbf{u}_0 \in N(\Omega)$, by means of the approximate equation

$$\dot{\mathbf{u}} + (\mathbf{u}_\varepsilon \cdot \operatorname{grad})\, \mathbf{u} = - \frac{1}{\rho} \operatorname{grad} p + \nu\Delta\mathbf{u}, \tag{2.2.24}$$

taking the initial condition $\mathbf{u}(x, 0) = (\mathbf{u}_0)_\varepsilon$, $\mathbf{u}_\varepsilon$ and $(\mathbf{u}_0)_\varepsilon$ being the mean functions of $\mathbf{u}$, respectively $\mathbf{u}_0$. A (global) existence and uniqueness theorem has been proved for the solution of (2.2.24) in the class of the functions from $C^\infty(\Omega \times (0, T))$ which belong to the intersection $C^\circ((0, T), N(\Omega)) \cap C^0((0, T), L^\infty(\Omega))$ [18].

Let $\mathbf{u}(t, \varepsilon)$ be the solution of Equation (2.2.24), corresponding to a certain value of ε; by means of the integrals

$$W(t, \varepsilon) = \int_\Omega |\mathbf{u}(t, \varepsilon)|^2 \, dx, \quad J^2(t, \varepsilon) = \int_\Omega |D\mathbf{u}|^2 \, dx,$$

[18] Here T may also take the value ∞.

we get

$$W(t, \varepsilon) \leqslant W(0) = \int_\Omega |\mathbf{u}_0^2| \, dx \tag{2.2.25}$$

and

$$\nu \int_0^T J^2(t, \varepsilon) \, dt + \frac{1}{2} W(t, \varepsilon) \leqslant \frac{1}{2} W(0). \tag{2.2.26}$$

Relation (2.2.25) expresses the property of boundedness of the mean energy of the solution $\mathbf{u}(t, \varepsilon)$ by the corresponding mean energy corresponding to the initial moment. (2.2.26) is the inequality of energy dissipation for Equation (2.2.24). In this way, for every t belonging to the complement of a negligible subset of $(0, T)$, $\{\mathbf{u}(t, \varepsilon)\}_{\varepsilon>0}$ and $\{D\mathbf{u}(t, \varepsilon)\}_{\varepsilon>0}$ are bounded sets of elements from $L^2(\Omega)$; then by Theorem 2.1.6 there exists a sequence $\varepsilon_n \to 0$ such that the corresponding sequence $\{\mathbf{u}_n(t)\}$ is weakly converging in $N(\Omega)$ towards an element $\mathbf{u}(t) \in N^1(\Omega)$. In this way, we obtain an element $\mathbf{u} \in L^2((0, \infty), N^1(\Omega)) \cap L^\infty((0, \infty), N(\Omega))$ satisfying Equation (2.2.15) Hence, $\mathbf{u}$ is a turbulent solution of the problem (1.1.1)—(1.1.4)*.

For bounded domains from R^3, the method used by Kiselev and Ladyzhenskaya [44] and subsequently by Shinbrot [86] differs from Leray's method by the fact that instead of (2.2.24), it approximates Equation (1.1.1) by the linear steady equation

$$\frac{1}{\varepsilon}(\mathbf{u} - \mathbf{f}) + (\mathbf{f} \cdot \operatorname{grad}) \mathbf{u} - \nu \Delta \mathbf{u} = -\frac{1}{\rho} \operatorname{grad} p, \tag{2.2.27}$$

with $\mathbf{f} \in N(\Omega)$ and $\varepsilon > 0$. This can be written as

$$\mathbf{u} + \varepsilon(\mathbf{f} \cdot \operatorname{grad}) \mathbf{u} - \varepsilon\nu \Delta \mathbf{u} = -\frac{\varepsilon}{\rho} \operatorname{grad} p + \mathbf{f}. \tag{2.2.28}$$

Hence, it is of the form (2.2.6) with $A\mathbf{u} = \mathbf{u} + \varepsilon(\mathbf{f} \cdot \operatorname{grad})\, \mathbf{u} - \varepsilon\nu\Delta\mathbf{u}$. For every $\mathbf{u} \in N^1$ and $\mathbf{v} \in \mathfrak{N}$, we have

$$(\mathbf{u}, A^+ \mathbf{v}) = (\mathbf{u}, \mathbf{v}) + \varepsilon((\mathbf{f} \cdot \operatorname{grad}) \, \mathbf{u}, \mathbf{v}) + \varepsilon\nu(\operatorname{grad} \mathbf{u}, \operatorname{grad} \mathbf{v}).$$

Therefore, assuming that $\mathbf{f} \in \mathfrak{N}$ and denoting $M = \|\mathbf{f}\|_\infty$, we get

$$|(\mathbf{u}, A^+\mathbf{v})| \leqslant |\mathbf{u}| \, |\mathbf{v}| + \varepsilon M |\operatorname{grad} \mathbf{u}| \cdot |\mathbf{v}| + \varepsilon\nu |\operatorname{grad} \mathbf{u}| \cdot |\operatorname{grad} \mathbf{v}|.$$

Applying inequality (2.1.19), we obtain

$$|(\mathbf{u}, A^+\mathbf{v})| \leqslant c \|\mathbf{u}\| \|\mathbf{v}\|.$$

On the other hand, since $((\mathbf{f} \cdot \operatorname{grad}) \, \mathbf{v}, \mathbf{v}) = 0$, we have

$$(\mathbf{v}, A^+ \mathbf{v}) = (\mathbf{v}, \mathbf{v}) + \varepsilon\nu((\mathbf{v}, \mathbf{v})),$$

whence

$$(\mathbf{v}, A^+ \mathbf{v}) \geqslant \varepsilon\nu \|\mathbf{v}\|^2.$$

By Theorem 2.2.4, Equation (2.2.28) therefore admits at least one weak solution $\mathbf{u}(\varepsilon) \in N^1(\Omega)$ corresponding to a given $\mathbf{f} \in \mathcal{N}(\Omega)$. It can be shown that this solution exists also for $\mathbf{f} \in N$. To this purpose, consider a sequence $\{\mathbf{f}_i\}$ of functions from $\mathcal{N}$, converging to the element $\mathbf{f} \in N$ and let $\mathbf{u}_i \in N^1$ be the weak solution corresponding to the data $\mathbf{f}_i$. For every $\mathbf{v} \in \mathcal{N}$(or even from N^1), $\mathbf{u}_i$ satisfies the equation

$$(2.2.29) \qquad (\mathbf{u}_i, \mathbf{v}) + \varepsilon((\mathbf{f}_i \ \mathrm{grad})\, \mathbf{u}_i, \mathbf{v}) + \varepsilon\nu((\mathbf{u}_i, \mathbf{v})) = (\mathbf{f}_i, \mathbf{v}).$$

Taking $\mathbf{v} = \mathbf{u}_i$, we obtain

$$(\mathbf{u}_i, \mathbf{u}_i) + \varepsilon\nu((\mathbf{u}_i, \mathbf{u}_i)) = (\mathbf{f}_i, \mathbf{u}_i)$$

and therefore

$$|\mathbf{u}_i|^2 + \varepsilon\nu \| \mathbf{u}_i \|^2 \leqslant |\mathbf{f}_i|\,|\mathbf{u}_i| \leqslant \frac{1}{2}(|\mathbf{f}_i|^2 + |\mathbf{u}_i|^2),$$

or

$$|\mathbf{u}_i|^2 + 2\varepsilon\nu \| \mathbf{u}_i \|^2 \leqslant |\mathbf{f}_i|^2.$$

It follows that the sequence $\{\mathbf{u}_i\}$, being bounded in N^1, contains a subsequence $\{\mathbf{u}_j\}$ weakly converging to an element $\mathbf{u} \in N^1$. By the corollary to Theorem 2.1.4, this sequence is strongly converging in N. Then, for $j \to \infty$, $(\mathbf{u}_j, \mathbf{v}) \to (\mathbf{u}, \mathbf{v})$, $((\mathbf{u}_j, \mathbf{v})) \to ((\mathbf{u}, \mathbf{v}))$, $(\mathbf{f}_j, \mathbf{v}) \to$ $\to (\mathbf{f}, \mathbf{v})$, and for $((\mathbf{f}_j \cdot \mathrm{grad})\, \mathbf{u}_j, \mathbf{v})$ we have the estimate

$$|((\mathbf{f}_j \cdot \mathrm{grad})\, \mathbf{u}_j, \mathbf{v}) - ((\mathbf{f} \cdot \mathrm{grad})\, \mathbf{u}, \mathbf{v})| \leqslant$$
$$\leqslant |((\mathbf{f}_j - \mathbf{f})\, \mathrm{grad}\, \mathbf{u}, \mathbf{v})| + |((\mathbf{f} \cdot \mathrm{grad})\, (\mathbf{u}_j - \mathbf{u}), \mathbf{v})|.$$

Hence

$$|((\mathbf{f}_j \cdot \mathrm{grad})\, \mathbf{u}_j, \mathbf{v}) - ((\mathbf{f} \cdot \mathrm{grad})\, \mathbf{u}, \mathbf{v})| \leqslant$$
$$|\mathbf{f}_j - \mathbf{f}| \cdot \| \mathbf{u}_i \| \cdot M_1 + |((\mathbf{f} \cdot \mathrm{grad})\, (\mathbf{u}_j - \mathbf{u}), \mathbf{v})|,$$

with $M_1 = \sup_{x \in \Omega} |\mathbf{v}(x)|$, where the right-hand side tends to zero owing to the convergence $\mathbf{u}_j \to \mathbf{u}$ which is weak in N^1 and strong in N. Passing to the limit in (2.2.29), we see that $\mathbf{u} = \mathbf{u}(\varepsilon)$ is a weak solution of Equation (2.2.28) corresponding to the data $\mathbf{f} \in N^1$.

Let $\mathbf{u}_1(\varepsilon)$ be the solution of Equation (2.2.8) corresponding to the data $\mathbf{f} = \mathbf{u}_0$ ($\mathbf{u}_0$ being the initial date of problem (1.1.1)—(1.1.4)*) and, by recurrence, let $\mathbf{u}_k(\varepsilon)$ be the solution corresponding to the data $\mathbf{f} = \mathbf{u}_{k-1}(\varepsilon)$. Further, for every $t > 0$, we can apply the last part of Leray's approach to the family $\{\mathbf{u}(\varepsilon, t)\}_{\varepsilon > 0}$, where $\mathbf{u}(\varepsilon, t)$ is an element of $N^1(\Omega)$ defined by

$$\mathbf{u}(\varepsilon, t) = \mathbf{u}_k(\varepsilon), \ (\varepsilon k \leqslant t \leqslant \varepsilon(k+1), \ k = 0, 1, \ldots),$$

and satisfying the relations

$$\| \mathbf{u}(\varepsilon, t) \|^2 + 2\varepsilon\nu \| \mathrm{D}\mathbf{u}(\varepsilon, t) \|^2 \leqslant \| \mathbf{u}_0 \|^2, \ (t \geqslant \varepsilon),$$

and

$$\frac{1}{\varepsilon}\int_{\varepsilon}^{\infty} \| \mathbf{u}(\varepsilon, t) - \mathbf{u}(\varepsilon, t-\varepsilon) \|^2 \, dt + 2\nu \int_{\varepsilon}^{\infty} \| \mathbf{Du}(\varepsilon, t) \|^2 \, dt \leqslant \| \mathbf{u}_0 \|^2,$$

analogous to (2.2.25) and (2.2.26).

Note that both theorems are of global type with respect to t. Hence the asymptotic stability problem of the corresponding solutions makes sense.

Galerkin-Faedo-Hopf's method is useful for the proof of the local (with respect to t) theorem of existence of strong solutions of the problem (1.1.1) — (1.1.4)* projected on $L^2((0, T), N(\Omega))$ in the form

$$\frac{\mathrm{d}\mathbf{u}}{\mathrm{d}t} + \mathbf{Au} = \mathbf{Ru} + \mathbf{f}. \tag{2.2.22}$$

In order to obtain a priori estimates for the solutions of this equation, we use the following lemmas

Lemma 2.2.1 [76]. *If* $\mathbf{u}$ *is a solution of* (2.2.22), *then*

$$\int_0^T \left(\frac{\mathrm{d}\mathbf{u}}{\mathrm{d}t}, \mathbf{Au}(t) \right) \mathrm{d}t = \frac{1}{2} \{ \| \mathbf{u}(T) \|^2 - \| \mathbf{u}(0) \|^2 \}. \tag{2.2.30}$$

This lemma follows by straightforward calculations using the mean functions introduced in Section 2.1.2.

Lemma 2.2.2 [76]. *For every* $\mathbf{u} \in \mathrm{N}^2(\Omega)$, *inequality*

$$| \mathbf{Ru} |^2 \leqslant k \| \mathbf{u} \|^3 \, | \mathbf{Au} |, \tag{2.2.31}$$

holds, where the constant k *depends only on the two- or three-dimensional domain* Ω.

Proof. Using the inequality $| \mathbf{Pu} | \leqslant | \mathbf{u} |$ and the generalized Hölder inequality

$$\int_{\Omega} | \mathbf{u\,v\,w} | \, \mathrm{d}x \leqslant \left(\int_{\Omega} | \mathbf{u} |^3 \, \mathrm{d}x \right)^{1/3} \left(\int_{\Omega} | \mathbf{v} |^2 \right)^{1/2} \left(\int_{\Omega} | w |^6 \right)^{1/6},$$

we get

$$| R\mathbf{u} |^2 = | P(\mathbf{u} \cdot \mathrm{grad})\, \mathbf{u} |^2 \leqslant | (\mathbf{u} \cdot \mathrm{grad})\, \mathbf{u} |^2 \leqslant \int_{\Omega} | \mathbf{u} |^2 \, (| \mathbf{Du} |^2)^{1/2}$$

$$(| \mathbf{Du} |^2)^{1/2} \, \mathrm{d}x \leqslant \left(\int_{\Omega} | \mathbf{u} |^6 \, \mathrm{d}x \right)^{1/3} \left(\int_{\Omega} | \mathbf{Du} |^2 \, \mathrm{d}x \right)^{1/2} \left(\int_{\Omega} | \mathbf{Du} |^6 \, \mathrm{d}x \right)^{1/6} =$$

$$= \| \mathbf{u} \|_6^2 \, | \mathbf{Du} | \, \| \mathbf{Du} \|_6 .$$

On the other hand, if $\mathbf{v} \in \mathring{W}^{1,2}$, then we have an inequality of the form $\| \mathbf{v} \|_6 \leqslant C_1 \| \mathbf{v} \|_{1,2}$. For $n = 2$ this follows from Theorem 2.1.2, and for $n = 3$ from inequality (2.1.16). Hence, owing to the equivalence between the norm $\|\cdot\|_{1,2}$ and $\|\cdot\|$, we obtain the inequality $\|\mathbf{v}\|_6 \leqslant C_2\|\mathbf{v}\|$. Then, for $\mathbf{v} = \mathrm{D}\mathbf{u}$, $\mathbf{u} \in N^2$, it follows that $\|\mathrm{D}\mathbf{u}\|_6 \leqslant C_2\|\mathrm{D}\mathbf{u}\| = C_2|\mathrm{D}^2\mathbf{u}|$. Taking into account Cattabriga's inequality

$$\|\mathbf{u}\|_{N^2} \leqslant k_1|\mathbf{A}\mathbf{u}|, \tag{2.2.18$'$}$$

and the equivalence (Remark 2.2.4) between the norm $\|\mathbf{u}\|_{N^2}$ and $|\mathrm{D}^2\mathbf{u}|$ we obtain (2.2.31).

Remark. Using Young's inequality

$$ab \leqslant \frac{a^p}{p\varepsilon^p} + \frac{b^q\varepsilon^q}{q}, \tag{2.2.32}$$

which is valid for every positive numbers a, b, ε and for every pair of conjugate numbers (p, q), (2.2.23) implies, for $p = q = 2$,

$$|R\mathbf{u}|^2 \leqslant k\left(\frac{\|\mathbf{u}\|^6\varepsilon^2}{2} + \frac{|A\mathbf{u}|^2}{2\varepsilon^2}\right). \tag{2.2.33}$$

Multiplying Equation (2.2.22) by $A\mathbf{u}$, integrating on $(0, T)$ and taking into account (2.2.30), we get

$$\frac{1}{2}\{\|\mathbf{u}(T)\|^2 - \|\mathbf{u}(0)\|^2\} + \int_0^T |A\mathbf{u}(t)|^2\,\mathrm{d}t = \tag{2.2.34}$$

$$= \int_0^T (R\mathbf{u}, A\mathbf{u})\,\mathrm{d}t + \int_0^T (\mathbf{f}, A\mathbf{u})\,\mathrm{d}t.$$

It follows that almost everywhere in $[0, T]$, we have

$$\frac{1}{2}\frac{\mathrm{d}}{\mathrm{d}t}\|\mathbf{u}\|^2 + |A\mathbf{u}|^2 = (R\mathbf{u}, A\mathbf{u}) + (\mathbf{f}, A\mathbf{u}), \tag{2.2.35}$$

which, after applying Schwarz's inequality, becomes

$$\frac{1}{2}\frac{\mathrm{d}}{\mathrm{d}t}\|\mathbf{u}\|^2 + |A\mathbf{u}|^2 \leqslant |R\mathbf{u}|\cdot|A\mathbf{u}| + |\mathbf{f}|\cdot|A\mathbf{u}|. \tag{2.2.36}$$

Applying now Young's inequality to each term of the right-hand side of (2.2.36), we get

$$|\mathbf{f}|\,|A\mathbf{u}| \leqslant \frac{|\mathbf{f}|^2\varepsilon_1^2}{2} + \frac{|A\mathbf{u}|^2}{2\varepsilon_1^2},$$

$$|R\mathbf{u}|\,|A\mathbf{u}| \leqslant \frac{|R\mathbf{u}|^2\varepsilon_2^2}{2} + \frac{|A\mathbf{u}|^2}{2\varepsilon_2^2} \leqslant \frac{\varepsilon_2^2}{2}\left[k_1\left(\frac{\|\mathbf{u}(t)\|^6\varepsilon^2}{2} + \frac{|A\mathbf{u}|^2}{2\varepsilon^2}\right)\right] +$$

$$+ \frac{|A\mathbf{u}|^2}{2\varepsilon_2^2} = \frac{k_1\varepsilon^2\varepsilon_2^2}{4}\|\mathbf{u}\|^6 + \left(\frac{k_1\varepsilon_2^2}{4\varepsilon^2} + \frac{1}{2\varepsilon_2^2}\right)|A\mathbf{u}|^2,$$

where in the last estimate we used inequality (2.2.33). Taking into account (2.2.36), we obtain

$$\frac{1}{2}\frac{\mathrm{d}}{\mathrm{d}t}\|\mathbf{u}\|^2+|A\mathbf{u}|^2\leqslant\frac{k_1\varepsilon\varepsilon_2^2\|\mathbf{u}\|^6}{4}+\frac{\varepsilon_1^2}{2}|\mathbf{f}|^2+\left(\frac{k_1\varepsilon_2^2}{4\varepsilon^2}+\frac{1}{2\varepsilon_2^2}+\frac{1}{2\varepsilon_1^2}\right)|A\mathbf{u}|^2.$$

Choosing ε, ε_1 and ε_2 such that $\frac{k_1\varepsilon_2^2}{4\varepsilon^2}+\frac{1}{2\varepsilon_2^2}+\frac{1}{2\varepsilon_1^2}=1$, we deduce the inequalities

$$\frac{1}{2}\frac{\mathrm{d}}{\mathrm{d}t}\|\mathbf{u}\|^2\leqslant\frac{k_1\varepsilon^2\varepsilon_2^2}{4}\|\mathbf{u}\|^6+\frac{\varepsilon_1^2}{2}|\mathbf{f}|^2,$$

(2.2.37)
$$\frac{1}{2}\frac{\mathrm{d}}{\mathrm{d}t}\|\mathbf{u}\|^2\leqslant k_2(\|\mathbf{u}\|^6+|\mathbf{f}|^2)$$

$\left(\text{where } k_2=\max\left\{\frac{k_1\varepsilon^2\varepsilon_2^2}{4},\ \frac{\varepsilon_1^2}{2}\right\}\right)$. If $\mathbf{f}\in L^\infty((0,T),N(\Omega))$, and $c_f = \mathrm{ad}\max_{t\in(0,T)}|\mathbf{f}(t)|^{1/3}$, from (2.2.37) it follows that

(2.2.38)
$$\frac{1}{2}\frac{\mathrm{d}}{\mathrm{d}t}\|\mathbf{u}\|^2\leqslant k_2(\|\mathbf{u}\|^2+c_f^2)^3.$$

This inequality yields the following *a priori* estimate for the solutions of Equation (2.2.22)

(2.2.39)
$$\|\mathbf{u}(t)\|^2\leqslant\frac{\|\mathbf{u}(0)\|^2+c_f^2}{\sqrt{1-4k_2t(\|\mathbf{u}(0)\|^2+c_f^2)^{-2}}}-c_f^2,$$

which is valid in the time interval $(0, T)$, where

(2.2.40)
$$T=\frac{1}{4k_2}(\|\mathbf{u}(0)\|^2+c_f^2)^{-2}.$$

From (2.2.34), we obtain for $\int_0^T|A\mathbf{u}|^2\,\mathrm{d}t$ the estimate

$$\int_0^T|A\mathbf{u}|^2\,\mathrm{d}t\leqslant\int_0^T(R\mathbf{u},A\mathbf{u})\,\mathrm{d}t+\int_0^T(\mathbf{f},A\mathbf{u})\,\mathrm{d}t+\frac{1}{2}\|\mathbf{u}(0)\|^2.$$

Using as above Schwarz's and Young's inequalities, combined with (2.2.33), it follows that

$$\int_0^T|A\mathbf{u}|^2\,\mathrm{d}t\leqslant c_1(\varepsilon_1)\int_0^T|\mathbf{f}|^2\,\mathrm{d}t+c_2(k_1,\varepsilon^2,\varepsilon_2^2)\|\mathbf{u}\|^6+c_3(\varepsilon,\varepsilon_1,\varepsilon_2,k_1)\times$$
$$\times\int_0^T|A\mathbf{u}|^2\,\mathrm{d}t+\frac{1}{2}\|\mathbf{u}(0)\|^2.$$

Choosing ε, ε_1, ε_2 and k from c_3 such that $1 - c_3 > 0$ and relabelling the constants, we have

$$\int_0^T |A\mathbf{u}|^2\, dt \leqslant c_1 \int_0^T |\mathbf{f}|^2\, dt + c_2 \|\mathbf{u}\|^6 + \frac{1}{2}\|\mathbf{u}(0)\|^2.$$

Then, taking into account the estimate (2.2.39) on $\|\mathbf{u}\|$, we get an *a priori* estimate for $\int_0^T |A\mathbf{u}|^2\, dt$.

Finally, in order to obtain an *a priori* upper bound on the expression $\int_0^T \left|\frac{d\mathbf{u}}{dt}\right|^2 dt$, we multiply Equation (2.2.22) by $\frac{d\mathbf{u}}{dt}$, integrate and, using (2.2.30), we obtain the inequality

(2.2.41) $$\int_0^T \left|\frac{d\mathbf{u}}{dt}\right|^2 dt + \frac{1}{2}\{\|\mathbf{u}(T)\|^2 - \|\mathbf{u}(0)\|^2\} =$$

$$= \int_0^T \left(R\mathbf{u}, \frac{d\mathbf{u}}{dt}\right) dt + \int_0^T \left(\mathbf{f}, \frac{d\mathbf{u}}{dt}\right) dt.$$

We then use the same method as for the first two estimates as well as these estimates themselves.

By means of the estimates obtained above, the existence of the solution of Equation (2.2.22) can be proved as shown below. Let

$$0 < \lambda_1 \leqslant \lambda_2 \leqslant \ldots \leqslant \lambda_n \to \infty$$

be the spectrum of A, let $\{\boldsymbol{\omega}_k\}$ be an orthogonal system complete in $N(\Omega)$ and $N^1(\Omega)$ and normed in $N(\Omega)$, consisting of the corresponding eigenvectors, and let V_n be the linear space spanned by the first n vectors of the system ($n = 1, 2, \ldots$) (Appendix 1).

If $\mathbf{u}_n$ is the orthogonal projection of $\mathbf{u}$ on V_n, then the projection of Equation (2.2.22) on this space is

(2.2.42) $$\left(\frac{d\mathbf{u}_n}{dt}, \boldsymbol{\omega}_k\right) + (A\mathbf{u}_n, \boldsymbol{\omega}_k) = (R(\mathbf{u}_n), \boldsymbol{\omega}_k) + (\mathbf{f}, \boldsymbol{\omega}_k)$$

$$(k = 1, 2, \ldots, n).$$

Hence,

(2.2.43) $$\mathbf{u}_n = \sum_{k=1}^{n} C_{k,n}(t)\boldsymbol{\omega}_k,$$

where the coefficients $C_{k,n}$ may be determined from the system of ordinary differential equations

$$\frac{dC_{k,n}}{dt} - \sum_{i=0}^{n} a_{k,i} C_{i,n} + \sum_{i,s=0}^{n} a_{k,i,s} C_{i,n} C_{s,n} = f_k \tag{2.2.44}$$

$$(k = 1, 2, \ldots, n)$$

with the initial conditions

$$C_{k,n}(0) = (\mathbf{u}_0, \boldsymbol{\omega}_k) \quad (k = 1, 2, \ldots, n), \tag{2.2.45}$$

where $a_{k,i}$ and $a_{k,i,s}$ are constants and $f_k = (\mathbf{f}, \boldsymbol{\omega}_k)$.

Applying the operator A to (2.2.43) and taking into account the equality $A\boldsymbol{\omega}_k = \lambda_k \boldsymbol{\omega}_k$, we have

$$A\mathbf{u}_n(t) = \sum_{k=1}^{n} \lambda_k C_{k,n}(t) \boldsymbol{\omega}_k. \tag{2.2.46}$$

Then, multiplying (2.2.42) by $\lambda_k C_{k,n}(t)$ and summing with respect to k, we get for $\mathbf{u}_n$ the relation

$$\frac{1}{2} \frac{d}{dt} \| \mathbf{u}_n \|^2 + | A\mathbf{u}_n |^2 = (R\mathbf{u}_n, A\mathbf{u}_n) + (\mathbf{f}(t), A\mathbf{u}_n). \tag{2.2.47}$$

i.e. $\mathbf{u}_n$ satisfies Equation (2.2.35). From the above considerations, we obtain for $\| \mathbf{u}_n \|$, $\int_0^T | A\mathbf{u}_n |^2 \, dt$ and $\int_0^T \left| \frac{d\mathbf{u}_n}{dt} \right| dt$ estimates which do not depend on n. Then, from the sequence $\{\mathbf{u}_n\}$ a subsequence $\mathbf{u}_{n_k}$ can be extracted, such that $\mathbf{u}_{n_k}$ converges strongly in $L^2((0, T), N)$ towards a function $\mathbf{u}(t)$, and $A\mathbf{u}_{n_k}$, $\frac{d\mathbf{u}_{n_k}}{dt}$ and $R(\mathbf{u}_{n_k})$ are weakly convergent in $L^2((0, T), N)$ towards $A\mathbf{u}$, $\frac{d\mathbf{u}}{dt}$ and $R\mathbf{u}$ respectively. Consequently, $\mathbf{u}$ satisfies Equation (2.2.22), i.e.u $\mathbf{u}$ is a strong solution of the problem (1.1.1)—(1.1.4)*.

By the same method we can prove the existence of solutions to the equation

$$\frac{d\mathbf{v}}{dt} + A\mathbf{v} + M(\mathbf{u}^*, \mathbf{v}) = R(\mathbf{v}) + \mathbf{g}(t) \tag{2.2.23}$$

with the initial condition

$$\lim_{t \to +0} \mathbf{v}(t) = \mathbf{v}_0. \tag{2.2.21}$$

This yields strong solutions of problem (1.1.1)—(1.1.3) with the nonhomogeneous boundary condition (1.1.4)*. Since $\mathbf{u}^* \in W^{2,2}(\Omega)$, by Theorem 2.1.1 we have $\mathbf{u}^* \in C(\overline{\Omega})$ and $\sup_{x \in \Omega} | \mathbf{u}^*(x) | < c \| \mathbf{u} \|_{2,2}$.

Whence

$$|M(\mathbf{u}^*, \mathbf{v})| \leqslant |(\mathbf{u}^* \cdot \text{grad})\, \mathbf{v}| + |(\mathbf{v} \cdot \text{grad})\, \mathbf{u}^*| \leqslant$$

$$\leqslant \left\{\int_\Omega |\mathbf{u}^*(x)|^2 \left[\sum_{i,j} \left(\frac{\partial v_j}{\partial x_i}\right)^2\right] \mathrm{d}x\right\}^{\frac{1}{2}} +$$

$$+ \left\{\int_\Omega |\mathbf{v}(x)|^2 \left[\sum_{i,j} \left(\frac{\partial u_j^*}{\partial x_i}\right)^2\right] \mathrm{d}x\right\}^{\frac{1}{2}} \leqslant \sup_{x \in \Omega} |\mathbf{u}^*(x)|\, \|\mathbf{v}\| +$$

$$+ \left(\int_\Omega |\mathbf{v}(x)|^4 \,\mathrm{d}x\right)^{1/4} \left\{\int_\Omega \left[\sum_{i,j} \left(\frac{\partial u_j^*}{\partial x_i}\right)^2\right]^2 \mathrm{d}x\right\}^{\frac{1}{4}} \leqslant k \|\mathbf{u}^*\|_{2,2} \|\mathbf{v}\| = c_1 \|\mathbf{v}\|.$$

Hence, for a given $\mathbf{u}^*$, $M_{\mathbf{u}^*} = M(\mathbf{u}^*, \cdot)$ is a bounded linear operator from N^1 to N.

By scalar multiplication in N of Equation (2.2.23) by $A\mathbf{v}$ and by using the boundedness of $M_{\mathbf{u}^*}$, we get the inequality

$$\frac{1}{2} \frac{\mathrm{d}}{\mathrm{d}t} \|\mathbf{v}(t)\|^2 \leqslant k\{\|\mathbf{v}\|^6 + c_1^2 \|\mathbf{v}\|^2 + |\mathbf{g}(t)|^2\}.$$

As above, a local theorem for the existence and uniqueness of solutions of problem (2.2.23), (2.2.21) can be proved.

Remarks: i) Using the same method, we can prove the continuous dependence of the solutions from $C^\circ((0, T), N^1)$ on initial data in N^1.

ii) The Navier-Stokes equations, linearized about $\mathbf{u}^*$, obtained from (2.2.23) by putting $R(\mathbf{v}) = 0$, admit a solution in every time interval $[0, T]$. Hence, in this case, an existence theorem in the large can be proved.

§ 2.3. COMPLETENESS OF NORMAL MODES

2.3.1. Motions in bounded domains

Consider a stationary fluid flow $(\overline{\mathbf{u}}(\mathbf{x}), \bar{p}(\mathbf{x}))$, perturbed up to a nonstationary flow $(\mathbf{u}(\mathbf{x}, t), p(\mathbf{x}, t))$, which corresponds to the same body forces and boundary conditions as the stationary one. Then the infinitesimal perturbation $\mathbf{v}(\mathbf{x}, t)$, $q(\mathbf{x}, t))$, $(\mathbf{v} = \mathbf{u} - \overline{\mathbf{u}}$, $q = p - \bar{p})$ is a solution of the following linear initial-value problem

$$\frac{\partial \mathbf{v}}{\partial t} + (\overline{\mathbf{u}} \cdot \text{grad})\, \mathbf{v} + (\mathbf{v} \cdot \text{grad})\, \overline{\mathbf{u}} - \nu \Delta \mathbf{v} = -\frac{1}{\rho} \text{grad}\, q, \tag{2.3.1}$$

$$\mathbf{v}|_{t=0} = \mathbf{v}_0, \tag{2.3.2}$$

where $\mathbf{v}$ is a solenoidal vector field vanishing on $\partial\Omega$ and sufficiently smooth so that each term of (2.3.1) is a continuous function. Hence $\mathbf{v}$ belongs to the class (2.1.5).

Using the method of separation of variables, let us choose particular perturbations $(\mathbf{v}, q)$ in the form of normal modes

$$\mathbf{v}(\mathbf{x}, t) = e^{-\sigma t}\,\mathbf{v}_0(\mathbf{x}), \quad q(\mathbf{x}, t) = e^{-\sigma t}\,q_0(\mathbf{x}). \tag{2.3.3}$$

Putting $\mathbf{v}_0 = \boldsymbol{\omega}$, problem (2.3.1), (2.3.2) is reduced to the eigenvalue problem

$$-\sigma\boldsymbol{\omega} - \nu\Delta\boldsymbol{\omega} + (\overline{\mathbf{u}}\cdot\text{grad})\,\boldsymbol{\omega} + (\boldsymbol{\omega}\cdot\text{grad})\,\overline{\mathbf{u}} = -\frac{1}{\rho}\,\text{grad}\;q_0, \tag{2.3.4}$$

$$\boldsymbol{\omega}\,|_{\partial\Omega} = 0, \tag{2.3.5}$$

where the eigenvalues σ_n are complex numbers, and the eigenfunctions are just the initial perturbations $\boldsymbol{\omega}_n(=\mathbf{v}_0)$.

Note that the use of this method restricts the class S_1 of perturbations to a class of normal modes $\mathbf{v}_n = \mathbf{e}^{-\sigma_n t}\,\boldsymbol{\omega}_n(\mathbf{x})$, and the class S_2 of the initial perturbations to that of the eigenfunctions $\boldsymbol{\omega}_n$. For hydrodynamic stability purposes, the perturbations must be arbitrary in these classes. So we make the following hypothesis: the system $\{\boldsymbol{\omega}_n\}$ is complete in S_2. Hence every $\mathbf{v}_0 \in S_2$ is of the form

$$\mathbf{v}_0 = \sum_{n=1}^{\infty} c_n\boldsymbol{\omega}_n(\mathbf{x}), \tag{2.3.6}$$

and every $\mathbf{v} \in S_1$ is of the form

$$\mathbf{v} = \sum_{n=1}^{\infty} c_n e^{-\sigma_n t}\boldsymbol{\omega}_n(\mathbf{x}), \tag{2.3.7}$$

where c_n are constants. We shall analyse this hypothesis in the class of strong solutions for bounded Ω of class C^2.

The rest case $(\overline{\mathbf{u}} \equiv 0)$. In this case, problem (2.3.4), (2.3.5) becomes

$$\sigma\boldsymbol{\omega} + \nu\Delta\boldsymbol{\omega} = \frac{1}{\rho}\,\text{grad}\;q_0, \tag{3.2.8}$$

$$\boldsymbol{\omega}\,|_{\partial\Omega} = 0, \tag{2.3.5}$$

the strong solutions of which are (Section 2.2.2) pairs of the form $(\boldsymbol{\omega}, q)$, with $\boldsymbol{\omega} \in N^2$ and $\text{grad}\;q \in G(\Omega)$, which satisfy Equation (2.3.8). The projection of this equation on N is

$$-\sigma\boldsymbol{\omega} = \nu A\boldsymbol{\omega}, \tag{2.3.8)'$$

or, equivalently,

$$A^{-1}\boldsymbol{\omega} = \frac{\nu}{\sigma}\,\omega, \tag{2.3.8)''$$

where $A = -P\Delta$ and A^{-1} is a selfadjoint, completely continuous operator defined on the Hilbert space N (Theorem 2.2.3). In particular these considerations show the existence of a complete system of orthonormal eigenvectors $\{\boldsymbol{\omega}_n\}$, in N^1 such that every initial perturbation from rest is of the form

$$\mathbf{v}_0 = \sum_{n=1}^{\infty} ((\mathbf{v}_0, \boldsymbol{\omega}_n))\boldsymbol{\omega}_n. \tag{2.3.6$'$}$$

The perturbations $\mathbf{v}$, at every moment $t > 0$, admit the expansion

$$\mathbf{v} = \sum_{n=1}^{\infty} ((\mathbf{v}, \boldsymbol{\omega}_n))\,\boldsymbol{\omega}_n, \tag{2.3.7$'$}$$

whose Fourier coefficients $c_n = ((\mathbf{v}, \boldsymbol{\omega}_n))$ satisfy the system

$$c_n' + \sigma c_n = 0, \tag{2.3.9}$$

$$c_n(0) = ((\mathbf{v}_0, \boldsymbol{\omega}_n)). \tag{2.3.10}$$

The solutions of this system are $c_n(t) = c_n(0)\mathrm{e}^{-\sigma_n t}$. Every $\mathbf{v}_0 \in N^1$ and $\mathbf{v} \in C((0, T), N^1)$ may be expressed by (2.3.6) and respectively (2.3.7), where $c_n = c_n(0)$. Thus, in the particular case when $\bar{\mathbf{u}} = 0$, the completeness of the normal modes in N^1 is proved [45].

The general case ($\bar{\mathbf{u}} \not\equiv 0$). Assume again that Ω is bounded and of class C^2 and consider the strong solutions of the problem (2.3.4), (2.3.5), projected on N, in the form

$$-\sigma\boldsymbol{\omega} + \nu A\boldsymbol{\omega} + M(\bar{\mathbf{u}}, \boldsymbol{\omega}) = 0, \tag{2.3.11}$$

where $M(\bar{\mathbf{u}}, \boldsymbol{\omega}) = -P[(\bar{\mathbf{u}}\cdot\operatorname{grad})\boldsymbol{\omega} + (\boldsymbol{\omega}\cdot\operatorname{grad})\,\bar{\mathbf{u}}]$, and $M\bar{\mathbf{u}} = M(\mathbf{u},\cdot)$ is a bounded map from N^1 to N (Section 2.2.4). Note that the operator A, being the inverse of a continuous everywhere defined operator, is closed (Theorem 2.2.3) and $M_{\bar{\mathbf{u}}}$ is closed since it is bounded; it follows that $\tilde{A} \overset{\text{def}}{=} A + M_{\bar{\mathbf{u}}}$ is also closed and therefore its spectrum consists only of eigenvalues, which in general are complex numbers (A being generally nonsymmetric).

Applying the operator $(-A)^{-1}$ to Equation (2.3.11) we get

$$\omega = -\frac{1}{\nu}A^{-1}M_{\bar{\mathbf{u}}}\boldsymbol{\omega} + \frac{\sigma}{\nu}A^{-1}\boldsymbol{\omega}, \tag{2.3.11$'$}$$

where $A^{-1}M$ is completely continuous since it is the product of a completely continuous operator and a bounded one. Therefore, in the right-hand side of (2.3.11) we have a completely continuous operator, hence, its spectrum is discrete and of the form $\{\sigma_n\}$. This set is called the *stability spectrum* and depends continuously on the kinematic viscosity ν (or on the Reynolds number if the equations are written in a dimensionless form).

Equation (2.3.1) belongs to the class studied by Keldysh [41], from whose results we deduce that the following system of small oscillations

$$(2.3.12)\qquad \mathbf{v}_n = e^{-\sigma_n t}\left(\boldsymbol{\omega}_{jk}^{(n)} + \frac{t}{1!}\,\boldsymbol{\omega}_{j(k-1)}^{(n)} + \ldots + \frac{t^k}{k!}\,\boldsymbol{\omega}_{j_0}^{(n)}\right)$$

where $\boldsymbol{\omega}_{j_0}^{(n)}$ are eigenvectors and $\boldsymbol{\omega}_{jk}^{(n)}$ are their associated eigenvectors, corresponding to the eigenvalue σ_n of Equation (2.3.11)', is complete in the class of strong solutions of the problem (2.3.1), (2.3.2). In other words, every perturbation $\mathbf{v}$ has the form

$$(2.3.13)\qquad \mathbf{v} = \sum_{i=1}^{\infty} P_i(\mathbf{x}, t)e^{-\sigma_i t}\,\mathbf{v}_0(\mathbf{x}),$$

where $P_i(\mathbf{x}, t)$ is a polynomial of degree $k(i, j)$ with respect to t. Hence, around every stationary solution, the perturbations have the form (2.3.13), and the initial perturbations are of the form (2.3.6).

Remark 2.3.1. In the case of a stationary motion $\overline{\mathbf{u}} \neq 0$, the general form of the perturbations $\mathbf{v}$ is (2.3.13); and when the eigenvalues belonging to the stability spectrum are simple the form of these perturbations is given by (2.3.7). In any case, from (2.3.13) or from (2.3.7) it follows that, when every eigenvalue has positive real part, $\mathbf{v} \to 0$ as $t \to \infty$ and therefore in this case, the basic flow $\overline{\mathbf{u}}$ is linearly stable. Since the stability spectrum depends on the Reynolds number, it has positive real parts for certain Reynolds number $\mathrm{Re} < \mathrm{Re}_{cr}$ where Re_{cr} is the critical Reynolds number corresponding to the neutral stability (§ 2.5). Hence for $\mathrm{Re} < \mathrm{Re}_{cr}$ we have linear stability and for $\mathrm{Re} > \mathrm{Re}_{cr}$ we have linear instability. It is to be expected that the same should occur in the nonlinear theory where instead of Equation (2.3.1) we have a nonlinear equation which contains the term $(\mathbf{v} \cdot \mathrm{grad})\,\mathbf{v}$. Nevertheless, as will be seen in the next paragraph, the linearization principle ensures the nonlinear stability of $\overline{\mathbf{u}}$ for $\mathrm{Re} < \mathrm{Re}_{cr}$ and its stability for $\mathrm{Re} > \mathrm{Re}_{cr}$ only in the case when $\|\mathbf{v}_0\|$ is small enough.

2.3.2. Motions in unbounded domains

The main types of motion studied in hydrodynamic stability theory (i.e. plane Couette and Poiseuille flows or Couette and Poiseuille flows in tubes) take place in domains unbounded in one or two directions such that the perturbations are assumed periodic in those directions. In the case of plane parallel flows, Squire's theorem (Section 1.1.4) shows that the critical perturbations are the two-dimensional ones. For flows through pipes, they are assumed axi-symmetric. This is why, by the normal mode method, perturbations are sought in the form

(2.3.3)′ $$\mathbf{v}(\mathbf{x}, t) = e^{-\sigma t + i\alpha y}\boldsymbol{\omega}(x),$$

where α is the wave number in the x-direction. Introducing the function which describes the perturbation of velocity (and of the temperature) $e^{-\sigma t + i\alpha y}\boldsymbol{\varphi}(x)$, we obtain an eigenvalue problem for the non selfadjoint equation

(2.3.14) $$L\varphi = \sigma M\varphi,$$

with homogeneous conditions for $\boldsymbol{\varphi}$ and $\dfrac{d\boldsymbol{\varphi}}{dx}$ at $x = 0$ and $x = 1$, where L and M are ordinary differential operators and M is positive definite. Thus, we have to deal once more with the problem of the completeness of normal modes (2.3.3)′ in the class of all perturbations.

Equation (2.3.14) may be specialized in order to yield the equations corresponding to the linear stability problem of plane parallel flows having velocity $(\bar{\mathbf{u}}, 0, 0)$, to the Couette flow between rotating cylinders, and to the convection. For instance, for plane parallel flows, (2.3.14) reduces to the Orr-Sommerfeld equation (Section 1.1.3), where

$$L = \left(\frac{d^2}{dx^2} - \alpha^2\right)^2 - i\alpha\, \mathrm{Re}\left[\bar{\mathbf{u}}\left(\frac{d^2}{dx^2} - \alpha^2\right) - \bar{\mathbf{u}}''\right],\ M = -\left(\frac{d^2}{dx^2} - \alpha^2\right).$$

Di Prima and Habetler [15] considered Equation (2.3.14) in a Hilbert space H, which, under some additional hypothesis, is reduced to the equation

(2.3.14)′ $$T\boldsymbol{\varphi} = \sigma\boldsymbol{\varphi}$$

in the energy space H_M of the positive definite map M, where $T = M^{-1}L$ satisfies the hypotheses of the following theorem.

Theorem (*Naimark* [67]). *Let* T *be a linear closed operator defined on a dense subspace of the Hilbert space* H, *whose resolvent* $(\mathrm{T} - \mu\mathrm{I})^{-1}$ *is completely continuous for a certain* $\mu = \mu_0$, *and let* $\{\mathrm{r_k}\}$ *be a sequence of strictly positive numbers such that* a) *no eigenvalues of* T *exist on the circumferences of the circles* $\{\mathrm{C_k}\}$ *of radius* $\mathrm{r_k}$, *having the center at the origin*; b) $\lim_{k\to\infty} \mathrm{r_k} = \infty$; c) *lim sup* $\lim_{k\to\infty}\sup_{\mu\in C_k} \|(\mathrm{T} - \mu\mathrm{I})^{-1}\| = 0$. *If* $\boldsymbol{\varphi}_{k_1}, \ldots$ $\ldots, \boldsymbol{\varphi}_{kn_k}$ *are the eigenvectors and the associated eigenvectors corresponding to the eigenvalues of* T *from the domain* $\mathrm{r_{k-1}} < |\mu| < \mathrm{r_k}$ *for* $\mathrm{k} > 1$ *and* $|\mu| < \mathrm{r_1}$ *for* $\mathrm{k} = 1$, *then every* $\boldsymbol{\varphi} \in \mathrm{H_M}$ *admits the representation*

(2.3.15) $$\boldsymbol{\varphi} = \sum_{i=1}^{\infty}\sum_{j=1}^{\infty} a_{ij}\boldsymbol{\varphi}_{ij}$$

which is convergent in the norm of $\mathrm{H_M}$.

In this way, the formal procedure is justified, according to which, in the nonlinear stability theory the Fourier components of the perturbations periodic in space are expanded in terms of the eigenfunctions of the corresponding linear problem. This procedure is developed in general and applied to the case of plane parallel flows in ref. [16] and to Couette flow between rotating cylinders in ref. [14].

In the case of nonviscous fluids subjected to the action of magnetic forces, the problem of completeness of the normal modes has been studied in [18] and [17]. Using classical methods, expansion theorems in terms of the eigenfunctions of the linear stability problem of the plane parallel flows and of the flows in tubes are given in ref. [84].

The normal mode method, according to which the solution of the initial-value problem (2.3.1) and (2.3.2) is sought in the form (2.3.3), has been used in hydrodynamic stability theory for about one century (Chap. 1). This method relies on physical reasonings concerning the time dependence of the acoustical and mechanical perturbations which appear in this theory, as well as on the remark that Equation (2.3.1) is invariant with respect to time translations. As a consequence, instead of the initial-value problem expressed by relations (2.3.1) and (2.3.2) the eigenvalue problem (2.3.4), (2.3.5) is studied, regardless of the fact that their solutions do not always coincide. This leads to some ambiguities, paradoxes or omissions, as has been pointed out by Case in [3]—[7]. Thus, for instance, in the case of plane parallel flows between two plates, the solution of the Rayleigh equation has two branches, and in the linear frame no criterion of selecting one of them exists; for plane Couette flows, the Rayleigh equation has no solution which also satisfies the homogeneous boundary conditions on the plates (in this last case the continuous spectrum is not taken into account). Moreover, this method sometimes leads to an arbitrary choice of boundary conditions and arises the problem of proving the completeness of normal modes. All these difficulties are removed by the use of method of integral transforms when solving the initial value problem (2.3.1), (2.3.2). For plane parallel flows with velocity $\mathbf{\bar{u}} = (\bar{u}(y), 0, 0)$ between two plates located at $y = 0$ and $y = 1$, we take the Fourier transform with respect to x and Laplace transform with respect to t, i.e.

$$v_\alpha(y, t) = \int_{-\infty}^{+\infty} v_2(x, y, t)\, e^{-i\alpha x}\, dx,\ v_0(y) = \int_0^{\infty} v_\alpha(y, t)\, e^{\sigma t}\, dt,$$

where $\mathbf{v} = (v_1, v_2, v_3)$.
From (2.3.1) and (2.3.2) we get

$$(2.3.16)\quad (-\sigma + i\alpha\bar{u})\left(\frac{d^2}{dy^2} - \alpha^2\right)v_\sigma - i\alpha\bar{u}''v_\sigma - \nu\left(\frac{d^2}{dy^2} - \alpha^2\right)^2 v_\sigma = \\ = \left(\frac{d^2}{dy^2} - \alpha^2\right)v_\alpha(y, 0).$$

This becomes the Orr-Sommerfeld equation if $v_\alpha(y, 0) = 0$ and $\sigma = i\alpha c$. If $u'' = 0$ (i.e. the motion is plane Couette flow) (2.3.16) has a continuous spectrum which we consider in Section 2.4.3.

The normal mode method is suitable for the study of problem (2.3.1), (2.3.2) when the corresponding operators in the equation with respect to v_σ have a discrete spectrum. Its use for the case of singular operators or operators with a continuous spectrum is more difficult as it requires consideration of the initial-value problem (2.3.1), (2.3.2). Then the singularities at σ give rise to poles of the transform $v_\sigma\left(\text{hence } v_\sigma \sim \dfrac{1}{\sigma^k}\right)$, and these in their turn, through the inversion formula

$$v_\alpha(y, t) = \frac{1}{2\pi i}\int_C v_\sigma(y)\, e^{-\sigma t}\, d\sigma,$$

become poles in t. Eventually, $\mathbf{v}_2$ will no longer be purely exponential with respect to t (as in the normal mode method) and will have [6] the behaviour

$$v_2 \sim t^{-(k-1)}\, e^{-\sigma t}.$$

These considerations concern the nonviscous case to which the singular operator corresponds in (2.3.16).

§ 2.4. LINEARIZATION PRINCIPLE

2.4.1. The finite-dimensional case

Consider the system of differential equations

$$\text{(2.4.1)} \qquad \frac{dx}{dt} = f(x, \lambda),$$

where f is a function of class C^1 with respect to x, $f : R^{n+1} \to R^n$, $x \in R^n$, λ is a real parameter and $f = (f_1, \ldots, f_n)$. Consider a stationary solution (x_0, λ) of the system (2.4.1) and a nonstationary one $y(t) = x'(t) + x_0$, where $x'(t)$ is a perturbation. Then from the system (2.4.1) we obtain the system

$$\text{(2.4.1)}' \qquad \frac{dx'_j}{dt} = f_j(x' + x_0, \lambda_0) =$$

$$= f_j(x_0, \lambda_0) + \left.\frac{\partial f_j}{\partial x_i}\right|_{(x_0,\, \lambda)} \cdot x'_i + 0(|x'|^2),$$

which may be also written in the form

$$\frac{dx'}{dt} = Ax' + 0(|x'|^2),$$

where A is the constant square matrix $\left\| \frac{\partial f_j}{\partial x_i}\bigg|_{(x_0,\ \lambda)} \right\|$. Neglecting the nonlinear terms with respect to x' from (2.4.1)$'$ we obtain the linear system

$$\frac{dz'}{dt} = Az', \tag{2.4.2}$$

which is called the first variation, or the first approximation of the system (3.1.1). Let $z' = e^{tA}z_0$ be its solution and denote by $\sigma_k = (\sigma_r)_k(\lambda) + i(\sigma_i)_k(\lambda)$, $k = 1, ..., n$, the eigenvalues of the matrix A. Taking into account the expression of z', it may be easily seen that, in the frame of the linearized theory, the solution (x_0, λ_0) is asymptotically stable if $(\sigma_r)_k < 0$, $k = 1, ..., n$ and asymptotically unstable if at least one integer j, $1 \leqslant j \leqslant n$, exists, such that $(\sigma_r)_j > 0$.

Theorem 2.4.1 *(Liapunov). If* $(\sigma_r)_k < 0$, $k = 1..., n$, *then the solution* (x_0, λ_0) *of the system* (2.1.1) *is asymptotically stable; and if at least there exists one integer* j, $1 \leqslant j \leqslant n$, *such that* $(\sigma_r)_j > 0$, *this solution is asymptotically unstable.*

This theorem is the principle of linearized stability and shows that the linear theory is sufficient to determine nonlinear stability. The first part of this theorem is known as Perron's theorem and corresponds to the nonperiodic asymptotic behaviour (if $(\sigma_i)_k = 0$, $k = 1, ..., n$) of the perturbations, or to the periodic behaviour if there exists at least one integer r, $1 \leqslant r \leqslant n$, such that $(\sigma_i)_r \neq 0$.

In the neutral case, namely when some eigenvalues of the matrix A have vanishing real parts and the rest are negative, the linearization principle says nothing. Since Liapunov's theorem gives the stability of the first approximation, in the above mentioned case certain nonlinear terms must be taken into account. Thus, when $(\sigma_r)_l = 0$, the corresponding Jordan cell has dimension 1; and we have stability if at least one integer l, $1 \leqslant l \leqslant n$ exists, such that $(\sigma_r)_l = 0$. When the corresponding Jordan cell has dimension greater than 1, we have instability. If $(\sigma_r)_k = 0$, $k = 1, ..., n$, and the normal form of the matrix A is diagonal, then the solutions are bounded for every real t, and in general are almost periodic functions [28]. Let us now pass to the infinite dimensional case when instead of Equation (2.4.1) defined in the finite-dimensional Banach space R^n we have an evolution equation in an infinite-dimensional Banach space, namely Equation (2.2.22) in the space L^2 $((0, T), N)$.

2.4.2. Linearization principle in hydrodynamic stability

Consider the Navier-Stokes equations with the boundary condition $\mathbf{u}|_{\partial\Omega} = \mathbf{w}$, and with body forces $\mathbf{f} \in N$ which do not depend on time

$$\frac{d\mathbf{u}}{dt} + A\mathbf{u} = R(\mathbf{u}) + \mathbf{f}. \tag{2.2.22}$$

Assume also that the element $\mathbf{u}^* \in W^{2,2}$ which extends $\mathbf{w}$ to the interior of Ω (section 2.2.4) is a stationary solution of this equation, called basic solution.

Definition 2.4.1 We say that the solution $\mathbf{u}^*$ is *stable in the large* (or just *stable*) with respect to perturbations of the initial conditions if, for every strong solution $\mathbf{u}(t)$ of Equation (2.2.22) corresponding to an initial condition $\mathbf{u}_0 \in W^{1,2}$ with $\mathbf{u}^* - \mathbf{u}_0 \in N^1$ and for every $\varepsilon > 0$, there exists $\eta(\varepsilon) > 0$ such that for every $t > 0$, $\| \mathbf{u}^* - \mathbf{u}(t) \| < \varepsilon$ for $\| \mathbf{u}^* - \mathbf{u}_0 \| < \eta$; if, in addition, $\lim_{t\to\infty} \| \mathbf{u}^* - \mathbf{u}(t) \| = 0$, then $\mathbf{u}^*$ is called *asymptotically*[19] *stable in the large* (or just *asymptotically stable*).

Denoting $\mathbf{u}(t) = \mathbf{u}^* + \mathbf{v}(t)$, we show now that the perturbation $\mathbf{v}$ satisfies the equation

$$\frac{d\mathbf{v}}{dt} + A\mathbf{v} + M_{\mathbf{u}^*}(\mathbf{v}) = R(\mathbf{v}), \tag{2.4.3}$$

where $M_{\mathbf{u}^*}(\mathbf{v}) = P[(\mathbf{u}^* \cdot \text{grad})\mathbf{v} + (\mathbf{v} \cdot \text{grad})\mathbf{u}^*]$. Introducing the perturbation $\mathbf{v}$, the stability of $\mathbf{u}^*$ is reduced to the stability of the null solution of Equation (2.4.3); the solution $\mathbf{u}^*$ is asymptotically stable if, for every solution $\mathbf{v}(t)$ of Equation (2.4.3) corresponding to an initial condition $\mathbf{v}_0 \in N^1$, we have a) $\| \mathbf{v}(t) \| < \varepsilon$ for $\| \mathbf{v}_0 \| < \eta$ and b) $\lim_{t\to\infty} \| \mathbf{v}(t) \| = 0$.

Owing to the difficulty of the mathematical problems conected with the proof of stability for various types of motion in most papers devoted to this subject, small perturbations are considered at the initial moment and it is assumed that they remain small at every $t > 0$. Then, instead of (2.4.3), the perturbation $\mathbf{v}$ satisfies the equation obtained by linearization with respect to $\mathbf{u}^*$,

$$\frac{d\mathbf{v}}{dt} + A\mathbf{v} + M_{\mathbf{u}^*}(\mathbf{v}) = 0. \tag{2.4.4}$$

Hence, the problem is reduced to proving that the solution of Equation (2.4.4) satisfies Conditions a) and b). In other words, we

[19] The norm $\|\cdot\|$ is that of N^1.

must prove that $\mathbf{u}^*$ is linearly asymptotically stable. Hence, the linear problem of hydrodynamic stability is reduced to the proof of Conditions a) and b) for the solution $\mathbf{v}$ of the initial-value problem for Equation (2.4.4). Now, looking for these solutions in the form of normal modes, this problem becomes an eigenvalue problem

$$(2.4.4)' \qquad (A + M_{\mathbf{u}^*})\,\boldsymbol{\omega} = \sigma\boldsymbol{\omega}.$$

Using the completeness of the normal modes (Remark 2.3.1), we can see that if all the real parts of σ are strictly positive, then $\|\mathbf{v}\| \to 0$ for $t \to \infty$ and therefore $\mathbf{u}^*$ is linearly stable. If there exists a σ with $\sigma_r < 0$, then $\mathbf{u}^*$ is linearly unstable. Now, it remains to prove that the results of the linear theory are still valid for the nonlinear theory. For this reason, an important role is played (as in the case of ordinary differential equations) by the proof, for a given motion, of a theorem called the linearization principle, having the same statement as Liapunov's theorem from Section 2.4.1. The only difference is that σ are the eigenvalues of the operator $-(A + M_{\mathbf{u}^*})$ which is the Fréchet differential at the point $\mathbf{u}^*$ of the nonlinear operator $-(A + R)$.

The first part of the linearization principle in the class of the strong solution, for stationary basic montions in bounded domains, was proved by Prodi in 1962, and its second parts by Yudovich [36] (1965). Subsequently, Sattinger [82] proved both parts of the linearization principle in the class of Hopf's solutions. A deep study of this principle is done in Iooss' papers [26], [33]. For particular motions in unbounded domains, the linearization principle has been proved by Kirchgässner and Sorger [42] fot the Couette flow when the outer cylinder is at rest, and by Romanov [78] for the plane Couette flow. For time periodic basic motions corresponding to periodic body forces, this principle has been proved by Yudovich [37], [38] in the general case for bounded domains, and by Herfort [29] for the case of the Couette flow when the outer cylinder has axial oscillations.

Theorem 2.4.2 (*Prodi* [76]). *If Ω is a two- or three dimensional domain whose boundary is of class C^2, and the spectrum of the map $A + M_{\mathbf{u}^*}$ has only positive real parts, then the solution $\mathbf{u}^*$ of the Navier-Stokes equation is stable with respect to sufficiently small initial perturbations.*

We shall prove this theorem by means of some lemmas concerning the operators of the form $\widetilde{A} = A + M$, where $M\colon N^1 \to N$ is a bounded map; we shall assume that the spectrum of $\widetilde{A}$ is located in the right half-plane and that c_1 is a constant such that

$$(2.4.5) \qquad |M\mathbf{u}| \leqslant c_1\|\mathbf{u}\|, \quad (\mathbf{u} \in N^1).$$

Lemma 2.4.1. *Under the above hypotheses, the map* $-\tilde{A} = -A - M$ *defined on* N^2 *and taking values in* N, *is the generator of a* *semigroup of operators* $\exp\{-\tilde{A}t\}$ *strongly continuous in* N, *which satisfies the inequality*

$$\|\exp\{-\tilde{A}t\}\|_{N\to N}\leqslant\exp\left(-\frac{c_1^2}{4}t\right). \tag{2.4.6}$$

Proof. Assuming σ is real, taking the scalar product of $\mathbf{u}$ with the relation $\tilde{A}\mathbf{u}-\sigma\mathbf{u}=\mathbf{g}$ in N and taking into account property (2.2.15) of A, we get

$$\|\mathbf{u}\|^2+(\mathrm{M}\mathbf{u},\mathbf{u})-\sigma|\mathbf{u}|^2=(\mathbf{g},\mathbf{u}). \tag{2.4.7}$$

Using (2.4.5), it follows that

$$\|\mathbf{u}\|^2-\sigma|\mathbf{u}|^2\leqslant c_1\|\mathbf{u}\|\cdot|\mathbf{u}|+|\mathbf{g}|\cdot|\mathbf{u}|\leqslant\|\mathbf{u}\|^2+\frac{|\mathbf{u}|^2c_1^2}{4}+|\mathbf{g}|\cdot|\mathbf{u}|,$$

or

$$\left(-\sigma-\frac{1}{4}c_1^2\right)|\mathbf{u}|\leqslant|\tilde{A}\mathbf{u}-\sigma\mathbf{u}|.$$

By the hypothesis concerning the spectrum of A, the resolvent $(\tilde{A}-\sigma I)^{-1}$ exists for every $\sigma<0$, and from this inequality, it follows that

$$\|(\tilde{A}-\sigma I)^{-1}\|_{N\to N}\leqslant\left(-\sigma-\frac{1}{4}c_1^2\right)^{-1},\ \text{for}\ \sigma<-\frac{1}{4}c_1^2.$$

Putting $-\sigma=\mu$, for $\mu>\dfrac{c_1^2}{4}$ it follows that

$$\|(-\tilde{A}-\mu I)^{-1}\|_{N\to N}\leqslant\left(\mu-\frac{c_1^2}{4}\right)^{-1},$$

and the map $-\tilde{A}$ is closed (Section 2.3.1). Finally, $D(-\tilde{A})=N^2$, hence $D(-\tilde{A})$ is dense in N. The assertions in the lemma follow then from the Hille-Yosida theorem (Appendix 2).

Lemma 2.4.2. *The restriction of* $-\tilde{A}$ *to the subspace* $\tilde{\mathrm{N}}=\{\mathbf{u}\in\mathrm{N}^2\mid\tilde{A}\mathbf{u}\in\mathrm{N}^1\}$ *is the generator of a semigroup of maps strongly continuous in* N^1, *which satisfies the inequality*

$$\|\exp\{-\tilde{A}t\}\|_{N^1\to N^1}\leqslant\exp\left(-\frac{c_1^2}{4}t\right). \tag{2.4.6$'$}$$

Proof. Taking the scalar product of the expression $\tilde{A}\mathbf{u}-\sigma\mathbf{u}=\mathbf{g}$ with $A\mathbf{u}$ where $\mathbf{u}\in\tilde{N}$, we get

$$|A\mathbf{u}|+(M\mathbf{u},A\mathbf{u})-\sigma\|\mathbf{u}\|^2=((\mathbf{g},\mathbf{u})),$$

whence

$$|A\mathbf{u}|^2 - \sigma\|\mathbf{u}\|^2 \leqslant \|\mathbf{g}\| \cdot \|\mathbf{u}\| + c_1\|\mathbf{u}\| \cdot |A\mathbf{u}|.$$

As above, it follows that

$$\|(\tilde{A} - \sigma I)^{-1}\|_{N \to N} \leqslant \left(-\sigma - \frac{1}{4} c_1^2\right)^{-1}$$

for $0 < -\sigma - \frac{1}{4} c_1^2$, and we can again apply the Hille-Yosida theorem.

Consider now the linear differential equation

$$\frac{d\mathbf{v}}{dt} + A\mathbf{v} + M\mathbf{v} = \mathbf{h}(t) \tag{2.4.8}$$

with $\mathbf{h}(t) \in L^2((0, T), N(\Omega))$ and $\mathbf{v}_0 \in N^1$. Its solution $\mathbf{v} \in C((0, T), N^1(\Omega))$, satisfying Conditions a) and b) from the definition of the strong solutions (Section 2.2.3), can be expressed in the form

$$\mathbf{v}(t) = \exp\{-t\tilde{A}\}\mathbf{v}_0 + \int_0^t \exp\{-\tau\tilde{A}\} \cdot \mathbf{h}(t-\tau)\, d\tau, \tag{2.4.9}$$

where $\exp\{-t\tilde{A}\}$ is the semigroup generated by $-\tilde{A}$ in N^1.

Lemma 2.4.3. *For* $0 \leqslant t \leqslant 1$, *the operator* $\exp\{-t\tilde{A}\}$ *from* N *into* N^1 *satisfies an inequality of the form*

$$\|\exp(-t\tilde{A})\|_{N \to N^1} \leqslant \frac{c_2}{t} \cdot \tag{2.4.10}$$

Proof. Since for $\mathbf{v}_0 \in N^1$, $\mathbf{v}(t) = \exp\{-t\tilde{A}\}\mathbf{v}_0$ is the solution of the homogeneous equation

$$\frac{d\mathbf{v}}{dt} + A\mathbf{v} + M\mathbf{v} = 0,$$

we obtain, after scalar multiplication in N by $\mathbf{v}(t)$,

$$\frac{1}{2}\frac{d}{dt}|\mathbf{v}|^2 + \|\mathbf{v}\|^2 = -(M\mathbf{v}, \mathbf{v}).$$

Therefore

$$\frac{1}{2}\frac{d}{dt}|\mathbf{v}|^2 \leqslant -\|\mathbf{v}^2\| + c_1\|\mathbf{v}\| \cdot |\mathbf{v}| \leqslant \frac{1}{2} c_1^2|\mathbf{v}|^2 - \frac{1}{2}\|\mathbf{v}\|^2,$$

whence

$$(2.4.11) \qquad |\mathbf{v}(t)|^2 \leqslant e^{c_1^2 t}\left[|\mathbf{v}_0|^2 - \int_0^t e^{-c_1^2\tau}\,\|\mathbf{v}(\tau)\|^2\,d\tau\right]^{20)}.$$

For $0 \leqslant t \leqslant 1$, it follows that

$$\int_0^t \|\mathbf{v}(\tau)\|^2\,d\tau \leqslant c_3|\mathbf{v}_0|^2;$$

integrating the inequality

$$|\mathbf{v}(t)|^2 \leqslant c_4\|\mathbf{v}(\tau)\|^2,$$

from 0 to t we get

$$t\|\mathbf{v}(t)\|^2 \leqslant c_4\int_0^t \|\mathbf{v}(\tau)\|^2\,d\tau \leqslant c_3c_4|\mathbf{v}_0|^2.$$

Since $\mathbf{v}_0$ is arbitrary in N^1, the inequality is immediate.

Lemma 2.4.4. *If we denote by* $\mathbf{v}_1(t) = \int_0^t \exp\{-\tau\tilde{A}\}\,\mathbf{h}(t-\tau)\,d\tau$ *the solution of Equation* (2.4.8) *corresponding to the initial value* $\mathbf{v}_0 = 0$, *then* $\mathbf{v}_1$ *satisfies the inequality*

$$(2.4.12) \qquad \|\mathbf{v}_1\|^2 \leqslant c_5\int_0^t |\mathbf{h}(\tau)|^2\,d\tau, \text{ for } 0 \leqslant t \leqslant 1.$$

Proof. After scalar multiplication of (2.4.8) by $A\mathbf{v}$, we get

$$\frac{1}{2}\frac{d}{dt}\|\mathbf{v}\|^2 + |A\mathbf{v}|^2 + (M\mathbf{v}, A\mathbf{v}) = (\mathbf{h}, A\mathbf{v}).$$

Applying Young's inequality (2.2.32) with $p = q = 2$ and $\varepsilon = \frac{1}{\sqrt{2}}$, we get

$$(2.4.13) \qquad \frac{1}{2}\frac{d}{dt}\|\mathbf{v}\|^2 + \frac{1}{2}|A\mathbf{v}|^2 \leqslant |\mathbf{h}|^2 + c_1^2\|\mathbf{v}\|^2.$$

---

[20] In order to deduce the last inequality, the following property [52] is used: if $y(t)$ is a non-negative absolutely continuous function satisfying for almost every $t \in [0, T]$ the inequality $\frac{dy(t)}{dt} \leqslant c_1(t)y(t) \pm c_2(t)$, where c_1 and c_2 are non-negative functions integrable on $[0, T]$, then y satisfies the inequality

$$y(t) \leqslant \left\{\exp\left[\int_0^t c_1(\tau)\,d\tau\right]\right\}\left[y(0) \pm \int_0^t c_2(\tau)\exp\left\{-\int_0^\tau c_1(\eta)\,d\eta\right\}d\tau\right].$$

The absolute continuity of the function $\|\mathbf{v}(t)\|^2$ folows from (2.2.34).

Hence,

$$\frac{1}{2}\frac{\mathrm{d}}{\mathrm{d}t}\|\mathbf{v}\|^2 \leqslant \frac{1}{2}(c_1^2\|\mathbf{v}\|^2 + |\mathbf{h}|^2).$$

Assuming that $\mathbf{v}_0 = 0$ and proceeding as in the derivation of (2.4.11) we obtain

$$\|\mathbf{v}\|^2 \leqslant \mathrm{e}^{c_1^2 t}\left[\int_0^t |\mathbf{h}(\tau)|^2 \mathrm{e}^{-c_1^2\tau}\,\mathrm{d}\tau\right] \leqslant c_5\int_0^t |\mathbf{h}(\tau)|^2\,\mathrm{d}\tau, \quad (0 \leqslant t \leqslant 1).$$

Lemma 2.4.5. *For* t > 0, *every solution* $\mathbf{v}$(t) *of Equation* (2.4.8) *satisfies the inequality*

$$\|\mathbf{v}(t)\|^2 + \int_0^t |A\mathbf{v}(\tau)|^2\,\mathrm{d}\tau \leqslant c_6\left\{\|\mathbf{v}_0\|^2 + \int_0^t |\mathbf{h}(\tau)|^2\,\mathrm{d}\tau\right\}. \tag{2.4.14}$$

Proof. Taking into account (2.4.9) and (2.4.6)′, we have the estimate

$$\|\mathbf{v}(t)\|^2 \leqslant 2(\|\mathbf{v}_0\|^2 \exp(-2\gamma t) + \|\mathbf{v}_1(t)\|^2 + \|\mathbf{v}_2(t)\|^2), \tag{2.4.15}$$

where

$$\gamma = \frac{c_1^2}{4},\quad \mathbf{v}_1(t) = \int_0^1 \exp\{-\tau\tilde{A}\}\mathbf{h}(t-\tau)\,\mathrm{d}\tau,\quad \mathbf{v}_2(t) = $$

$$= \int_0^t \exp\{-\tau\tilde{A}\}\,\mathbf{h}(t-\tau)\,\mathrm{d}\tau.$$

Applying Lemma 2.4.4 for $t = 1$, we obtain

$$\|\mathbf{v}_1(t)\|^2 \leqslant c_5\int_0^1 |\mathbf{h}(t-1+\tau)|^2\,\mathrm{d}\tau = c_5\int_0^1 |\mathbf{h}(t-\tau)|^2\,\mathrm{d}\tau. \tag{2.4.16}$$

On the other hand, for $t \geqslant 1$, we have

$$\|\exp\{-t\tilde{A}\}\|_{N\to N^1} \leqslant \|\exp\{-\tilde{A}\}\exp\{-(t-1)\tilde{A}\}\|_{N\to N^1} \leqslant$$

$$\leqslant \|\exp\{-\tilde{A}\}\|_{N\to N^1}\|\exp\{-(t-1)\tilde{A}\}\|_{N^1\to N^1}.$$

Using relations (2.4.6)′ and (2.4.10) for $t = 1$, it follows that

$$\|\exp\{-t\tilde{A}\}\|_{N\to N^1} \leqslant c_7\exp(-\gamma t), \text{ for } t \geqslant 1.$$

Then we have

$$\|\mathbf{v}_2(t)\| \leqslant \int_1^t \|\exp\{-\tau\tilde{A}\}\,\mathbf{h}(t-\tau)\|\,\mathrm{d}\tau \leqslant$$

$$\leqslant \int_1^t \|\exp\{-\tau\tilde{A}\}\|_{N\to N^1}|\mathbf{h}(t-\tau)|\,\mathrm{d}\tau \leqslant c_7\int_1^t \exp(-\gamma\tau)|\mathbf{h}(t-\tau)|\,\mathrm{d}\tau.$$

If γ' is such that $0 < \gamma' < \gamma$, then

$$\text{(2.4.17)}\quad \| \mathbf{v}_2(t) \|^2 \leqslant c_7^2 \left\{ \int_1^t \exp(-(\gamma - \gamma')\tau) \exp(-\gamma'\tau) \, | \mathbf{h}(t-\tau) | \, d\tau \right\}^2 \leqslant$$

$$\leqslant c_7^2 \int_1^\infty \exp(-2(\gamma - \gamma')\tau) \, d\tau \int_1^t \exp(-2\gamma'\tau) \, | \mathbf{h}(t-\tau) |^2 \, d\tau \leqslant$$

$$\leqslant c_8 \int_1^t \exp(-2\gamma'\tau) \, | \mathbf{h}(t-\tau) |^2 \, d\tau.$$

From the relations (2.4.15)—(2.4.17), we deduce that

$$\text{(2.4.18)}\qquad \| \mathbf{v}(t) \|^2 \leqslant c_9 \{ \exp(-2\gamma t) \| \mathbf{v}_0 \|^2 + \int_0^t \exp(-2\gamma'(t-\tau)) \, | \mathbf{h}(\tau) |^2 d\tau \}.$$

Integrating inequalities (2.4.18) and (2.4.13), we obtain the relations

$$\text{(2.4.19)}\qquad \int_0^t \| \mathbf{v}(\tau) \|^2 \, d\tau \leqslant (2\gamma)^{-1} c_9 \| \mathbf{v}_0 \|^2 + c_9 \int_0^t d\eta \int_0^\eta \exp(-2\gamma'(\eta - \tau)) \, | \mathbf{h}(\tau) |^2 \, d\tau =$$

$$= (2\gamma)^{-1} c_9 \| \mathbf{v}_0 \|^2 + (2\gamma')^{-1} c_9 \int_0^t | \mathbf{h}(\tau) |^2 \, d\tau$$

and

$$\text{(2.4.20)}\qquad \int_0^t | A\mathbf{v}(\tau) |^2 \, d\tau \leqslant \| \mathbf{v}_0 \|^2 + 2c_1^2 \int_0^t \| \mathbf{v}(\tau) \|^2 \, d\tau + 2 \int_0^t | \mathbf{h}(\tau) |^2 \, d\tau.$$

On the other hand, we note from (2.4.18) that

$$\text{(2.4.21)}\qquad \| \mathbf{v}(t) \|^2 \leqslant c_9 \left\{ \| \mathbf{v}_0 \|^2 + \int_0^t | \mathbf{h}(\tau) |^2 \, d\tau \right\}.$$

Relation (2.4.14) is then an immediate consequence of inequalities (2.4.19)—(2.4.21).

Now let δ be a number such that the spectrum of $\tilde{A}$ is contained in $\mathcal{R}e(\sigma) > \delta$. If in (2.4.8) we put $\mathbf{v}(t) = \mathbf{w}(t) \exp(-\delta t)$, we obtain an equation of the same type as (2.4.8), namely,

$$\text{(2.4.8)}'\qquad \frac{d\mathbf{w}}{dt} + \tilde{A}_1 \mathbf{w} = \mathbf{h}_1(t),$$

where $\tilde{A}_1 = \tilde{A} - \delta I = A + (M - \delta I)$, $\mathbf{h}_1(t) = \exp(\delta t) \, \mathbf{h}(t)$; and by the hypothesis concerning A, the spectrum of $\tilde{A}_1$ is contained in the right halfplane. Applying Lemma 2.4.5 to Equation (2.4.8)′, we obtain for the solution of Equation (2.4.8) the inequality

(2.4.22) $$\|\mathbf{v}(t)\|^2 + \int_0^t \exp(-2\delta(t-\tau))\,|A\mathbf{v}(\tau)|\,\mathrm{d}\tau \leqslant$$
$$\leqslant c^* \left\{ \exp(-2\delta t)\,\|\mathbf{v}_0\|^2 + \int_0^t \exp(-2\delta(t-\tau)\,|\mathbf{h}(\tau)|^2\,\mathrm{d}\tau \right\}.$$

Using this inequality, we can prove now Theorem 2.4.2. In fact, Equation (2.4.3) is formally of the same type as (2.4.8), for $M = M_{\mathbf{u}^*}$ and $\mathbf{h} = R(\mathbf{v})$. Taking into account inequality (2.2.31), from (2.4.22) we obtain the estimate

$$\|\mathbf{v}(t)\|^2 + \int_0^t \exp(-2\delta(t-\tau))\,|A\mathbf{v}(\tau)|^2\,\mathrm{d}\tau \leqslant c^* \left\{ \exp(-2\delta t)\,\|\mathbf{v}_0\|^2 + \right.$$
$$\left. + k_3 \int_0^t \exp(-2\delta(t-\tau))\,|A\mathbf{v}(\tau)|\,\|v(\tau)\|^3\,\mathrm{d}\tau \right\}.$$

Applying Young's inequality, we further deduce that

(2.4.23) $$\|\mathbf{v}(t)\|^2 \leqslant c^* \exp(-2\delta t)\,\|\mathbf{v}_0\|^2 +$$
$$+ \frac{c^{*2}k_3^2}{4} \int_0^t \exp(-2\delta(t-\tau))\,\|\mathbf{v}(\tau)\|^6\,\mathrm{d}\tau.$$

By substitution $\|\mathbf{v}(t)\|^2 \exp(2\delta t) = \psi(t)$, we obtain for $\psi(t)$ the inequation

$$\psi(t) \leqslant c^*\|\mathbf{v}_0\|^2 + \frac{c^{*2}k_3^2}{4} \int_0^t \exp(-4\delta\tau)\,\psi^3(\tau)\,\mathrm{d}\tau.$$

Hence, for every $t \geqslant 0$ we have $\psi(t) \leqslant \varphi(t)$, where $\varphi(t)$ is the solution of the differential equation

(2.4.24) $$\frac{\mathrm{d}\varphi}{\mathrm{d}t} = \frac{c^{*2}k_3^2}{4} \exp(-4\delta t)\,\varphi^3,$$

corresponding to the initial data $\varphi(0) = c^*\|\mathbf{v}_0\|^2$. Solving Equation (2.4.24) and considering again the variable $\mathbf{v}$, we obtain the estimate

(2.4.25) $$\|\mathbf{v}(t)\|^2 \leqslant \frac{c^*\|\mathbf{v}_0\|^2\,\mathrm{e}^{-2\delta t}}{\left(1 - \dfrac{k_3^2 c^{*4}}{8\delta}\,(1-\mathrm{e}^{-4\delta t})\,\|\mathbf{v}_0\|^4\right)^{1/2}}, \quad (0 \leqslant t < \infty),$$

which holds for

(2.4.26) $$\|\mathbf{v}_0\|^4 < \frac{8\delta}{k_3^2 c^{*4}}.$$

Inequality (2.4.25) ensures the asymptotic stability of the null solution of Equation (2.4.3) corresponding to an initial value $\mathbf{v}_0$ which satisfies the relation (2.4.26), and so Prodi's theorem is proved.

For the sake of simplicity, we have assumed that $\nu = 1$; in the general case, instead of relation (2.4.26), we would have obtained an inequality containing ν. Passing to dimensionless coordinates, this inequality defines in the plane (δ, Re) the domain of validity of the linearization principle.

Let us now state the second part of the linearization principle in the class of strong solutions for the two- or three- dimensional domain of motions of class C^2.

Theorem 2.4.3 *(Yudovich* [36]). *If* $\mathrm{A} + \mathrm{M}_{\mathbf{u}^*}$ *has at least one eigenvalue whose real part is negative, then the solution* $\mathbf{u}^*$ *is asymptotically unstable.*

In this way, by the linearization principle, the proof of the stability or instability of a basic flow with respect to small initial perturbations is reduced to the complete determination of the spectrum of the linear problem (2.4.4). The particular cases for which this proof has been given are: the Kolmogorov flow [60], Couette flow at large Re [49], plane Poiseuille flow [50] and plane Couette flow [78], the first three being unstable and the last one — analysed in detail in Section 2.4.3 — being stable.

Remark 2.4.1. If in (2.4.7) we assume that σ is an arbitrary complex number and separate the real and imaginary part of both sides, then, applying Young's inequality, we obtain the relation

$$\left[-\mathcal{R}e(\sigma) + \frac{1}{4c_1^2}\mathcal{I}m^2(\sigma) - c_1^2\right]|\mathbf{u}|^2 \leqslant \frac{1}{c_1^2}|\tilde{A}\mathbf{u} - \sigma\mathbf{u}|^2.$$

We deduce that if M is a continuous operator, whose norm is bounded by the constant c_1, then the spectrum of the map $\tilde{A} = A + M$ is located in the interior of the parabola whose equation is $\mathcal{R}e(\sigma) = \frac{1}{4c_1^2}\mathcal{I}m(\sigma) - c_1^2$. In the particular case $M = M_{\mathbf{u}^*}$, it follows that the assertion is true for the spectrum of the hydrodynamic problem linearized with respect to $\mathbf{u}^*$.

2.4.3. Stability of plane Couette flows

Consider plane Couette flow in the domain $\Omega = \{(x, y, z) \mid |x| \leqslant 1, \ |y| \leqslant \infty, \ |z| \leqslant \infty\}$, having the velocity profile $\mathbf{v}_0 = (0, 0, x)$. Owing to the extreme simplicity of this profile, the study of the linear stability problem for the corresponding motion was initiated as early as the beginning of this century. This study was reduced to the determination of the eigenvalues of the Orr-Sommerfeld equation

$$(2.4.27) \qquad (D^2 - \alpha^2)^2 \varphi = i\alpha R_1(x - c)(D^2 - \alpha^2)\varphi,$$

$$\left(-1 \leqslant x \leqslant 1,\ D = \frac{d}{dx}\right),$$

under the boundary conditions

$$(2.4.28) \qquad \varphi(\pm 1) = \varphi'(\pm 1) = 0,$$

the eigenvalues being $c = c_r + ic_i$ and the eigenfunctions are $\varphi(x)$. As in Section 1.1.3, α is the wave number, R_1 the Reynolds number, c_r the ratio between the propagation speed of the perturbation and the velocity of the upper plate, and c_i is the amplification factor. The eigenvalues c depend on α and R_1 and, in general, c_i takes various signs at various points of the plane (α, R_1). The domain consisting of the points (α, R_1) such that $c_i < 0$ is called the stability domain, its complementary set with respect to the first quadrant being the domain of instability.

The partial results obtained in the study of the problem (2.4.27), (2.4.28) [100], [96] converge to the assertion that plane Couette flow is linearly stable for every $\alpha \geqslant 0$ and $R_1 > 0$; the proof of this fact has been given by Romanov [77], [78] and is based on many of the already mentioned old results. Let us quote some of them. Synge's method (Chap. 1) allows the proof of the following

Lemma 2.4.6. *For every $\alpha > 0$ and $R_1 > 0$, we have*

$$(2.4.29) \qquad -1 < c_r < 1,$$

and for

$$(2.4.30) \qquad R_1^2 < 8\alpha^2(\alpha^2 + 1),$$

the plane Couette flow is stable.

Relation (2.4.29) shows that the velocity of propagation of the perturbation lies between the velocity of deplacement of the plates, and inequality (2.4.30) defines the domain of stability in the plane (α, Re_1). A larger domain of stability is given by the following lemma obtained by Joseph's method

Lemma 2.4.7. *If*

$$(2.4.31) \qquad \alpha R_1 \leqslant 2\sqrt{(2 - \eta_0)(1 - \eta_0)}\,\alpha^3 + \frac{9}{16}\pi^3(1 - \eta_1),$$

then

$$(2.4.32) \qquad \alpha R_1 c_i \leqslant -\alpha^2\eta_0 - \frac{\pi^4}{\pi^2 + 4\alpha^2}\eta_1, \text{ (where } 0 < \eta_i < 1,\ i = 0,1).$$

Finally a stronger result is Orr's stability criterion, which asserts that the plane Couette flow is stable for $R_1 < 44.3$ and every $\alpha > 0$.

Putting $\varphi'' - \alpha^2\varphi = f$, problem (2.4.27), (2.4.28) becomes

$$f'' - \alpha^2 f = \mathrm{i}\alpha R_1(x - c)f, \tag{2.4.33}$$

$$\int_{-1}^{+1} \mathrm{e}^{\alpha x} f \,\mathrm{d}x = 0, \int_{-1}^{+1} \mathrm{e}^{-\alpha x} f \,\mathrm{d}x = 0. \tag{2.4.34}$$

Problem (2.4.33), (2.4.34) has been analysed for the neutral case ($c_i = 0$) by von Mises [62], [63] who approximated Equation (2.4.33) by a sequence of equations with finite differences. A rigorous treatment is due to Hopf, who reduced the problem (2.4.27), (2.4.28) to an eigenvalue problem for a second order differential equation by means of the transformation $z = (\alpha^2 - \mathrm{i}\alpha R_1(x - c))\,(\lambda R_1)^{2/3}$. In this way, Equation (2.4.27) turns into a Bessel equation

$$f'' + zf = 0, \tag{2.4.35}$$

and Conditions (2.4.28) are replaced by (2.4.34). Two linearly-independent solutions of Equation (2.4.35) expressed by means of Hankel functions H are $f_1 = z^{1/2} H^{(1)}_{1/3}\left(\frac{2}{3} z^{3/2}\right)$ and $f_2 = z^{1/2} H^{(2)}_{1/3}\left(\frac{2}{3} z^{3/2}\right)$. The function φ satisfies the equation

$$\frac{\mathrm{d}^2\varphi}{\mathrm{d}z^2} + k^2\varphi = Af_1 + Bf_2, \left(k = \left(\frac{\alpha^2}{R_1}\right)^{2/3}\right),$$

where A and B are arbitrary constants. Multiplying this relation by $\sin kz$, and respectively by $\cos kz$, integrating between z_1 and z_2 (corresponding to $x = +1$ and $x = -1$) and eliminating A and B, we obtain the equation

$$\int_{z_1}^{z_2}\int_{z_1}^{z_2} \sin k(z' - z'')\, \varphi_1(z')\, \varphi_2(z'')\, \mathrm{d}z'\, \mathrm{d}z'' = 0.$$

This was solved by Hopf only for small or large values of the argument $\frac{2}{3} z^{3/2}$. For these cases, it follows that $c_i < 0$.

Introduce now, instead of the Hankel functions, those of Airy, whose advantages are well known in asymptotic methods. Instead of z we choose the independent variable $\xi = (x - \mathrm{i}d)\varepsilon$, where $d = -\mathrm{i}c + \frac{\alpha}{R_1}$ is the eigenvalue of problem (2.4.33), (2.4.34), and $\varepsilon = (\alpha R_1)^{-1/3}$. Then, instead of Equation (2.4.35) we obtain

$$\frac{\mathrm{d}^2 f}{\mathrm{d}\xi^2} = \mathrm{i}\xi f. \tag{2.4.35$'$}$$

Adding to this equation the boundary conditions (2.4.34), we get an eigenvalue problem whose two linearly-independent solutions are $Ai(e^{\frac{i\pi}{6}}\xi)$ and $Ai(e^{\frac{5\pi i}{6}}\xi)$, where $Ai(\xi)$ is the Airy function which satisfies the equation $Ai''(\xi) = \xi Ai(\xi)$. Then the characteristic equation is

$$\int_{-1}^{+1} e^{-\alpha x} Ai\left(e^{\frac{i\pi}{6}} \frac{x - id}{\varepsilon}\right) dx \int_{-1}^{+1} e^{\alpha x} Ai\left(e^{\frac{5i\pi}{6}} \frac{x - id}{\varepsilon}\right) dx -$$

$$-\int_{-1}^{+1} e^{\alpha x} Ai\left(e^{\frac{i\pi}{6}} \frac{x - id}{\varepsilon}\right) dx \int_{-1}^{+1} e^{-\alpha x} Ai\left(e^{\frac{5i\pi}{6}} \frac{x - id}{\varepsilon}\right) dx = 0, \tag{2.4.36}$$

from which we must deduce d (and hence c) in terms of α and ε. From this equation, we can derive the following

Lemma 2.4.8. *If for all* $\mathfrak{Im} z \leqslant \delta$ *the relation*

$$e^{\alpha x} \left| 1 - e^{-2\alpha} w\left(z, \frac{2}{\varepsilon}\right) - \alpha\varepsilon \int_0^{\frac{2}{\varepsilon}} e^{-\alpha\varepsilon x} w(z, x)\, dx \right| > \left| 1 - \right. \tag{2.4.37}$$

$$\left. - e^{2\alpha} w\left(z, \frac{2}{\varepsilon}\right) + \alpha\varepsilon \int_0^{\frac{2}{\varepsilon}} e^{\alpha\varepsilon x} w(z,x)\, dx \right|,$$

holds, where

$$w(z, x) = \frac{A_0(z + x)}{A_0(z)} = \exp\left(\int_0^x \frac{A_0'(z + t)}{A_0(z + t)}\, dt\right), \tag{2.4.38}$$

then Equation (2.4.36) *has no roots in the halfplane* $\mathfrak{Re}\, d \geqslant -\delta\varepsilon$, $\delta \in [0, \delta_0)$.

The determination of δ_0 follows from

Lemma 2.4.9. *The function* $A_0(z) = \int_{ze^{i\pi/6}}^{+\infty} Ai(t)dt$ *has no roots in the sector* $-\pi \leqslant \arg z \leqslant \frac{2\pi}{3}$.

Using this lemma and the asymptotic expansion of Wasow [100] for $A_0(z)$, it follows that the function $A_0(z)$ has no roots in a certain halfplane $\mathfrak{Im}\, z < \delta_0$, $\delta_0 > 0$. This δ_0 is the constant which appeared in Lemma 2.4.8.

For $\mathfrak{Im}\, z \leqslant \delta$, we have

$$|w(z, x)| = \exp\left(\int_0^x \mathfrak{Re} \frac{A_0'(z + t)}{A_0(z + t)}\, dt\right) \leqslant e^{-\alpha x}, \; x \geqslant 0,$$

where

$$a = a(\delta) = -\max_{\mathfrak{Im}\, z \leqslant \delta} \mathfrak{Re} \{A_0'(z)/A_0(z)\}, \; \delta \in [0, \delta_0). \tag{2.4.39}$$

Lemma 2.4.10 *In every sector* $|\arg(z\, e^{\frac{i\pi}{6}})| \leqslant \pi - \varepsilon$, $\varepsilon > 0$, *the formula*

$$A_0'(z)/A_0(z) = -e^{\frac{i\pi}{6}}(z\, e^{\frac{i\pi}{6}}) + 0(z^{-1})$$

holds for each $\delta \in [0, \delta_0)$. *The value of the maximum of* (2.4.39) *is reached on a certain interval* $z = i\delta + r$, $-r_0(\delta) \leqslant r \leqslant r_0(\delta)$, *the function* $a(\delta)$ *being continuous on* $\delta \in [0, \delta_0)$.

Lemma 2.4.11. *With the above notation, the following inequalities*

$$(2.4.40) \quad \mathcal{Re}\{A_0'(z)/A_0(z)\,|_{z\leqslant -5}\} < -\frac{1}{2}; \qquad \mathcal{Re}\{A_0'(z)/A_0(z)\,|_{z\geqslant 2,5} < -\frac{1}{2}$$

hold.

Lemma 2.4.10 implies that the maximum from (2.4.39) is reached on the real axis. Taking into account Lemma 2.4.11 Romanov calculated numerically the expression $\mathcal{Re}\{A_0'(z)/A_0(z)\}$ on the segment $[-10; 5]$, taking for $A_0(z)$ [100],

$$(2.4.41)' \qquad A_0(z) = \frac{1}{3} - \int_0^{ze^{\frac{i\pi}{6}}} Ai(t)\, dt$$

and using for $A_0'(z)$ the asymptotic expansion

$$Ai(z) = \frac{1}{3^{2/3}\,\Gamma\left(\frac{2}{3}\right)} \sum_{n=0}^{\infty} \frac{1 \cdot 4 \ldots (3n-2)}{(3n)!} z^{3n} -$$

$$(2.4.41) \qquad - \frac{1}{3^{1/3}\Gamma\left(\frac{1}{3}\right)} \sum_{n=0}^{\infty} \frac{2 \cdot 5 \ldots (3n-1)}{(3n+1)!} z^{3n+1}.$$

He also calculated the same expression by numerical integration of the differential equation $y'' + 3y'y + y^3 = izy$, which is satisfied by $y(z) = A_0'(z)/A_0(z)$, using as initial values those given by (2.4.41) and (2.4.41)′. In both calculations he obtained $\max\limits_{-5\leqslant z\leqslant 2,5} \mathcal{Re}\{A_0'(z)/A_0(z)\} = -0.48432715$, reached at $z = -3.0185988$.

It follows that

$$(2.4.42) \quad \max_{\mathcal{Im} z\leqslant 0} \mathcal{Re}\{A_0'(z)/A_0(z)\} = \max_{-5\leqslant z\leqslant 2.5} \mathcal{Re}\{A_0'(z)/A_0(z)\} = -0.48432715.$$

Using this equality and the continuity of $a(\delta)$, we see that $a(\delta)$ is continuous and positive for $\delta \in [0, \delta_1]$ for some positive δ_1.

From inequality (2.4.37), we can deduce one more Lemma which, together with the Lemma 2.4.7 shows that for every point (α, R_1) from the first quadrant, the spectrum of problem (2.4.27), (2.4.28) has a vanishing imaginary part. Hence, plane Couette flow is linearly stable for every Reynolds number.

Lemma 2.4.12. *If*

$$\alpha\varepsilon \leqslant \frac{a}{2}\left(e^{\frac{2a}{\varepsilon}} - 5\right), \quad \varepsilon \leqslant \frac{a}{\ln 3}, \tag{2.4.43}$$

then

$$\alpha R_1 c_i \leqslant -\alpha^2 - \delta(\alpha R_1)^{2/3}, \quad (\delta \in [0, \delta_1)). \tag{2.4.44}$$

The points from the first quadrant of the plane (α, R_1) are in one-to-one correspondence with those of the first quadrant of the plane (α, ε) through the relation defining ε. The domain of the first quadrant of the plane (α, ε), where inequality (2.4.43) holds, includes the domain (containing the origin)

$$\alpha\varepsilon \leqslant 2a, \quad \varepsilon \leqslant \frac{a}{\ln 3}\cdot \tag{2.4.45}$$

On the other hand, the domain (2.4.31) contains in the first quadrant a domain lying on the opposite side of the origin

$$[2\sqrt{(2-\eta_0)(1-\eta_0)}]^{-1/3} \leqslant \alpha\varepsilon, \left[\frac{9}{16}\pi^3(1-\eta_1)\right]^{-1/3} \leqslant \frac{a}{\ln 3}\cdot \tag{2.4.46}$$

Hence, in order that the union of the domains (2.4.45) and (2.4.46) cover the first quadrant $(\alpha \geqslant 0,\ R_1 > 0)$, or, equivalently, $\alpha \geqslant 0$, $\varepsilon > 0$, it is sufficient to take $a(\delta)$ such that

$$[2\sqrt{(2-\eta_0)(1-\eta_0)}]^{-1/3} \leqslant 2a, \left[\frac{9}{16}\pi^3(1-\eta_1)\right]^{-1/3} = \frac{a}{\ln 3}\cdot$$

Hence, we obtain $a = a(\delta) > \max\left(2^{-3/2}, \frac{\ln 3}{\pi}\left(\frac{4}{3}\right)^{2/3}\right) \approx 0.423$, this inequality is satisfied for $\delta \in [0, \delta_1)$.

Therefore, Lemmas 2.4.7 and 2.4.12 show that for every $\alpha \geqslant 0$ and $R_1 > 0$, the flow is stable, namely,

$$\alpha R_1 c_i < -\widetilde{c}, \tag{2.4.47}$$

where

$$\widetilde{c} = \max\left\{\alpha^2\eta_0 + \frac{\pi^4}{\pi^2 + 4\alpha^2}\eta_1,\ \alpha^2 + \delta(\alpha R_1)^{2/3}\right\}\cdot$$

In this way, the linear stability problem for plane Couette flow was completely solved.

According to the linearization principle, the motion will be nonlinearly stable for initial data having small norm in $N^1(\Omega)$. In fact, consider the equation $\widetilde{A}\mathbf{u} = \lambda\mathbf{u} + \mathbf{h}$, $\mathbf{h} \in N(\Omega)$, where $\widetilde{A}$ is the operator from Section 2.4.2 in the particular case $\mathbf{u}_0 = \mathbf{v}_0$. In order to determine the eigenvalue λ, a Fourier transform with respect to variables y and z is applied to both sides of this equality. Then the spectrum of $\widetilde{A}$ is continuous and consists of $\lambda_n(\xi_2, \xi_3, \mathrm{Re})$, where the independent parameters ξ_2 and ξ_3 run over the real axis, and λ_n runs over the discrete set of eigenvalues of the problem

$$(2.4.27)' \quad \frac{1}{\mathrm{Re}}(\mathrm{D}^2 - \xi_2^2 - \xi_3^2)^2\varphi = (\mathrm{i}\xi_2 x + \lambda)(\mathrm{D}^2 - \xi_2^2 - \xi_3^2)\varphi, -1 \leqslant x \leqslant 1,$$

$$(2.4.28)' \quad \varphi(\pm 1) = \varphi'(\pm 1) = 0,$$

and over the set of eigenvalues λ' of the problem

$$\varphi'' = (\xi_2^2 + \xi_3^2 + i\xi_2 \mathrm{Re} x + \lambda' \mathrm{Re})\varphi, \ \varphi(\pm 1) = 0.$$

In order to estimate $\mathcal{R}e\lambda$, we consider $\xi_2 \geqslant 0$ (without loss of generality, since $\overline{\lambda(\xi_2)} = \lambda(-\xi_2)$, $\overline{\varphi(\xi_2)} = \varphi(-\xi_2)$). Denoting $\alpha = \sqrt{\xi_2^2 + \xi_3^2}$, $R_1 = \xi_2\mathrm{Re}/\alpha$, $\lambda = -i\xi_2 c$, Problem (2.4.27)′, (2.4.28)′ turns into problem (2.4.27), (2.4.28), $m(\mathrm{Re})$ being the maximum of the real parts of the spectrum of $\widetilde{A}$, i.e.

$$m(\mathrm{Re}) = \max\left\{\sup_n \sup_{\alpha \geqslant 0, 0 \leqslant R_1 \leqslant Re} \frac{\alpha R_1}{\mathrm{Re}}(c_i)_n(\alpha, R_1), -\frac{\pi^2}{4\mathrm{Re}}\right\}.$$

We have

$$(2.4.48) \quad m(\mathrm{Re}) \leqslant -\frac{c}{\mathrm{Re}}.$$

Theorem 2.4.4. *For every* $\mathrm{Re} > 0$ *(where* Re *is the Reynolds number),* $\mu \in (0, -\mathrm{m}(\mathrm{Re}))$ *a* $\delta = \delta(\mathrm{Re}, \mu) > 0$ *exists such that for all initial data from the sphere* $\|\mathbf{u}_0\|_{1,2} \leqslant \delta$ *the Navier-Stokes equations for plane Couette flow admit a strong solution* $\mathbf{u}(t)$ *and we have the estimates*

$$\|\mathbf{u}(t)\|_{1,2} \leqslant C(\mathrm{Re}, \mu)\, \mathrm{e}^{-\mu t} \|\mathbf{u}_0\|_{1,2}, \quad 0 \leqslant t \leqslant \infty,$$

$$\|\mathbf{u}(t)\|_{1,p} \leqslant C(\mathrm{Re}, \mu)\, t^{-\beta} \mathrm{e}^{-\mu t} \|\mathbf{u}_0\|_{1,2}, \quad 0 < t < \infty,$$

where $2 < \mathrm{p} < 6$, $\frac{3}{4} - \frac{3}{2\mathrm{p}} < \beta < \frac{1}{2}$, *and* C *is a constant.*

§ 2.5. THE PRINCIPLE OF EXCHANGE OF STABILITIES (P.E.S.)

2.5.1. The neutral state and P.E.S.

In this paragraph the Reynolds number is denoted by λ, and by definition $\xi_0(\lambda) = \inf_{i \in N} \mathcal{R}e\{\sigma_i(\lambda)\}$, where $\{\sigma_i(\lambda)\}$ is the spectrum of the linearized stability problem (§ 2.3). It can be proved [34] that $\xi_0(\lambda)$ is a decreasing function of λ. We define the critical Reynolds number as the least positive value of λ for which $\xi_0(\lambda) = 0$. The state of the fluid corresponding to $\lambda = \lambda_{cr}$ is called the *neutral* or *marginal* state.

Remark 2.5.1. Let $\eta_0(\lambda)$ be the imaginary part of the eigenvalue of the stability spectrum whose real part is $\xi_0(\lambda)$. Then $\sigma = \xi_0(\lambda) + i\eta_0(\lambda)$ is the eigenvalue of the dimenssionless equation

$$(2.3.11)' \qquad -\sigma \mathbf{v} + \lambda^{-1} A\mathbf{v} + M(\overline{\mathbf{u}}, \mathbf{v}) = 0$$

corresponding to (2.3.11); in the neutral case, since $\xi_0(\lambda_{cr}) = 0$, (2.3.11)′ becomes

$$(2.3.11)'' \qquad -i\eta_0(\lambda)\mathbf{v} + \lambda^{-1}A\mathbf{v} + M(\overline{\mathbf{u}}, \mathbf{v}) = 0.$$

For every assigned η_0, $\lambda_{cr}(\eta_0)$ is the least eigenvalue of this equation.

Taking into account the general form (2.3.13) of the linear perturbations, we may deduce estimates for $\mathbf{v}$. Hence the basic flow $\overline{\mathbf{u}}$ is linearly stable for $\xi_0(\lambda) > 0$, and linearly unstable for $\xi_0(\lambda) < 0$. We shall assume now that

a) $\xi_0(0) > 0$

and

b) the function $\xi_0(\lambda)$ changes sign as λ crosses λ_{cr}. Hence for λ slightly supercritical, we have $\xi_0(\lambda) < 0$; then for $\lambda < \lambda_{cr}$, all the stability spectrum lies in the right halfplane. In particular, for $\lambda < \lambda_{cr}$ $\xi_0(\lambda) > 0$, and, according to the above assertions, $\overline{\mathbf{u}}$ is lineary stable. Similarly for $\lambda > \lambda_{cr}$ it follows that $\overline{\mathbf{u}}$ is lineary unstable. Thus, the neutral state separates the domain of linear stability of $\overline{\mathbf{u}}$ from that of instability, λ_{cr} being *the point of loss of linear stability* for the basic solution.

From the fact that the stability spectrum lies in the interior of a parabola (Remark 2.4.1.) and since it has no finite accumulation points (§ 2.3), it follows that for $\lambda = \lambda_{cr}$ only a finite number of points of this spectrum lie on the imaginary axis, the rest of the spectrum belonging to the right halfplane, i.e. $\delta > 0$ exists such that $\mathcal{R}e\{\sigma(\lambda)\} > \delta$. Hence, in the neutral case, some normal modes (with $\mathcal{R}e\{\sigma(\lambda)\} > \delta$), are damped out, whereas others hawing $\mathcal{R}e\{\sigma(\lambda)\} = 0$ and $\mathcal{I}m\{\sigma(\lambda)\} \neq 0$ are oscillatory: for $t \to \infty$, they are amplified if σ is a multiple eigenvalue, or their amplitude remains unchanged if σ is a simple eigenvalue. Finally, the normal mode $\sigma(\lambda_{cr}) = 0$ (if it exists) does not oscillate but is nonstationary and amplified (if 0 is a multiple eigenvalue). It follows that in general nothing can be said about the stability of $\overline{\mathbf{u}}$ at $\lambda = \lambda_{cr}$. The stability spectrum is decomposed into two parts: one on the right of $\mathcal{R}e\{\sigma\} = \delta$ and one on the real axis. While the first one cannot induce instability since the corresponding normal modes are damped out, $\overline{\mathbf{u}}$ is lineary unstable at $\lambda = \lambda_{cr}$ if at least one σ on the imaginary axis is multiple. If all the stability spectrum has the rank 1, then we say that $\overline{\mathbf{u}}$ is *neutrally stable*. This only implies that the perturbations are not amplified, and the possibility exists that their amplitude is constant. The principle of exchange of stabilities is a hypothesis which, as will be seen, simplifies the distribution of the stability spectrum.

Definition, We say that *the principle of exchange of stabilities* (P.E.S.), holds, if every perturbation which is not damped out does not oscillate with respect to time. In the framework of linear stability theory, the general from (2.3.13) of perturbations shows that the P.E.S. holds if $\mathcal{I}m\{\sigma(\lambda)\} \neq 0$ implies that $\mathcal{R}e\{\sigma(\lambda)\} > 0$. It follows that for $\lambda = \lambda_{cr}$, the eigenvalue with vanishing real part has also vanishing imaginary part. Therefore the single point of the stability spectrum located on the imaginary axis is at the origin, the rest of this spectrum lying to the right of $\mathcal{R}e\{\sigma\} = \delta$. In other words, the most unstable mode, corresponding to the eigenvalue 0, *does not oscillate* and it is nonstationary when 0 is multiple and stationary when 0 is a simple eigenvalue of the stability spectrum.

Remark 2.5.2. When P.E.S. holds, we have $\eta_0 = 0$, and therefore λ_{cr} is the least eigenvalue of the problem

$$\lambda^{-1}A\mathbf{v} + M(\overline{\mathbf{u}}, \mathbf{v}) = 0. \qquad (2.3.11)'''$$

From P.E.S. and the assumption that $\xi_0(\lambda)|_{\lambda>\lambda_{cr}} < 0$, it follows that for $\lambda > \lambda_{cr}$, the stability spectrum is real. This is why P.E.S. holds for all flows whose stability spectrum is real. In particular, a class of basic flows for which this principle holds is that in which the operator $\widetilde{A}$ (§ 2.3) is selfadjoint. This is the reason why the first proof of P.E.S. [71] was obtained for Bénard convection, where $\widetilde{A}$ is selfadjoint.

2.5.2. **Proof of P.E.S. for particular motions**

Consider a horizontal layer of fluid at rest, heated from below, lying between two rigid parallel planes $z = 0$ and $z = 1$. The solution of the hydrodynamic equations corresponding to this state of static equilibrium is called the thermal conduction. Assume that we have perturbed this equilibrium with perturbations whose velocity and temperature are normal modes $\mathbf{v} = \mathbf{V} \exp[\mathrm{i}(k_x x + k_y y) - \sigma t]$ and respectively $\theta = \Theta \exp[\mathrm{i}(k_x x + k_y y) - \sigma t]$, where $\mathbf{V} = (U, V, W)$. Then, in the classical frame, the linear stability of the thermal conduction state is described by the equation

$$\left(\frac{\mathrm{d}^2}{\mathrm{d}z^2} - \alpha^2 - \mathrm{Pr}\,\sigma\right) F = -\mathrm{Ra}\,\alpha^2 W, \tag{2.5.1}$$

and the boundary conditions

$$\Theta = W = DW = 0 \quad \text{for} \quad z = 0 \text{ and } z = 1, \tag{2.5.2}$$

where $F = \left(\frac{\mathrm{d}^2}{\mathrm{d}z^2} - \alpha^2\right)\left(\frac{\mathrm{d}^2}{\mathrm{d}z^2} - \alpha^2 + \sigma\right) W$, $\alpha = \sqrt{k_x^2 + k_y^2}$, Pr is the Prandtl number and Ra denotes the Rayleigh number.

Multiplying (2.5.1) by the conjugate $\bar{F}$ of F, integrating by parts on [0, 1] and taking into account (2.5.2) we get

$$(\mathfrak{Im}\,\{\sigma\}) \left\{ \mathrm{Pr} \int_0^1 |F|^2\,\mathrm{d}z + \mathrm{Ra}\,\alpha^2 \int_0^1 \left[\left| \frac{\mathrm{d}W}{\mathrm{d}z} \right|^2 + \alpha^2 |W|^2 \right] \mathrm{d}z \right\} = 0.$$

Now since $\mathrm{Pr} > 0$ and $\mathrm{Ra} > 0$, it follows that $\mathfrak{Im}\,\sigma = 0$, and hence P.E.S. holds. The above used method is called the definite integrals method.

There are numerous studies which show that the spectrum of $\tilde{A}$ for certain basic flows (most of them being convective ones) is real. To prove this result, the definite integrals method or variational principles are mainly used [103], [95], [85], [81]. Sometimes, from the knowledge of the eigenfunctions of $\tilde{A}$, criteria for the validity of P.E.S. can be given [9], [24].

A much investigated motion is Couette flow between rotating cylinders. Experiments and numerical calculations show that P.E.S. holds for this flow. In this case, $\tilde{A}$ is not selfadjoint, σ depending on λ as well as on the wave number of the perturbation, which implies the difficulty of proving that P.E.S. holds. So far, exist only partial results, based on perturbation methods [12] or conditions equivalent with P.E.S., or numerical calculations which assert the validity of P.E.S. [99].

2.5.3. Branching (bifurcation) of solutions of the hydrodynamic equations

Consider, in the Banach space B, the equation

$$\mathbf{u} = F_1(\mathbf{u}, \lambda), \tag{2.5.3}$$

where λ is a positive parameter and F_1 is a nonlinear operator; let $\bar{\mathbf{u}}(\lambda)$ be a solution of this equation. We look for solutions of Equation (2.5.3) of the form

$$\mathbf{u} = \bar{\mathbf{u}} + \mathbf{v},$$

where $\mathbf{v}$ satisfies the equation

$$\mathbf{v} = F(\mathbf{v}, \lambda) \tag{2.5.4}$$

and F is a compact nonlinear operator. By construction, Equation (2.5.4) admits the solution $\mathbf{v} = 0$ for every λ. We say that λ_0 is a *branch (bifurcation) point* for solutions of Equation (2.5.3) if, for every $\varepsilon > 0$ and $\eta > 0$, an eigenvalue λ and a corresponding eigenvector $\mathbf{v}$ of $F - I$ exist such that $|\lambda - \lambda_0| < \varepsilon$ and $\|\mathbf{v}\|_B < \eta$.

Assume that for small values of λ the solution $\mathbf{v} \equiv 0$ is unique. Then the first branch point corresponding to the least eigenvalue λ_1 is a point of *loss of uniqueness* of the trivial (vanishing) solution and of the appearance of nonvanishing solutions of Equation (2.5.4). The point $(\lambda_1, 0)$ in the $\lambda - \|\mathbf{v}\|$ plane will be called the first branch point.

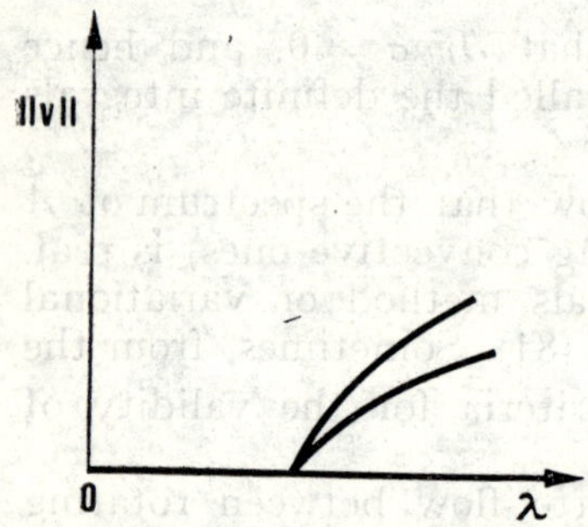

Fig. 2.5.1

The branching process around it is represented in fig. 2.5.1: for every $\lambda \in [0, \lambda_1)$ Equation (2.5.4) has the unique solution $\mathbf{v} \equiv 0$; for $\lambda > \lambda_1$ in a certain neighbourhood of $(\lambda_1, 0)$, besides the trivial solution, there exist also nonvanishing solutions belonging to continuous curves (called branches) which appear at $(\lambda_1, 0)$. As we shall see in Chap. 3, the branch points λ are found among the real eigenvalues λ_i of the linear completely continuous operator $A(\lambda)$ where $A(\lambda)$ is the *Fréchet differential* of F at the point 0. Real eigenvalues λ_i with odd multiplicity are always branch points.

Let $\mathbf{u}$ and $\mathbf{v}$ be respectively a stationary solution of the Navier-Stokes equations and a stationary perturbation, which is a strong solution of equation

$$\lambda^{-1} A\mathbf{v} + M\mathbf{v} = R(\mathbf{v}). \tag{2.5.5}$$

If M^{-1} exists, then (2.3.5) can be put in the form

$$\mathbf{v} = -\lambda^{-1}M^{-1}A\mathbf{v} + M^{-1}R(\mathbf{v}), \tag{2.5.5$'$}$$

which corresponds to the linear equation

$$\mathbf{v} = -\lambda^{-1}M^{-1}A\mathbf{v}, \tag{2.5.6}$$

where $-\lambda M^{-1}A$ is the Fréchet differential at the point $\mathbf{v} \equiv 0$ of the operator $-\lambda M^{-1}A + M^{-1}R$. Hence, the branch points of the solution $\mathbf{v} \equiv 0$ of (2.5.5) (and therefore of the solution $\bar{\mathbf{u}}$ of the Navier-Stokes equations) are the real eigenvalues of odd multiplicity of the operator $M^{-1}A$. That is why the spectrum $\{\lambda_i\}$ of this Fréchet differential is called the *branching spectrum*.

From the above considerations it follows that the first branch point is the least positive eigenvalue λ_1 if its multiplicity is odd. But Equation (2.5.6) is equivalent to Equation (2.3.11)'' ad by Remark 2.5.2, we have

$$\lambda_{cr} = \lambda_1. \tag{2.5.7}$$

Consequently, when P.E.S. holds, if ξ_0 satisfies hypotheses a) and b) from section 2.5.1, and the multiplicity of the least eigenvalue of the branching spectrum is odd, the point of loss of linear stability coincides with the point of loss of uniqueness of the basic solution, due to be the appearance of new solutions called *secondary solutions.* The hypotheses that the P.E.S. holds is essential, otherwise the possibility would exist that λ_{cr}, depending on η_0, does not attain its minimum at $\eta_0 = 0$.

In Chapter 3 we shall see that if the P.E.S. holds in the case of the Navier-Stokes equations, then two branches of stationary solutions appear at λ_{cr}. Hence as λ crosses λ_{cr}, the motion passes from a stable stationary state ($\bar{\mathbf{u}}$) to another state corresponding to a stable stationary secondary solution, which justifies the name of "exchange of stabilities".

The above considerations point out the particular importance of *solving the linear eigenvalue problem of hydrodynamic stability*. The fact that the operator corresponding to this problem is, as a rule, not selfadjoint, makes the treatement of this problem very difficult in general; this is why only few studies in this connection exist, and they concern classes of motions such as, for instance, that of the plane parallel flows [25], [79]. So, there is the necessity of a *numerical* investigation of the problem for various particular cases of motions. From the knowledge of the stability spectrum and of the multiplicity of the eigenvectors as well as of the associated vectors, we derive the validity of the linearization principle and the completeness of the normal modes. The knowledge of the branching spectrum shows the location of the points where new solutions appear.

By virtue of P.E.S., for $\lambda > \lambda_1$ the nonstationary perturbations contain the factor $\exp[-\mathcal{Re}\{\sigma\}t]$ and, as we have seen in § 2.3, these perturbations are the generalized solutions of a nonlinear initial-value problem. Using the theory of the semigroups of operators Iooss [34], [36] proved that, in certain conditions, as $t \to \infty$, the perturbations tend towards one of the secondary solutions from branches which appear at λ_1. For many motions, the linear theory indicates the stability of both secondary solutions. It follows that the most unstable perturbations are finite and can only be treated in the framework of the nonlinear theory. Consequently, *the hydrodynamic stability problem is completely solved for every motion, only after the asymptotic behaviour of the corresponding nonlinear initial value problem has been determined.*

An important problem which was not treated in this chapter is that of the regularity of generalized solutions. In this respect, it is worth mentioning that for most configurations of motion, body forces and initial and boundary conditions in the laminar regime, the generalized solutions are also classical [52]. It seems that in advanced stages of transition, and consequently in the stability problem of the secondary solutions, the physical phenomenon is suitably described by generalized solutions only. Moreover, the complete solution of the eigenvalue problem which appears in the study of stability can be done only in the generalized frame. Finally, by reducing the mixed problem for the Navier-Stokes equations to equations in Hilbert spaces, the proof of existence, uniqueness and stability of generalized solutions is easily done by standard methods of the theory of these spaces.

§ 2.6. UNIVERSAL CRITERIA OF HYDRODYNAMIC STABILITY

2.6.1. Stationary basic flows

Consider a fluid flow having the velocity $\bar{\mathbf{u}}$ in a bounded two-dimensional domain Ω; denote by d the maximal diameter of Ω, by V_{max} the maximum modulus of the velocity and by ν the kinematic viscosity. By definition, the dimensionless number is $\mathrm{Re} = \dfrac{V_{max}\, d}{\nu}$. Let $\mathbf{u} = \bar{\mathbf{u}} + \mathbf{v}$ be a perturbed flow, corresponding to the same body forces as $\bar{\mathbf{u}}$, and denote by $K(t) = \int_\Omega \frac{1}{2} |\mathbf{v}|^2$ the mean kinetic energy of the perturbation $\mathbf{v}$. The basic flow $\bar{\mathbf{u}}$ is called *stable in the mean*, or simply *stable*, if $\lim_{t\to\infty} K(t) = 0$, that is, if $\lim_{t\to\infty} |\mathbf{v}(t)|^2 = 0$[21].

[21] As in the previous sections, $|\cdot|$ and $\|\cdot\|$ stand for the norm in L^2 and in N^1 respectively.

We note that if $\frac{dK}{dt} < 0$ for every $t > 0$, it follows that $|\mathbf{v}| \to 0$ as $t \to \infty$. The classical stability criteria assert that if Re is less than a certain number α then $\frac{dK}{dt} < 0$; hence, it follows that for given values of d and ν this motion is stable if the maximum velocity of the basic flow belongs to the interval $\left(0, \frac{\alpha\nu}{d}\right)$.

In the above, α is the constant which satisfies the isoperimetric inequality

$$\alpha d^{-2}\int_\Omega |\bar{\mathbf{u}}|^2 \, dx \leqslant \int_\Omega |\mathrm{grad}\, \mathbf{u}|^2 \, dx, \tag{2.6.1}$$

in $N^1(\Omega)$ (Section 1.3.1). Hence, $\sqrt{\alpha d^{-2}}$ is the constant of immersion of N^1 into N;

$$\alpha^{-\frac{1}{2}} d = \max_{\mathbf{u} \in N^1} \frac{|\mathbf{u}|}{\|\mathbf{u}\|}, \tag{2.6.1)'$$

and αd^{-2} is the least eigenvalue λ_1 of the problem

$$\begin{cases} -\Delta \mathbf{u} + \mathrm{grad}\, p = \lambda \mathbf{u}, \\ \mathrm{div}\, \mathbf{u} = 0, \\ \mathbf{u}|_{\partial\Omega} = 0. \end{cases} \tag{2.6.1)''$$

The numerical calculation of α is extremely important, since it represents the upper bound up to which we have certain stability. As we have seen in section 1.3.1, $5.302\pi^2 \leqslant \alpha \leqslant 5.312\pi^2$.

Assume now that the stationary basic motion corresponds to the turbulent solution $\bar{\mathbf{u}}$, which satisfies the equation

$$\int_0^T [-(\bar{\mathbf{u}}, \dot{\boldsymbol{\varphi}}) + \nu((\bar{\mathbf{u}}, \boldsymbol{\varphi})) + ((\bar{\mathbf{u}} \cdot \mathrm{grad})\, \bar{\mathbf{u}}, \boldsymbol{\varphi})] \, dt = (\bar{\mathbf{u}}, \boldsymbol{\varphi}(0)) + \int_0^T (\mathbf{f}, \boldsymbol{\varphi}) \, dt, \tag{2.2.18}$$

(where the body force $\mathbf{f} \in L^2(\Omega)$ does not depend on t), and the perturbed motion corresponds to the turbulent solution $\mathbf{u} = \bar{\mathbf{u}} + \mathbf{v}$, which satisfies the equation

$$\int_0^T [-(\mathbf{u}, \dot{\boldsymbol{\varphi}}) + \nu((\mathbf{u}, \boldsymbol{\varphi})) + ((\mathbf{u} \cdot \mathrm{grad})\, \mathbf{u} \cdot \boldsymbol{\varphi})] \, dt = \tag{2.2.18)'$$

$$= (\bar{\mathbf{u}} + \mathbf{v}_0, \boldsymbol{\varphi}(0)) + \int_0^T (\mathbf{f}, \boldsymbol{\varphi}) \, dt,$$

where $\mathbf{f}$ is the above body force and $\mathbf{v}_0 \in N(\Omega)$. Under these conditions, we can prove [52] the existence of the generalized derivative $\dot{\mathbf{v}} \in L^2(\Omega \times (0, T))$ and, consequently,

$$-\int_0^T (\mathbf{v}, \dot{\varphi})\, dt = \int_0^T (\dot{\mathbf{v}}, \varphi)\, dt. \tag{2.6.2}$$

Subtracting (2.2.18) from (2.2.18)′, using (2.6.2) and taking the arbitrary function φ to be the perturbation $\mathbf{v}$, we get

$$\begin{aligned}\int_0^T [(\dot{\mathbf{v}}, \mathbf{v}) + \nu \|\mathbf{v}\|^2 + ((\mathbf{v}\cdot \operatorname{grad})\, \bar{\mathbf{u}}, \mathbf{v}) + ((\bar{\mathbf{u}} \cdot \operatorname{grad})\, \mathbf{v}, \mathbf{v}) +\\ + ((\mathbf{v}\cdot \operatorname{grad})\, \mathbf{v}, \mathbf{v})]\, dt = |\mathbf{v}(0)|^2.\end{aligned} \tag{2.6.3}$$

Integrating by parts, we obtain

$$\int_0^T ((\mathbf{v}\cdot \operatorname{grad})\, \mathbf{v}, \mathbf{v})\, dt = \int_0^T ((\bar{\mathbf{u}}\cdot \operatorname{grad})\, \mathbf{v}, \mathbf{v})\, dt = 0,$$

$$\int_0^T ((\mathbf{v}\cdot \operatorname{grad})\, \bar{\mathbf{u}}, \mathbf{v})\, dt = -\int_0^T ((\mathbf{v}\cdot \operatorname{grad})\, \mathbf{v}, \bar{\mathbf{u}})\, dt.$$

Hence differentiation of (2.6.3) yields the energy relation

$$\frac{1}{2}\frac{d}{dt}|\mathbf{v}|^2 + \nu\|\mathbf{v}\|^2 = ((\mathbf{v}\cdot \operatorname{grad})\, \mathbf{v}, \bar{\mathbf{u}}). \tag{2.6.4}$$

Taking into account the generalized Hölder inequality of indices $\frac{1}{4}, \frac{1}{2}$ and $\frac{1}{4}$, and the Ladyzhenskaya inequality (2.1.14) together with (2.6.1), the right-hand side of (2.6.4) admits the following estimates

$$\begin{aligned}|((\mathbf{v}\cdot \operatorname{grad})\, \mathbf{v}, \bar{\mathbf{u}})| &\leqslant |\mathbf{v}|_4 \|\mathbf{v}\|\, |\bar{\mathbf{u}}|_4 \leqslant 2^{1/4} |\mathbf{v}|^{1/2} \|\mathbf{v}\|^{1/2} \|\mathbf{v}\| 2^{1/4} |\bar{\mathbf{u}}|^{1/2} \|\mathbf{u}\|^{1/2} =\\ &= \sqrt{2}\, |\mathbf{v}|^{1/2} \|\mathbf{v}\|^{3/2} |\bar{\mathbf{u}}|^{1/2} \|\mathbf{u}\|^{1/2} \leqslant \sqrt{2}\, d^{1/2} \alpha^{-\frac{1}{4}} \|\mathbf{v}\|^2 |\bar{\mathbf{u}}|^{1/2} \|\bar{\mathbf{u}}\|^{1/2}.\end{aligned}$$

Consequently

$$\frac{d}{dt}\left(\frac{1}{2}|\mathbf{v}|^2\right) \leqslant (-\nu + \sqrt{2}\, d^{1/2} \alpha^{-1/4} |\bar{\mathbf{u}}|^{1/2} \|\bar{\mathbf{u}}\|^{1/2}) \|\mathbf{v}\|^2, \tag{2.6.5}$$

and hence a family of universal stability criteria can be deduced in the form

$$\mathrm{Re}_\gamma < \frac{\alpha^{(1/4+\gamma/2)}}{\sqrt{2}}, \quad \left(0 \leqslant \gamma \leqslant \frac{1}{2}\right), \tag{2.6.6}$$

here Re_γ is the generalized Reynolds number

$$\mathrm{Re}_\gamma = \frac{d^{1/2+\gamma}\,|\bar{\mathbf{u}}|^{1/2}\,\|\bar{\mathbf{u}}\|^{1/2+\gamma}}{\nu}.$$

The criteria (2.6.6) lie between the Ladyzhenskaia condition [51] corresponding to $\gamma = \frac{1}{2}$

$$\text{(2.6.7)} \qquad \frac{d\,\|\bar{\mathbf{u}}\|}{\nu} < \frac{\sqrt{\alpha}}{\sqrt{2}}$$

and that from [22] corresponding to $\gamma = 0$

$$\text{(2.6.8)} \qquad \frac{d^{1/2}\,|\bar{\mathbf{u}}|^{1/2}\,\|\bar{\mathbf{u}}\|^{1/2}}{\nu} < \sqrt[4]{\frac{\alpha}{4}}.$$

Finally, applying again (2.6.1) in (2.6.6) we find the criterion

$$\text{(2.6.9)} \qquad \mathrm{Re}' = \frac{|\bar{\mathbf{u}}|}{\nu} < \frac{1}{\sqrt{2}}.$$

Relations (2.6.9), (2.6.7) and (2.6.6) give respectively an upper bound for the mean kinetic energy $\frac{1}{2}\,|\bar{\mathbf{u}}|^2$ of the basic motion, for the mean gradient of its velocity, and for one of their combinations, below which the basic flow is stable. That is why criteria corresponding to generalized solutions are useful in those cases when it is necessary to know the domain of variation for the kinetic energy and the mean gradient of the velocity, in order that the basic flow remains stable.

We note that for particular domains Ω, we can determine the corresponding parameter α. In this way particular stability criteria which ensure the stability for larger domains of variation for Re_γ are obtained.

The case of the three-dimensional domain Ω can be treated analogously [22], the inequality (2.1.15) being used instead of (2.1.14).

2.6.2. Nonstationary basic flows

Consider a nonstationary basic flow whose velocity is the single turbulent solution $\mathbf{u}_1(t) \in L^2_{loc}((T,\infty), N^1) \cap L^\infty_{loc}((T,\infty), N)$ which satisfies the equation

$$\text{(2.6.10)} \int_T^\infty [-(\mathbf{u}_1, \dot{}\,) + \nu((\mathbf{u}_1, \boldsymbol{\varphi})) + ((\mathbf{u}_1\cdot \mathrm{grad})\,\mathbf{u}_1, \boldsymbol{\varphi})]\,\mathrm{d}t = \int_T^\infty (\mathbf{f}_1, \boldsymbol{\varphi})\,\mathrm{d}t$$

for every $\varphi \in C_0((T, \infty), N^1)$, $\dot{\varphi} \in L^2_{loc}((T, \infty), N)$, $\mathbf{f}_1(t) \in L^\infty((T, \infty), L^2)$ and corresponds to the initial condition $\lim_{t\to T+0} \mathbf{u}_1(t) = \mathbf{u}_0 \in N$. Denote by $\mathbf{u} = \mathbf{u}_1 + \mathbf{v}$ another solution corresponding to the body force $\mathbf{f}$; then for every linear operator $B \geqslant 0$, which is bounded in N and commutes with A (§ 2.4), the perturbation $\mathbf{v}$ satisfies [74] the following energy relation

$$(2.6.11) \quad \frac{(\mathbf{v}(t), B\mathbf{v}(t) - (\mathbf{v}(s), B\mathbf{v}(s))}{2} - \int_s^t [\nu((\mathbf{v}(\tau), B\mathbf{v}(\tau))) +$$

$$+ b(\mathbf{u}(\tau), \mathbf{v}(\tau), B\mathbf{v}(\tau)) + b(\mathbf{v}(\tau), \mathbf{u}_1(\tau), B\mathbf{v}(\tau))] \,\mathrm{d}\tau = \int_s^t (\mathbf{h}(\tau), B\mathbf{v}(s)) \,\mathrm{d}\tau,$$

where $T < s \leqslant t < \infty$, $\mathbf{h}(t) = \mathbf{f}(t) - \mathbf{f}_1(t)$, $b(\mathbf{u}, \mathbf{v}, \mathbf{w}) = \int_\Omega (\mathbf{u} \cdot \operatorname{grad}) \mathbf{v} \cdot \cdot \mathbf{w} \mathrm{d}x$. If $B = I$, $\mathbf{f} = \mathbf{f}_1$, and $s = T = 0$, then (2.6.11) reduces to (2.6.4).

Using Galerkin's method, this relation allows, for domains Ω of class C^2, one to prove the existence and uniqueness of a turbulent solution $\mathbf{u} \in L^\infty_{loc}((T, \infty), N^1)$ with $A\mathbf{u} \in L^2_{loc}((T, \infty), N)$. Choosing $B = -AP_n$, where P_n is the orthogonal projection of N^1 on V^n, $P_n\mathbf{v} = (\mathbf{v}, \mathbf{w}_1)\mathbf{w}_1 + \ldots + (\mathbf{v}, \mathbf{w}_n)\mathbf{w}_n$ and $\mathbf{w}_k$ are the eigenvectors of problem (2.6.1)'', relation (2.6.11) becomes

$$(2.2.35) \quad \frac{1}{2}\frac{\mathrm{d}}{\mathrm{d}t} \| \mathbf{u}(t) \|^2 + \nu | A\mathbf{u}|^2 = -b(\mathbf{u}(t), \mathbf{u}(t), A\mathbf{u}(t)) + (\mathbf{f}(t), A\mathbf{u}(t)).$$

As in Section 2.2.4, we can obtain estimates for $\|\mathbf{u}(t)\|$. Every turbulent solution $\mathbf{u}$ is [20] a continuous function defined on (T, ∞) with values in N^1, bounded on every interval $[s, \infty) \subset (T, \infty)$; and there exists a constant $\mathcal{C}$ depending on Ω, ν and $\|\mathbf{f}\|_\infty$ such that

$$(2.6.12) \quad \limsup_{t\to\infty} \| \mathbf{u}(t) \| \leqslant \mathcal{C},$$

and $\mathcal{C} \to 0$ as $\frac{1}{\nu}$ or $\|\mathbf{f}\|_\infty$ tend to zero.

Now let $B = \mathbf{E}$, where E is an orthogonal projection in N and $\mathbf{h} = 0$ (hence the perturbation $\mathbf{v}$ is due only to the perturbation $\mathbf{v}_0$). By differentiating relation (2.6.11) with respect to t we get

$$(2.6.11)' \quad \frac{1}{2}\frac{\mathrm{d}}{\mathrm{d}t} |E\mathbf{v}(t)|^2 + \nu \|\mathbf{E}\mathbf{v}(t)\|^2 + b(\mathbf{u}(t), \mathbf{v}(t), \mathrm{E}\mathbf{v}(t)) +$$

$$+ b(\mathbf{v}(t), \mathbf{u}_1(t), E\mathbf{v}(t)) = 0.$$

As for the estimate of the term $((\mathbf{v}\cdot \text{grad})\,\mathbf{v}, \overline{\mathbf{u}})$ from Section 2.6.1, we have

$$|b(\mathbf{u}, \mathbf{v}, E\mathbf{v})| = |b(\mathbf{u}, (I-E)\,\mathbf{v}, E\mathbf{v})| =$$

$$= |b(\mathbf{u}, E\mathbf{v}, (I-E)\,\mathbf{v})| \leqslant \sqrt{2}\,|\mathbf{u}|^{1/2}\|\mathbf{u}\|^{1/2}\|E\mathbf{v}\|\,|(I-E)\mathbf{v}|^{1/2}\|(I-E)\mathbf{v}\|^{1/2},$$

$$b(\mathbf{v}, \mathbf{u}_1, E\mathbf{v})| \leqslant |b(E\mathbf{v}, \mathbf{u}_1, E\mathbf{v}) + |b((I-E)\mathbf{v}, \mathbf{u}_1, E\mathbf{v})| \leqslant \sqrt{2}\|\mathbf{u}_1\|\,|E\mathbf{v}|\,\|E\mathbf{v}\| +$$

$$+ \sqrt{2}\,|\mathbf{u}_1|^{1/2}\|\mathbf{u}_1\|^{1/2}\cdot\|E\mathbf{v}\|\,|(\mathbf{I}-E)\mathbf{v}|^{1/2}\|(I-E)\mathbf{v}\|^{1/2}.$$

Let $\mathcal{C}_\sigma$ be a constant such that $\|\mathbf{u}(t)\| \leqslant \mathcal{C}_\sigma$, $\|\mathbf{u}_1(t)\| \leqslant \mathcal{C}_\sigma$ for $\sigma \leqslant t < \infty$ and $\lim_{\sigma\to\infty} \mathcal{C}_\sigma = \mathcal{C}$.

Then for $t \geqslant \sigma$, (2.6.11)$'$ becomes

$$\text{(2.6.13)} \qquad \frac{1}{2}\frac{\mathrm{d}}{\mathrm{d}t}|E\mathbf{v}^2| + \nu\|P\mathbf{v}\|^2 \leqslant 2\sqrt{2\alpha^{-1/2}d}\,\mathcal{C}_\sigma|(I -$$

$$- E)\,\mathbf{v}|^{1/2}\|(I-E)\,\mathbf{v}\|^{1/2}\|E\mathbf{v}\| + \sqrt{2}\mathcal{C}_\sigma|E\mathbf{v}|\,\|E\mathbf{v}\|.$$

Applying Young's inequality to each term in the right-hand side, we get

$$\text{(2.6.14)} \qquad 2\sqrt{2\alpha^{-1/2}d}\,\mathcal{C}_\sigma|(I-E)\mathbf{v}|^{1/2}\|(I-E)\,\mathbf{v}\|^{1/2}\|E\mathbf{v}\| \leqslant$$

$$\leqslant \frac{8\alpha^{-1/2}d\mathcal{C}_\sigma^2}{\nu}\,x^2|\,(I-E)\,\mathbf{v}|\;\|(I-E)\,\mathbf{v}\| + \frac{\nu}{4x^2}\|\mathbf{E}\mathbf{v}\|^2,$$

$$\text{(2.6.15)} \qquad \sqrt{2}\mathcal{C}_\sigma|E\mathbf{v}|\,\|E\mathbf{v}\| \leqslant 4y^2\frac{\mathcal{C}_\sigma^2}{\nu}|E\mathbf{v}|^2 + \frac{\nu}{8y^2}\|E\mathbf{v}\|^2,$$

whence

$$\text{(2.6.16)} \qquad \frac{1}{2}\frac{\mathrm{d}}{\mathrm{d}t}|E\mathbf{v}|^2 + \nu\|E\mathbf{v}\|^2 \leqslant \frac{8\alpha^{-1/2}d\mathcal{C}_\sigma^2}{\nu}\,x^2\,|(I -$$

$$- E)\,\mathbf{v}|\,\|(I-E)\,\mathbf{v}\| + 4y^2\frac{\mathcal{C}_\sigma^2}{\nu}|E\mathbf{v}|^2 + \left(\frac{\nu}{4x^2} + \frac{\nu}{8y^2}\right)\|E\mathbf{v}\|^2.$$

Taking $E = E_n$, where $E_n\mathbf{u} = (\mathbf{u}, \mathbf{w}_{n+1})\,\mathbf{w}_{n+1} + (\mathbf{u}, \mathbf{w}_{n+2})\,\mathbf{w}_{n+2} + \ldots$, and using the identity

$$|\,(I-E_n)\,\mathbf{v}\,| = |\,\mathrm{pr}_n\mathbf{v}|, \quad \|E_n\mathbf{v}\|^2 \geqslant \lambda_{n+1}|\,E_n\mathbf{v}|^2,$$

$$\|(I-E_n)\,\mathbf{v}\|^2 = \sum_{k=1}^{n}\lambda_k\,|(\mathbf{v}, \mathbf{w}_k)|^2 \leqslant \lambda_n\,|\mathrm{pr}_n\mathbf{v}|^2,$$

where $\mathrm{pr}_0\mathbf{v} = 0$, $\mathrm{pr}_n\mathbf{v} = ((\mathbf{v}, \mathbf{w}_1), \ldots, (\mathbf{v}, \mathbf{w}_n)) \in R^n$, we get from relation (2.6.16)

$$\frac{1}{2}\frac{\mathrm{d}}{\mathrm{d}t}|E_n\mathbf{v}|^2 + \frac{\nu\lambda_{n+1}(8x^2y^2 - 2y^2 - x^2)}{8x^2y^2}|E_n\mathbf{v}|^2 \leqslant$$

$$\leqslant \frac{8\alpha^{-1/2}d\mathcal{C}_\sigma^2\sqrt{\lambda_n}}{\nu}\,|\mathrm{pr}_n\mathbf{w}|^2x^2 + \frac{4y^2\mathcal{C}_\sigma^2}{\nu}|E_n\mathbf{v}|^2,$$

for $8x^2y^2 - 2y^2 - x^2 > 0$. It follows that

$$\frac{1}{2}\frac{\mathrm{d}}{\mathrm{d}t}|E_n\mathbf{v}|^2 + \left[\frac{\nu\lambda_{n+1}(8x^2y^2 - 2y^2 - x^2)}{8x^2y^2} - \frac{4y^2\mathcal{C}_\sigma^2}{\nu}\right]|E_n\mathbf{v}|^2 \leqslant$$

$$\leqslant \frac{8\alpha^{-\frac{1}{2}} d\mathcal{C}_\sigma^2\sqrt{\lambda_n}}{\nu}|\mathrm{pr}_n\mathbf{v}|^2x^2.$$

This last relation leads to the estimate

$$(2.6.17) \qquad |E_n\mathbf{v}(t)|^2 \leqslant |E_n\mathbf{v}(s)|^2\,\mathrm{e}^{-a(t-s)} + b\int_s^t \mathrm{e}^{-a(t-s)}|\mathrm{pr}\mathbf{v}(\tau)|^2\,\mathrm{d}\tau,$$

where we have put

$$a = \frac{\nu\lambda_{n+1}(8x^2y^2 - 2y^2 - x^2)}{8x^2y^2} - \frac{4y^2\mathcal{C}_\sigma^2}{\nu}, \quad \mathrm{b} = \frac{8\alpha^{-\frac{1}{2}} d\mathcal{C}_\sigma^2\sqrt{\lambda_n}}{\nu}x^2.$$

If $a > 0$, and $|\mathrm{pr}_n\mathbf{v}(t)| \to 0$ as $t \to \infty$, then from (2.6.17) it follows that $|E_n\mathbf{v}(t)|^2 \to 0$. Using $|\mathbf{v}(t)|^2 = |\mathrm{pr}_n\mathbf{v}(t)|^2 + |E_n\mathbf{v}(t)|^2$, it can be seen that $|\mathbf{v}(t)|^2 \to 0$ as $t \to \infty$. Consequently, the solution $\mathbf{u}_1$ is stable in the mean with respect to perturbations of the initial conditions. The sufficient conditions for stability were $|\mathrm{pr}_n\mathbf{v}(t)| \to 0$ as $t \to \infty$ and $a > 0$. Letting $\sigma \to \infty$, the last inequality may be written [22] as

$$\lambda_{n+1} > \frac{32x^2y^4\mathcal{C}^2}{\nu^2(8x^2y^2 - 2y^2 - x^2)},$$

the right-hand side taking its minimal value as $x \to \infty$ and $y = \frac{1}{2}$.

Thus, we obtain the criterion [22]

$$(2.6.18) \qquad \lambda_{n+1} > \frac{2\mathcal{C}^2}{\nu^2},$$

which for $n = 0$ becomes

$$(2.6.18)' \qquad \frac{\limsup\limits_{t\to\infty}\|\mathbf{u}_1\|\,d}{\nu} < \sqrt{\frac{\alpha}{2}},$$

i.e. the analog of the Ladyzhenskaya criterion for the nonstationary case.

Note that in the nonstationary case, to (2.6.18) we must add the condition $|\mathrm{pr}_n\mathbf{v}(t)| \to 0$ as $t \to \infty$; if $n = 0$ then this restriction

[22]) In [20] instead of (2.6.18) the weaker criterion $\lambda_{n+1} > \dfrac{16\mathcal{C}^2}{\nu^2}$ is proved.

is absent. Since λ_n is an increasing sequence, it follows that as n is increased, better criteria are obtained; but it appears the restriction that for $t \to \infty$ the projection of the perturbation on a finite dimension space of increasing dimension vanishes.

In the above, the perturbations have been supposed to concern only the initial conditions; if the body forces are also perturbed, then the calculations may be performed analogously, and hence we obtain the same criteria as above, but with the additional restriction that $\mathbf{f}(t) \to \mathbf{f}_1(t)$ as $t \to \infty$.

A stability criterion for the intermediate solutions, analogous to (2.6.18) is deduced in ref. [73].

REFERENCES *

[1] Adams, R. A., *The Rellich-Kondrachov theorem for unbounded domains*, "Arch. Rational Mech. Anal.", **29**, 5, 1968, 390—394.

[2] Birih, R. V., *On the spectrum of the small perturbations of the plane parallel Couette flow*, "Prikl. Mat. Meh.", **29**, 4, 1965, 798—800. (In Russian).

[3] Case, K. M., *Stability of inviscid plane Couette flow*, "Phys. Fl.", **3**, 1960, 143—148.

[4] Case, K. M., *Edge effects and the stability of plane Couette flow*, "Phys. Fl.", **3**, 1960, 432—435.

[5] Case, K. M., *Stability of an indealized Atmosphere I. Discussion of results*, "Phys. Fl.", **3**, 1960, 149—154.

[6] Case, K. M., *Taylor instability of an inversed atmosphere*, "Phys. Fl.", **3**, 1960, 366—368.

[7] Case, K. M., *Hydrodynamic stability and the initial value problem*, "Proc. Symposia in appl. math." vol. XIII, Hydrodynamic instability, ed. G. Birkhoff, R. Bellman, 1962, 25—33.

[8] Cattabriga, L., *Su un problema al contorno relativo al sistema di equazioni di Stokes*, "Rend. Sem. Mat. Padova", **31**, 1961, 308—340.

[9] Chadrasekhar, S., *Hydrodynamic and hydromagnetic stability*, Oxford, Clarendon Press, 1961.

[10] Cristescu, R., *Analiză funcțională*, Ed. Didactică și Pedagogică, București, 1979.

[11] Daletzkii, Yu. L., M. G. Krein, *Stability of the solutions of the differential equations in Banach spaces*, Moscow, Nauka, 1970. (In Russian).

[12] Davis, S. H., *On the principle of exchange of stabilities*, "Proc. Roy. Soc.", A, **310**, 1969, 341—358.

[13] Dikii, L. A., *On the stability of the plane parallel Couette flow*, "Prikl. Mat. Meh.", **38**, 2, 1964, 389—392. (In Russian).

[14] Di Prima, R. C., *Vector eigenfunction expansion for the growth of Taylor vortices in the flow between rotating cylinders*, "Nonlinear Partial Differential Equations", New York, Acad. Press., 1967, 19—42.

[15] Di Prima, R. C., G. J. Habetler, *A completeness theorem for nonselfadjoint eigenvalue problem in hydrodynamic stability*, "Arch. Rational Mech. Analysis", **34**, 3, 1969, 218—227.

[16] Eckhaus, W., *Studies in nonlinear stability theory*, Berlin (occidental) — Heidelberg — New York, Springer, 1965.

* For periodical references (quoted between " "). The figures indicate first the colume, then the number (if it exists), the year, the pages. This is valid also for references in the Ch. 3 and 4.

[17] Eisenfeld, J., *Completeness theorems in hydrodynamic stability*, Ph. D. Thesis, Univ. of Chicago, 1966.
[18] Eisenfeld, J., *Normal mode expansion and stability of Couette flow*, "Quart. Appl. Math.", **26**, 3, 1968, 433—440.
[19] Friedman, A., *Partial differential equations*, Holt, Rinehart and Winston, Inc., 1969.
[20] Foiaş, C., G. Prodi, *Sur le comportement global des solutions non stationnaires des équations de Navier-Stokes en dimension 2*, "Rend. Sem. Mat. Univ. Padova", **39**, 1967, 1—34.
[21] Gallagher, A. P., A. Mc. D. Mercer, *On the behaviour of small disturbances in plane Couette flow*, "J. Fluid, Mech", **13**, 1962, 91—100.
[22] Georgescu, A., *Universal criteria of hydrodynamic stability*, "Rev. Roum. Math. Pures et Appliquées", **20**, nr. 3 (1976), 287—302.
[23] Ghika, Al., *Functional analysis*, Ed. Acad., Bucharest, 1967. (In Romanian).
[24] Gibson, R. D., *Overstability in the magnetohydrodynamic Bénard problem at large Hartmann numbers*, "Proc. Cambr. Phil. Soc.", **62**, 1966, 287—299.
[25] Grohne, D., *Über das Spektrum bei Eigenschwingungen ebener Larminarströmungen*, "ZAMM", **34**, 1954, 344—357.
[26] Guiraud, J. P., G. Iooss, *Sur la stabilité des écoulements laminaires*, "C.R. Acad. Paris", **266**, 1960, 1283—1286.
[27] Haimovici, A., *The equations of the mathematical physics and elements of calculus of variations*, Ed. did. şi ped., Bucharest, 1966. (In Romanian).
[28] Halanay, A., *Qualitative theory of the differential equations*, Ed. Acad. R.P.R., Bucharest, 1963. (In Romanian).
[29] Herfort, P., *Linear stability analysis for the Taylor problem with oscillating outer cylinder*, X^{th} Symp. Fluid. Dyn. Rynia, Poland, 1971.
[30] Hopf, E., *Uber die Anfangswertaufgabe für die hydrodynamischen Grundgleichungen*, "Math. Nachrichten", 4, 1950—1951, 213—231.
[31] Hopf, L., *Der Verlauf kleiner Schwingungen auf einer Strömung reibender Flüssigkeit*, "Annalen der Physik", **44**, 4, 1914, 1—60.
[32] Iooss, G., *Application de la théorie des semigroupes à l'étude de la stabilité des écoulements laminaires*, "Journal de Mécanique", **8**, 4, 1967, 477—507.
[33] Iooss, G., *Sur la théorie de la stabilité non linéaire des écoulements laminaires*, "C.R. Acad. Sc. Paris", **26**), 1969, 333—336.
[34] Iooss, G., *Théorie non linéaire de la stabilité des écoulements laminaires*, Thèse, Paris, 1971.
[35] Iooss, G., *Non linear stability theory of laminar flows in the case of exchange of stabilities*, X^{th} Biennal Fluid Dynamics Symposium, Rynia, (Poland), 1971.
[36] Yudovich, V. I., *On the stability of stationary flow of viscous incompressible fluids*, "Dokl. Akad. Nauk", **161**, 5, 1965, 1037—1040. (In Russian).
[37] Yudovich, V. I., *On the stability of forced oscillations of the fluids*, "Dokl. Akad. Nauk", **195**, 2, 1970, 292—296. (In Russian).
[38] Yudovich, V. I., *On the stability of the self oscillations of the fluid*, "Dokl. Akad. Nauk", **195**, 3, 1970, 574—576. (In Russian)
[39] Kaniel, Sh., M. Shinbrot, *Smoothness of weak solutions of the Navier-Stokes equations*, "Arch. Rational Mech. Anal.", **24**, 1967, 302—324.
[40] Kato, T., H. Fujita, *On the nonstationary Navier-Stokes system*, "Rend. Sem. Mat. Padova", **32**, 1962, 243—260.
[41] Keldysh, M. V., *On the eigenvalues and eigenfunctions of some classes of nonselfadjoint equations*, "Dokl. Akad. Nauk," **77**, 1, 1951, 11—14. (In Russian).
[42] Kirchgässner, K., P. Sorger, *Stability analysis of branching solutions of the Navier-Stokes equations*, Proc. 12th Int. Congr. Appl. Mech., Stanford Univ., Aug. 1968, 257—268.
[43] Kirchgässner, K., P. Sorger, *Branching analysis for the Taylor problem*, "Quart. J. Mech. Appl. Math.", **22**, 2, 1969, 183—209.
[44] Kiselev, A. A., O. A. Ladyzhenskaya, *On the existence and uniqueness of the solution of nonstationary problem for viscous incompressible fluids*, "Izv. A.N. S.S.S.R.", **21**, 1957, 655—680. (In Russian).

[45] Krein, S. G., *On functional properties, of the maps of the vector analysis and of the hydrodynamics,* "Dokl. Akad. Nauk", **93**, 6, 1953, 969–972. (In Russian).
[46] Krein, S. G., *Differential equations in Banach spaces and their applications to hydrodynamics,* "Uspehi Mat. Nauk", **12,** 1. 1957, 208–211. (In Russian).
[47] Krasnosel'skii, M. A., S. G. Krein, P. E. Sobolevskii, *On differential equations with unbounded maps in Hilbert spaces,* "Dokl. Akad. Nauk", **112,** 6, 1957, 990–993. (In Russian).
[48] Kranosel'skii, M. A., S. G. Krein, *On differential equations in Banach spaces,* Proceedings of the third Mathematical Congress of Soviet Union, "Izv. Akad. Nauk", S.S.S.R. Moscow, **3,** 1958, 73–80. (In Russian).
[49] Krylov, A. L., *The proof of the instability of a certain flow of viscous incompressible fluid,* "Dokl. Akad. Nauk", **153,** 4, 1963, 787–789. (In Russian).
[50] Krylov, A. L., *On the stability of Poiseuille flow in plane channel,* "Prikl. Mat. Meh.", **30,** 4, 1966, 679–687. (In Russian).
[51] Ladyzhenskaya, O. A., *Solution 'in the large' of the nonstationary boundary value problem for the Navier-Stokes system with two space variables,* "Comm. Pure Appl. Math.", **12,** 3, 1959, 427–433.
[52] Ladyzhenskaya, O. A., *Mathematical problems of the dynamics of viscous incompressible fluids,* Nauka, Moscow, 1970. (In Russian).
[53] Leray, J., *Étude de diverses équations intégrales non linéaires et de quelques problèmes que pose l'Hydrodynamique,* "J. Math. Pures Appl.", **12,** 1933, 1–82.
[54] Leray, J., *Sur le mouvement d'un liquide visqueux remplissant l'espace,* "Acta Matematica", **63,** 1934, 193–248.
[55] Leray, J., *Essai sur le mouvement plan d'un liquide visqueux que limitent des parois,* "J. Math. Pures Appl.", série 9, **13,** 1934, 331–418.
[56] Lions, J. L., *Sur l'existence de solutions des équations de Navier-Stokes,* „C.R. Paris", **248,** 20, 1959, 2847–2849.
[57] Lions, J. L., *Quelques remarques sur les equations de Navier-Stokes,* "Bull. Math. de la Soc. Sci. Math. Phys. de la R.P.R.", **4** (52), 3–4, 1960, 81–88.
[58] Lions, J. L., E. Magenes, *Problèmes aux limites non homogènes et applications,* vol. I, Dunod, Paris, 1968.
[59] Lions, J. L., *Quelques méthodes de résolution des problèmes aux limites non linéaires,* Dunod, Gauthier-Villars, Paris, 1969.
[60] Meshalkin, L. D., Ya. G. Sinai, *Investigation of the stability of the stationary solution of the equation governing the plane flow of the viscous incompressible fluids,* "Prikl. Mat. Meh.", **25,** 6, 1961, 1140–1143. (In Russian).
[61] Mihlin, S. G., *Variational methods in mathematical physics,* Nauka, Moscow, 1970. (In Russian).
[62] von Mises, R., *Kleine Schwingungen und Turbulenz,* Jahresbericht der Deutschen Mathematiker Vereinigung, **21,** 1912, 241–248.
[63] von Mises, R., *Beitrag zum Oszillationsproblem,* Festschrift Heinrich Weber, Leipzig und Berlin, 1912, 252–282.
[64] Maurin, K., *Hilbert space methods,* Mir, Moscow, 1965. (In Russian).
[65] Morrey, C. B. Jr., *Multiple integrals in the calculus of variations,* Springer-Verlag, Berlin, 1966.
[66] Nagumo, M., *Lessons on the modern theory of the partial differential equations,* Mir. Moscow, 1967. (In Russian).
[67] Naimark, M. A., *On some characteristics of the completeness of the system of eigenvectors and associate vectors of linear maps in Hilbert spaces,* "Dokl. Akad. Nauk", **98**, 5, 1954, 727–730. (In Russian).
[68] Nikol'skii, S. M., *Approximation of the functions of several variables and the imbedding theorems,* Moscow, Mir, 1969. (In Russian).
[69] Oseen, S. W., *Über die Stokessche Formel und über eine verwandte Aufgabe in der Hydrodynamik,* "Arkiv för mathematik, astronomi och fysik", **6,** 29, 1910.
[70] Pascali, D., *Sobolev-Kondrashev's imbedding theorems,* "St. Cerc. Mat.", **17,** 8, 1965, 1231–1261. (In Romanian).
[71] Pellew, A., R. V. Southwell, *On maintained convection motion in a fluid heated from below,* "Proc. Roy. Soc. A", **176,** 1940, 312–343.
[72] Pomarenko, Yu, B., *On the stability of plane Couette flow,* "Prikl. Mat. Meh.", **32,** 4, 1968, 606–614. (In Russian).

[73] Pop, D., *Etude qualitative des solutions des équations de Navier-Stokes en dimension* 3, "Rend. Sem. Mat. Padova", **44,** 1970, 273—297.

[74] Prodi, G., *Qualche risultato riguardo alle equazioni de Navier-Stokes nel caso bidimensionale,* "Rend. Sem. Mat. Padova" **30,** 1960, 1—15.

[75] Prodi, G., *Résultats recents et problèmes anciens dans la théorie des équations de Navier-Stokes,* Colloques Intern. du CNRS, nr. 117, Les équations aux derivées partielles, Paris, 1962, 181—196.

[76] Prodi, G., *Teoremi di tipo locale per il sistema di Navier-Stokes e stabilità delle soluzioni stazionarie,* "Rend. Seminario Mat. Padova", **32,** 1962, 374—397.

[77] Romanov, V. A., *Stability of the plane parallel Couette flow,* 1/1971, Inst. Probl. Meh., Akad. Nauk S.S.S.R., Moscow. (In Russian).

[78] Romanov, V. A., *Stability of the plane parallel Couette flow,* „Funktzionalnyi Analiz i ego Prilojenia", **7,** 2, 1973, 62—73. (In Russian).

[79] Rosencrans, S. I., D. H. Sattinger, *On the spectrum of an operator occurring in the theory of hydrodynamic stability,* "J. Math. Phys.", **45,** 3, 1966, 289—300.

[80] Riis, E., *The stability of Couette flow in non-stratified and stratified viscous fluids,* "Geophys publ.", Oslo, **23,** 1962, 4.

[81] Sani, R. L., Ph. D. Thesis, *Department of Chemical Engineering,* Univ. of Minnesota, 1963.

[82] Sattinger, D. H., *The mathematical problem of hydrodynamic stability,* "J. Math. Mech.", **19,** 9, 1970, 797—817.

[83] Savulescu, St. N., *Transition from the laminar to the turbulent flow,* Ed. Acad., Bucharest, 1968. (In Romanian).

[84] Schensted, I. V., *Contributions to the theory of hydrodynamic stability,* Ph. D. Thesis, Univ. of Michigan, Ann Arbor, 1960.

[85] Sherman, M., S. Ostrach, *On the principle of exchange of stabilities for the magnetohydrodynamic thermal stability problem in completely confined fluids,* "J. Fluid Mech.", **24,** 4, 1966, 661—671.

[86] Shinbrot, M., *Lectures on fluid mechanics,* Gordon & Breach, London, 1973.

[87] Smirnov, V. I. *A course in higher mathematics,* vol. 5, Ed. Tehnică, 1963, București.

[88] Sobolev, S. L., *Some applications of the functional analysis to the mathematical physics,* Leningrad, 1950. (In Russian).

[89] Sobolev, S. L., *On a new problem for the systems of partial differential equations,* "Dokl. Akad. Nauk", **81,** 6, 1951, 1007—1009. (In Russian).

[90] Sobolev, S. L., *The equations of the mathematical physics,* Ed. tehnică, Bucharest, 1955. (In Romanian).

[91] Sobolevskii, P. E., *On nonstationary equations of the hydrodynamics of viscous fluids,* "Dokl. Akad. Nauk", **128,** 1, 1959, 45—48. (In Russian).

[92] Sobolevskii, P. E., *On the application of fractional powers of selfadjoint maps to the investigation of some nonlinear differential equations in Hilbert space,* "Dokl. Akad. Nauk", **130,** 2, 1960, 272—275. (In Russian).

[93] Sobolevskii, P. E., *On the application of method of the fractional powers of the maps to the investigation of Navier-Stokes equations,* "Dokl. Akad. Nauk", **155,** 1, 1964, 50—53. (In Russian).

[94] Sobolevskii, P. E., *The investigation of the Navier-Stokes equation by means of the methods of the theory of parabolic equations in Banach spaces,* "Dokl. Akad. Nauk.", **156,** 4, 1964, 745—748. (In Russian).

[95] Spiegel, E. A., "Astrophys. J", **132,** 1960, 250.

[96] Shtern, V. I., *Stability of the plane Couette flow,* "J. Prikl. Meh. i Teh. Phys.", **5,** 1969, 117—119. (In Russian).

[97] Volevich, L. R., B. I. Paneyah, *Some spaces of generalized functions and the imbedding theorems,* "Uspehi Mat. Nauk.", **20,** 1, 1965, 3—74. (In Russian).

[98] Vorovich, I., V. I. Yudovich, *Stationary flow of viscous incompressible fluids,* „Matem. Sbornik", **53,** 1961, 393—428. (In Russian).

[99] Velte, W., *Stabilität und Verzweigung stationärer Losungen der Navier-Stokesschen Gleichungen beim Taylorproblem,* "Arch. Rational Mech. Analysis", **22**, 1, 1966, 1—14.

[100] Wasow, W., *On small disturbances of plane Couette flow,* „J. Res. Nat. Bur. Standards", **51**, 1953, 195—202.

[101] Weinberger, H. F., *Exchange of stability in Couette flow,* "Bifurcation theory and nonlinear eigenvalue problems", ed. J. B. Keller & S. Antman, Benjamin, Amsterdam, 1969, 395—409.

[102] Weyl, H., *The method of orthogonal projection in potential theory,* "Duke Math. J." **7**, 1940, 411—444.

[103] Yih, C. S., *Thermal instability of viscous fluids,* "Quart. Appl. Math.", **17**, 1959, 25—42.

[104] Yosida, K., *Functional analysis,* Springer Verlag, Berlin (occidental), 1965.

Chapter 3

BRANCHING AND STABILITY OF SOLUTIONS OF THE NAVIER-STOKES EQUATIONS

Experimental evidence, the principle of exchange of stabilities and numerical computations by Ritz's and Galerkin's methods suggested that some steady flows become unstable owing to the appearance of new steady or time periodic flows—called secondary flows. This is why branching theory of solutions of the Navier-Stokes equations is closely connected with hydrodynamic stability theory.

After a presentation of Leray-Schauder's topological degree (§ 3.1), Liapunov-Schmidt's (§ 3.3) and Joseph-Sattinger's (§ 3.5) methods, we apply these methods (in §§ 3.2, 3.4 and 3.5 respectively) to the bifurcation of weak solutions of the Navier-Stokes equations. These problems are formulated as nonlinear eigenvalue problems for completely continuous maps in the stationary case and as evolution equations in a Banach space is the nonstationary case. We also consider the stability of secondary flows.

§ 3.1. TOPOLOGICAL DEGREE METHOD FOR NONLINEAR EQUATIONS IN BANACH SPACES (LERAY-SCHAUDER METHOD)

3.1.1. The finite-dimensional case

Let ω be a bounded open set of Euclidean space R^n, let $\partial\omega$ be its boundary and let $f\colon \overline{\omega} \to R^n$ be a map of class C^1, where $\overline{\omega}$ stands for the closure of $\omega(\overline{\omega} = \omega \cup \partial\omega)$. For every $p \in R^n$ which does not belong to $f(\partial\omega)$, we define $\varepsilon = \inf\limits_{x \in \partial\omega} |f(x) - p|$. Let $f_\varepsilon(y)$ be a positive function defined on $[0, \infty)$, which vanishes in some neighbourhood of $y = 0$ and for $y \geqslant \varepsilon$, and which satisfies

$$\int_{R^n} f_\varepsilon(|x|)\, \mathrm{d}x = 1.$$

By definition, *the topological degree* of f at the point p with respect to the set ω is

$$d(f, \omega, p) = \int_\omega f_\varepsilon(|f(x) - p|) J_f \, dx, \tag{3.1.1}$$

where J_f is the Jacobian of f, $J_f = \det \left\| \frac{\partial f_i}{\partial x_j}(x) \right\|$.

If to the hypothesis $p \notin f(\partial\omega)$ we add the hypothesis that p does not belong to the image under f of the set $\{x \in \omega \mid J_f(x) = 0\}$, it may be proved that the equation $f(x) = p$ has in ω only a finite number of solutions; and the analytic definition (3.1.1) coincides with the following algebraic definition

$$d(f, \omega, p) = n_1 - n_2, \tag{3.1.1$'$}$$

where

$$n_1 = \text{card}\, \{x \in f^{-1}(p) \mid J_f(x) > 0\},$$

$$n_2 = \text{card}\, \{y \in f^{-1}(p) \mid J_f(x) < 0\}.$$

In other words

$$d(f, \omega, p) = \sum_{\xi \in f^{-1}(p)} \text{sgn}\, J_f(\xi). \tag{3.1.2}$$

To do the effective calculus of the degree when $p \in f(\partial\omega)$, we choose a sequence of points $p_n \to p$ such that $p_n \notin f(\partial\omega)$, and then we have [38]

$$d(f, \omega, p) = \lim_{n \to \infty} d(f, \omega, p_n). \tag{3.1.1$''$}$$

The main properties of the topological degree are [46], [7], [47].

i) The topological degree is an integer; the degree of the identical map is equal to $+1$.

ii) If $d(f, \omega, p) \neq 0$, then the equation $f(x) = p$ has at least one solution in ω. (Thus the main advantage of the topological degree method is that it provides an existence theorem even when uniqueness is absent).

(iii) Additivity with respect to ω: if ω_1 and ω_2 are two disjoint open sets of R^n, for which the degree at the point p can be defined, then we have

$$d(f, \omega_1 \cup \omega_2, p) = d(f, \omega_1, p) + d(f, \omega_2, p).$$

iv) Invariance to homotopy: Let $f(x, \lambda)$ be a continuous function on $\omega \times (a, b)$ uniformly with respect to λ and let $p \notin f(\partial\omega)$ for $\lambda \in \in [a, b]$. Then $d(f(x, \lambda), \omega, p)$ is independent of λ.

v) Let $\xi \in \omega$ be a solution of the equation $f(x) = p$ and let ω_0 be a neighbourhood of ξ which does not contain any other solution of this equation. The number

$$i[f, \xi] = d(f, \omega_0, p) \tag{3.1.3}$$

is called the index of f with respect to the isolated solution ξ. If $J_f(\xi) \neq 0$. then from the relations (3.1.2), we derive the equality

$$i[f, \xi] = \text{sgn det} \left\| \frac{\partial f_i}{\partial x_j}\bigg|_{x=\xi} \right\|. \tag{3.1.4}$$

This equality reduces the study of the nonlinear problem to that of Jacobian, which is a linear operator.

vi) If the equation $f(x) = p$ has a finite number of solutions $\xi_1, \ldots, \xi_k$ in ω and $J_f(\xi_j) \neq 0$, then formula (3.1.2) can be written as

$$d(f, \omega, p) = \sum_{j=1}^{k} i[f, \xi_j]. \tag{3.1.2$'$}$$

It can be proved that this equality holds also when the Jacobian $J_f(x)$ vanishes at some solutions of the equation $f(x) = p$.

In the following we shall show that under the hypothesis of the principle of exchange of stabilities, the existence of branching solutions and in some cases their stability can be derived from a knowledge of the spectrum of the Jacobian.

Let the nonlinear nonstationary equation

$$\frac{dx}{dt} = f(x, \lambda), \tag{3.1.5}$$

where $f : R^n \times [a, b] \to R^n$ is a continuous function of class C^1 with respect to x. Denote by $x_0(\lambda)$ a stationary solution of Equation (3.1.5), for $\lambda \in [a, b]$. Denote by $f_i(x, \lambda)$ ($i = 1, \ldots, n$) the components of the vector function $f(x, \lambda)$ and by $A(\lambda)$ the square matrix $\left(\frac{\partial f_i}{\partial x_j}\bigg|_{(x_0(\lambda), \lambda)}\right)$. Then the stability of the solution $x_0(\lambda)$ with respect to perturbations $y(t)$ is governed by the linear equation

$$\frac{dy}{dt} = A(\lambda) y. \tag{3.1.6}$$

Now consider stationary solutions of Equation (3.1.5) of the form $x' + x_0$, which satisfy the nonlinear stationary equation

$$f(x' + x_0, \lambda) = 0. \tag{3.1.5$'$}$$

If in Equation (3.1.5)$'$ we neglect the nonlinear terms in x' we obtain the linear stationary equation

$$A(\lambda)\, x' = 0. \tag{3.1.6$'$}$$

Since the matrix $A(\lambda)$ is real, it follows that its eigenvalues $\sigma_k(\lambda) = (\sigma_r)_k(\lambda) + i(\sigma_i)_k(\lambda)$ are either real or complex-conjugate and they generate continuous curves in the plane (σ_r, σ_i) as λ varies. In particular these curves can be located on the real axis if the corresponding eigenvalue is real. Let λ_0 be the point where the solution $x_0(\lambda)$ loses its stability. By Liapunov's theorem, for $\lambda < \lambda_0$ we have $(\sigma_r)_k(\lambda) < 0$, $k = 1, 2, \ldots, n$ and for $\lambda > \lambda_0$ there exists at least one integer k such that $(\sigma_r)_k(\lambda) > 0$. At $\lambda = \lambda_0$ the following two situations may occur as the axis $\sigma_r = 0$ is crossed by a) a real simple eigenvalue $\sigma_1(\lambda)$, b) a pair of simple complex conjugate eigenvalues.

a) From the above-mentioned assumptions, it follows that $\sigma_1(\lambda_0) = 0$, $(\sigma_r)_k < 0$, $k = 2, \ldots, n$. Therefore, the principle of exchange of stabilities holds and $\left.\dfrac{d\sigma_1}{d\lambda}\right|_{\lambda=\lambda_0} > 0$. We shall prove that λ_0 is also a branching point of new stationary solutions, i.e. at the point λ_0 the uniqueness is lost. Without loss of generality, we may suppose that $x_0(\lambda) \equiv 0$. Let ω_0 be a sufficiently small neighbourhood of the point $x_0 \equiv 0$ such that for $\lambda < \lambda_0$ Equation (3.1.5)′ has in ω_0 the unique solution $x_0 \equiv 0$. Assuming that this is a solution for every λ, from the above properties and taking $p = 0$, we obtain

$$d(f, \omega_0, 0)\,|_{\lambda>\lambda_0} = d(f, \omega_0, 0)\,|_{\lambda<\lambda_0} =$$

$$= i[f, 0] = \operatorname{sgn} \det \left\| \left(\frac{\partial f_i}{\partial x_j} \right)_{(0,\lambda)} \right\| = \operatorname{sgn} \sigma_1(\lambda) \ldots \sigma_n(\lambda).$$

From the hypotheses concerning the spectrum of A, it follows that as λ crosses λ_0, then $\operatorname{sgn} \sigma_1(\lambda) \ldots \sigma_n(\lambda)$ changes its sign as well as $i[f, 0]$. Suppose that for $\lambda < \lambda_0$ it is equal to -1. Then for $\lambda > \lambda_0$ we have $i[f, 0] = +1$, and this means that besides the solution $x_0 \equiv 0$, some additional solutions $\xi_1, \ldots, \xi_k$ must exist so that the sum of their indices and $i[f, 0]$ is again equal to -1. For simplicity in the use of the topological degree theory we suppose that none of the solutions of (3.1.5)′ crosses the boundary $\partial\omega$. Note that this method shows the branching at the point $\lambda = \lambda_0$ but it does not yield the number of branches. By analytical methods, it can be proved that the point $(0, \lambda_0)$ is a bifurcation point, i.e. two branches may appear there: both of them supercritical $(\lambda > \lambda_0)$, both of them subcritical $(\lambda < \lambda_0)$, or one subcritical and the other supercritical. The supercritical branches are stable and the subcritical ones unstable. By branches we mean curves in the plane (x, λ), whose points are stationary solutions $(\xi(\lambda), \lambda)$ of Equation (3.1.5)′.

In case b) at the point $(0, \lambda_0)$ there appears, up to phase shifts, a single branch of solutions periodic with respect to time. Again the supercritical solutions are stable and the subcritical ones are unstable. Therefore, in case b) at the point λ_0 now appear nonstationary solutions without loss of uniqueness of the stationary solution.

Suppose now that at the bifurcation point $(0, \lambda_0)$ there appear supercritical branches $\xi_1(\lambda)$ and $\xi_2(\lambda)$ and let $\tilde{\sigma}_1(\lambda), \ldots, \tilde{\sigma}_n(\lambda)$ be the eigenvalues of A corresponding to $\xi_1(\lambda)$ and let us determine by topological degree the stability or instability of ξ_1 and ξ_2. For $\lambda > \lambda_0$, in a neighbourhood of λ_0, we have $\tilde{\sigma}_2, \ldots, \tilde{\sigma}_n$ which do not differ too much from $\sigma_2, \ldots, \sigma_n$ and these eigenvalues have negative real parts. Then sgn $\sigma_2 \ldots \sigma_n =$ sgn $\tilde{\sigma}_2 \ldots \tilde{\sigma}_n$ [44]. We have seen in the above that $i[f, \xi_1] = i[f, \xi_2] = -1$. Therefore

$$i[f, 0]\,|_{\lambda > \lambda_0} = +1 = -\,i[f, \xi_1],$$

$$i[f, \xi_1] = \operatorname{sgn} \det \left\| \frac{\partial f_i}{\partial x_j}\bigg|_{\xi_1, \lambda} \right\| = \operatorname{sgn} \tilde{\sigma}_1 \cdot \operatorname{sgn} (\tilde{\sigma}_2 \ldots \tilde{\sigma}_n),$$

$$i[f, 0] = \operatorname{sgn} \det \left\| \frac{\partial f_i}{\partial x_j}\bigg|_{0, \lambda} \right\| = \operatorname{sgn} \sigma_1 \cdot \operatorname{sgn} (\sigma_2 \ldots \sigma_n),$$

whence sgn $\tilde{\sigma}_1 = -$sgn $\sigma_1 = -1$. As $\tilde{\sigma}_2, \ldots, \tilde{\sigma}_n$ have negative real parts and sgn $\tilde{\sigma}_1 = -1$ this means that all the eigenvalues $\tilde{\sigma}_1, \ldots, \tilde{\sigma}_n$ have a negative real part and therefore, by the linearization principle, $\xi_1(\lambda)$ consists of stable solutions. Similarly [45], [46] it is proved that if two solutions, (x', λ') and (x'', λ''), from a neighbourhood of a critical point $(0, \lambda_0)$ have indices of opposite sign, then they have opposite stabilities.

3.1.2. The infinite-dimensional case (Leray-Schauder method).

Let ω be a bounded open set in a Banach space $\mathcal{B}$ and let $F : \overline{\omega} \to \mathcal{B}$ be a compact map (i.e. F is continuous and maps bounded closed sets onto compact sets) (see Appendix 1); and let $\Phi = I - F$, where I is the identity map. We assume that the point $p = 0$ does not belong to the set $\Phi(\partial\omega)$, or, equivalently, that F has no fixed points on the boundary $\partial\omega$, and we define the topological degree $d(\Phi, \omega)$ as equal to the topological degree $d(\Phi_\varepsilon, \omega_\varepsilon, 0)$, where ω_ε is an open set in a finite dimensional Banach subspace $\mathcal{B}_\varepsilon \subset \mathcal{B}$, and Φ_ε is a map which approximates Φ in a sense specified below.

Since $F(\partial\omega)$ is a compact set, and since the equation $x = F(x)$ has no solution on $\partial\omega$, it follows that $\inf\limits_{x \in \partial\omega} \|x - F(x)\| > 0$; let us denote this number by ε. The compactness of $F(\overline{\omega})$ implies the existence of a finite number of points $y_1, \ldots, y_k \in F(\overline{\omega})$, such that for every $x \in F(\overline{\omega})$ there exists at least one point y_i which satisfies the inequality $\|x - y_i\| \leqslant \frac{\varepsilon}{2}\cdot$ We can then define a continuous map T_ε

which maps the set $F(\overline{\omega})$ into the finite-dimensional subspacei $\mathcal{B}_\varepsilon \subset \mathcal{B}$ spanned by the points $y_i, \ldots, y_k$ and satisfies the inequality

$$\|T_\varepsilon(x) - x\| < \varepsilon \text{ [1]}. \tag{3.1.7}$$

Replacing, if necessary, $\mathcal{B}_\varepsilon$ by the subspace spanned by itself and by an arbitrary point of ω, we may assume that $\omega \cap \mathcal{B}_\varepsilon$ is not empty. We put $\omega_\varepsilon = \omega \cap \mathcal{B}_\varepsilon$, which, by construction, is a bounded nonempty open set of the finite-dimensional subspace $\mathcal{B}_\varepsilon$. Writing

$$\Phi_\varepsilon(x) = x - T_\varepsilon(F(x)),$$

we obtain a continuous map $\Phi_\varepsilon : \overline{\omega}_\varepsilon \to \mathcal{B}_\varepsilon$, which, by (3.1.7,) approximates Φ up to $\frac{\varepsilon}{2}$

$$\|\Phi_\varepsilon(x) - \Phi(x)\| \leqslant \frac{\varepsilon}{2}. \tag{3.1.8}$$

The definition of ε and inequality (3.1.8) yield, for $x \in \partial\omega_\varepsilon = \partial\omega \cap \mathcal{B}_\varepsilon$, the relation

$$\|\Phi_\varepsilon(x)\| \geqslant \|\Phi(x)\| - \|\Phi(x) - \Phi_\varepsilon(x)\| \geqslant \varepsilon - \frac{\varepsilon}{2} = \frac{\varepsilon}{2}, \tag{3.1.9}$$

which expresses the fact that the distance from the point $p = 0$ to the set $\Phi_\varepsilon(\partial\omega_\varepsilon)$ is strictly positive. In $\mathcal{B}_\varepsilon$, there exists then the finite-dimensional degree $d(\Phi_\varepsilon, \omega_\varepsilon, 0)$. It can be proved that it does not depend on the choice of Φ_ε and $\mathcal{B}_\varepsilon$; consequently we can define

$$d(\Phi, \omega) = d(\Phi_\varepsilon, \omega_\varepsilon, 0).$$

We list the following properties of the Leray-Schauder topological degree ([46], [31]), which can be deduced from the properties of finite-dimensional topological degree:

i) $d(I - F, \omega)$ is an integer defined when the completely continuous map F has no fixed point on $\partial\omega$; $d(I, \omega) = +1$.

ii) If $d(I - F, \omega) \neq 0$, then the map F admits at least one fixed point in ω.

---

[1] It is sufficient to take

$$T_\varepsilon(x) = \frac{\sum_{i=1}^{k} \mu_i(x)\, y_i}{\sum_{i=1}^{k} \mu_i(x)},$$

where

$$\mu_i(x) = \begin{cases} \varepsilon - \|x - y_i\| & \text{if } \|x - y_i\| < \varepsilon, \\ 0 & \text{if } \|x - y_i\| \geqslant \varepsilon. \end{cases}$$

iii) $d(I-F, \omega)$ is an additive function of ω.

iv) Let $F(x, \lambda)$ be a map uniformely continuous with respect to λ, which for every $\lambda \in [a, b]$ defines a compact map $F(\cdot, \lambda): \bar{\omega} \to \mathcal{B}$, with no fixed points on the boundary $\partial\omega$. Then $d(\Phi(\cdot, \lambda), \omega)$ does not depend on λ.

v) Let $x_0 \in \omega$ be an isolated fixed point of the map F and let $\omega_0 < \omega$ be a neighbourhood of x_0 which does not contain any other fixed points of F. Then the number

$$i[I-F, x_0] = d(I-F, \omega_0)$$

is called the index of F at the point x_0 and is independent of the choice of ω_0. Let F'_{x_0} be the Fréchet differential of F at the point x_0. If $I - F'_{x_0}$ is invertible, then x_0 is an isolated fixed point of $F(x)$ and we have

$$i[I-F, x_0] = d(I-F'_{x_0}, \omega) = \pm 1. \tag{3.1.10}$$

This property reduces the calculus of the index of a nonlinear map to that of a linear map, namely the Fréchet differential at the point under consideration [2)].

vi) If the map F has a finite number of fixed points $x_1, \ldots, x_k$ in ω, then

$$d(I-F, \omega) = \sum_{i=1}^{k} i[I-F, x_i]. \tag{3.1.11}$$

This reduces the calculus of the topological degree to that of indices.

Under the hypothesis that the point $x = 0$ is interior to ω, we apply the above properties to the study of the equation

$$x = F(x, \lambda), \tag{3.1.12}$$

where in addition to (iv), F satisfies the condition $F(0, \lambda) = 0$. We first prove the following

Lemma. *If* A *is a linear completely continuous map which has no characteristic value in the interval* $[0, \mu]$ *then*

$$d(I-\mu A, \omega) = +1. \tag{3.1.13}$$

Proof. By hypothesis, for every $\mu' \in [0, \mu]$, the linear map $I - \mu' A$ is invertible. Therefore the only fixed point of the map $\mu' A$ is $x = 0$, which does not belong to the boundary $\partial\omega$. By property iv), we have $d(I - \mu A, \omega) = d(I - \mu' A, \omega)$ which for $\mu' = 0$ gives the relation (3.1.13).

---

2) The Fréchet differential of a compact map F at a point x_0 is a completely continuous linear map F'_{x_0} which, assuming that the maps are defined on finite-dimensional Banach spaces, corresponds to the map associated with the matrix $\left(\frac{\partial f_i}{\partial x_j}\bigg|_{x_0}\right)$. Since the finite-dimensional degree of the map associated with an invertible matrix A is equal to sgn (det A), it is easy to see the analogy with relation (3.1.4).

Denoting by $A(\cdot, \lambda)$ the Fréchet differential of the map $F(\cdot, \lambda)$ at the point $x = 0$, we have

Theorem 3.1.1. *If for a particular value λ_0 of λ the map $A(\cdot, \lambda_0)$ has no characteristic value in the interval* [0, 1], *then*

$$i[I - F(\cdot, \lambda), 0] = +1.$$

Proof. By property v), we have $i\,[I-F(\cdot, \lambda_0), 0] = d(I-A(\cdot, \lambda_0), \omega)$ and by the lemma applied for $\mu = 1$ and $A = A(\cdot, \lambda_0)$, the right-hand side of this equality is equal to $+1$.

Corollary. *Under the same hypothesis, if the unique solution of the equation* $x = F(x, \lambda_0)$ *is* $x = 0$, *then for every* $\lambda \in [a, b]$, *we have*

$$d(I - F(\cdot, \lambda), \omega) = +1. \tag{3.1.14}$$

Proof. By property iv), we have $d(I-F(\cdot, \lambda), \omega) = d(I-F(\cdot, \lambda_0), \omega)$ and by hypothesis and property vi), it follows that $d(I-F(\cdot,\lambda), \omega) = i\,[I - F(\cdot, \lambda_0), 0]$, which, by virtue of Theorem 3.1.1, is equal to $+1$.

Now let $A(\cdot, \lambda)$ be of the form $A_1 + \lambda A_2$, where A_1 and A_2 are two linear completely continuous maps. If the positive eigenvalues λ_i of the problem

$$x = A(x, \lambda), \tag{3.1.15}$$

are $0 < \lambda_1 < \ldots < \lambda_i < \lambda_{i+1} < \ldots$, then $d(I - A(\cdot, \lambda), \omega) = i\,[I-F(\cdot, \lambda), 0]$ is constant on every interval $(\lambda_i, \lambda_{i+1})$ and is multiplied by $(-1)^{m_i}$ when λ crosses an eigenvalue of multiplicity equal to m_i.

Theorem 3.1.2. *Consider the nonlinear equation*

$$x = F(x, \lambda) \tag{3.1.12}$$

with the following properties: 1° $F(\cdot, \lambda)$ *is a map compact on* $\overline{\omega}$ *and uniformly continuous with respect to* λ; 2°) $F(0, \lambda) = 0$ *for every* λ, $F(x, 0) = 0$ *admits only the solution* $x = 0$; 3°) *the Fréchet differential* $F'_0(\cdot, \lambda) = A(\cdot, \lambda)$ *is of the form* $A_1 + \lambda A_2$ *and satisfies the hypotheses of Theorem 3.1.1 with* $\lambda_0 = 0$. *Then every positive eigenvalue* λ_i *of odd multiplicity on the linear equation (3.1.15) such that the nonlinear equation (3.1.12) has only the solution* $x = 0$ *for* $\lambda = \lambda_i$ *and has no solution of the boundary* $\partial\omega$ *for* $\lambda < \lambda_i$, *is a point of supercritical or subcritical branching of the solutions of this equation according as* $i\,[I - F(\cdot, \lambda), 0]$, *for* $\lambda < \lambda_i$ *is positive or negative.*

Proof. By the corollary of the theorem 3.1.1 it follows that $d(I - F(\cdot, \lambda), \omega) = 1$ for $0 \leqslant \lambda \leqslant \lambda_i$. Owing to the complete continuity, uniform continuity and uniquencess of the solution for $\lambda = \lambda_i$ there exists $\delta > 0$ such that for $\lambda \in [\lambda_i, \lambda_i + \delta]$ Equation (3.1.12) has no solution on the boundary $\partial\omega$ and therefore $d(I - F(\cdot, \lambda), \omega) \equiv 1$ for $\lambda_i < \lambda < \lambda_i + \delta$. We choose δ sufficiently small such that in the interval $[\lambda_i - \delta, \lambda_i + \delta]$ no other eigenvalue of the linear problem may exist. Then $i[I - F(\cdot, \lambda), 0]$ will be constant on each of the intervals $[\lambda_i - \delta, \lambda_i)$, $(\lambda_i, \lambda_i + \delta]$ and, since λ_i is of odd multiplicity, this index will change sign as λ crosses λ_i. Assume, for instance, that for $\lambda_i - \delta \leqslant \lambda < \lambda_i$ we have $i[I - F(\cdot, \lambda), 0] > 0$. Then for $\lambda_i < \lambda \leqslant \lambda_i + \delta$, this index being negative, it follows that $x = 0$ cannot be the unique solution of the nonlinear equation, since, if so, then $d(I - F(\cdot, \lambda), \omega) = 1$ would be contradicted by property vi). As $\lambda \to \lambda_i$ from above, again from the complete continuity, uniform continuity and uniqueness of the solution for $\lambda = \lambda_i$ it may be seen that $\| u(\lambda) \|_{\mathfrak{B}} \to 0$, where $u(\lambda)$ are the secondary solutions of Equation (3.1.12) corresponding to the eigenvalue λ. Hence at λ_i a supercritical branching occurs. In particular, this is valid, for $i = 1$. The case of subcritical branching is treated in a similar way.

If the nonlinear Equation (3.1.12) has the form

$$x = \lambda F(x), \tag{3.1.12$'$}$$

then, instead of Theorem 3.1.2, we have

Theorem 3.1.3 *(Krasnosel'skii [27]). A sufficient condition for the real number λ_i to be a branch point for Equation (3.1.12)', where* F *is a compact map on $\overline{\omega}$ is that λ_i should be a characteristic value of odd multiplicity of the Fréchet differential of the operator* F *at its fixed point* x = 0. *At λ_i a continuous branch of eigenfunctions* x(λ) *of* F *appears.*

Theorem 3.1.4. *A necessary condition for the real number λ to be a branch point for Equation* (3.1.12)', *where* F *is a compact map on $\overline{\omega}$ is that λ should be a characteristic value of the Fréchet differential of the map* F *at its fixed point* x = 0.

The necessity follows by reductio ad absurdum [40].

In addition to the usefulness of the topological degree method for determining branch points, it can sometimes be used to prove the existence of solutions of the nonlinear Equation (3.1.12)' for an arbitrary λ belonging to the interval [a, b], when all the solutions $x_1, \ldots, x_k$ corresponding to the particular value λ_0 of λ are known. If λ_0 is not a characteristic value of the maps $F'_{x_i} (i = 1, \ldots, k)$, then the indices $i[I - \lambda_0 F, x_i]$ can be calculated by formula (3.1.10) and the equality

$$d(I - \lambda_0 F'_{x_i}, \omega) = (-1)^{m(\lambda_0)}, \tag{3.1.16}$$

where $m(\lambda_0)$ is the sum of the multiplicities of the eigenvalues $\lambda_i < \lambda_0$, of the linear equation

$$x = \lambda F'_{x_i}(x). \tag{3.1.15$'$}$$

By formula (3.1.11), we obtain $d(I - \lambda_0 F, \omega)$. If this number is non zero and if for $\lambda \in [a, b]$ Equation (3.1.12)' has no solution on $\partial\omega$, then by property iv) it follows that $d(I - \lambda F, \omega) \neq 0$ and property ii) implies the existence of a solution of the equation (3.1.12)′ for every λ from the interval [a, b].

§ 3.2. BRANCHING OF SOLUTIONS OF THE NAVIER-STOKES EQUATIONS BY THE LERAY-SCHAUDER METHOD

Under the hypothesis that the principle of exchange of stabilities holds, topological degree theory can be applied to study the branching of stationary solutions of the Navier-Stokes system, for some particular types of flow. To this purpose we consider the system $x = F(x, \lambda)$ where F is the Navier-Stokes nonlinear map and the parameter λ is either the Reynolds or the Rayleigh number. To the map F is associated its Fréchet differential F_0' at the point $x = 0$, which is the Stokes linear map.

In order to apply the results from the previous paragraph, we must prove each time the compactness of the map F with respect to x and its uniform continuity with respect to λ and determine the eigenvalues of F_0' which have an odd multiplicity. This last problem is the major difficulty of the Leray-Schauder method.

The nonlinear eigenvalue problem for the Navier-Stokes map is considered in the generalized sense in a subspace of a Sobolev space $W^{l,2}$ (which is a Hilbert space). In all cases we find $m = 1$, so each real and positive characteristic value λ_i of the map F_0' is a branch point of new stationary solutions. Among these values we are interested only in the first (least) branch point (λ_1) which, by the above hypotheses, is the critical point of the neutral linear stability.

Generally, at the critical point λ_1, branches of stationary flows appear sub or supercritically from x_0. The supercritical secondary flows are stable whereas the subcritical ones are unstable. For some particular flows, we emphasize results of the following type: If $0 \leqslant \lambda \leqslant \lambda_1$, the basic stationary motion x_0 exists, is unique, and is stable for $0 \leqslant \lambda < \lambda_1$. As $\lambda \to \lambda_1$, the fluid motions come as close to the basic flow as is desired. After the loss of stability of the basic flow at the point λ_1, the nonstationary nonlinear perturbation grows and as $t \to \infty$ it tends to one of the stable secondary stationary motions.

3.2.1. Convective motions

In hydrodynamic stability theory as well as in the branching theory a frequently studied flow is thermal convection in the cellular form, which is also called the generalized Bénard flow [3], [4].

This flow sets in at the point at which the stability of the statical equilibrium of a fluid subjected to the effects of a temperature gradient is lost. The topological degree method was first used in hydrodynamics for a convection problem (in 1963) by Uhovskii and Yudovich [50] for a bounded domain, then (in 1964) by Velte [53] for a tube, in 1966 by Yudovich [17] for horizontal layers, and in 1967 by Babskii [1] for two spheres in a self-gravitational field.

Consider a Poiseuille type flow in a horizontal tube of cross section G in the plane (x, y), having the axis directed along the z-axis, the tube being heated from below [53]. In the Boussinesq approximation, if we denote by T the absolute temperature, by T_0 a reference temperature, and by Pr the Prandtl number, then the equations of motion, of heat transfer and of continuity become

$$(3.2.1)\quad \begin{cases} u_t + uu_x + vu_y + wu_z = -\dfrac{1}{\rho} p_x + \nu\Delta u, \\ v_t + uv_x + vv_y + wv_z = -\dfrac{1}{\rho} p_y + \nu\Delta v + g\alpha(T - T_0), \\ w_t + uw_x + vw_y + ww_z = -\dfrac{1}{\rho} p_z + \nu\Delta w, \\ T_t + uT_x + vT_y + wT_z = \dfrac{\nu}{\mathrm{Pr}} \Delta T, \\ u_x + v_y + w_z = 0, \end{cases}$$

where $\Delta \equiv \dfrac{\partial^2}{\partial x^2} + \dfrac{\partial^2}{\partial y^2} + \dfrac{\partial^2}{\partial z^2}$, g is the gravitational acceleration and α is the coefficient of expansion. Let G be inside the band $0 < y < h$ and $T^*(y) \equiv T_1 - (T_1 - T_0)\, y/h$, $T_1 > T_0$. Taking into account only solutions independent of z, we attach to system (3.2.1) the following initial and boundary conditions

$$(3.2.2)\quad \begin{gathered} u = u_0(x, y),\ v = v_0(x, y),\ w = w_0(x, y), \\ T = T_0(x, y) \text{ for } t = t_0 \text{ and } (x, y) \in G \end{gathered}$$

and

$$(3.2.3)\quad u = v = w = 0,\ T = T^* \text{ for } t \geqslant t_0 \text{ and } (x, y) \in \partial G.$$

Take $h, \dfrac{\nu}{h}, \dfrac{\rho\nu^2}{h^2}, \dfrac{h^2}{\nu}$ to be the characteristic length, velocity, pressure and time respectively. Denote by $\Theta = (T - T_0)(T_1 - T_0)^{-1}$ the dimensionless temperature, by $F = F(x, y, t)$ the stream function such that $u = F_y$, $v = -F_x$ and introduce the Grashof number

$Gr = g\alpha h^3(T_1 - T_0)\,\nu^{-2}$. Then problem (3.2.1)–(3.2.3) may be written as

$$(3.2.1)' \quad \begin{cases} \Delta\Delta F = \dfrac{\partial(\Delta F, F)}{\partial(x, y)} + \dfrac{\partial \Delta F}{\partial t} + \mathrm{Gr}\,\dfrac{\partial \Theta}{\partial x}, \\ \dfrac{1}{\mathrm{Pr}}\,\Delta\Theta = \dfrac{\partial(\Theta, F)}{\partial(x, y)} + \dfrac{\partial \Theta}{\partial t}, \\ \Delta w = \dfrac{\partial(w, F)}{\partial(x, y)} + \dfrac{\partial w}{\partial t} + \dfrac{\partial p}{\partial z}, \end{cases}$$

$$(3.2.2)' \quad F_y = u_0(x, y), \; F_x = -v_0(x, y), \; w = w_0(x, y)$$
$$\Theta = 0 \text{ for } t = t_0 \text{ and } (x, y) \in G,$$

$$(3.2.3)' \quad F = \frac{\partial F}{\partial x} = w = 0, \; \Theta = \Theta^* \equiv 1 - y \text{ for } t \geqslant t_0$$
$$\text{and } (x, y) \in \partial G.$$

We take $F_0 \equiv 0$, $\Theta_0 = 1 - y$ as the basic solution, which, in the case of a circular cylinder with $\dfrac{\partial p}{\partial z}$ = constant, is the Poiseuille flow, and we look for another solution in the form

$$(3.2.4) \quad F(x, y, t) = F_0 + \bar{f}(x, y, t), \; \Theta(x, y, t) = \Theta_0 + \mathrm{Pr}\,\bar{\theta}(x, y, t).$$

If $\bar{f}$ and $\bar{\theta}$ are nonstationary (i.e. normal modes)

$$(3.2.5) \quad \bar{f} = f(x, y)\exp(\gamma t), \; \bar{\theta} = \theta(x, y)\exp(\gamma t), \; (\gamma \text{ real}),$$

then the problem which governs the linear stability of the basic flow to infinitesimal perturbations of the form (3.2.5) is

$$(3.2.6) \quad \begin{cases} \Delta\Delta f = \mathrm{Pr}\mathrm{Gr}\;\theta_x + \gamma\Delta f, \\ \Delta\theta = f_x + \mathrm{Pr}\,\gamma\theta, \end{cases} \quad (x, y) \in G$$

$$(3.2.7) \quad f = \frac{\partial f}{\partial n} = \theta = 0, \; (x, y) \in \partial G.$$

By definition, the basic flow is linearly stable if $\gamma < 0$ and it is linearly unstable if $\gamma > 0$. In the neutral case $\gamma = 0$, putting $\lambda = \mathrm{Pr}\cdot\mathrm{Gr} = \mathrm{Ra}$ (Rayleigh number), we obtain

$$(3.2.8) \quad \begin{cases} \Delta\Delta f = \lambda\theta_x, \\ \Delta\theta = f_x, \end{cases} \quad (x, y) \in G$$

$$(3.2.9) \quad f = \frac{\partial f}{\partial n} = \theta = 0, \; (x, y) \in \partial G.$$

If G is a circle or a rectangle, then the weak solutions $f \in \mathring{W}^{2,2}(G)$, $\theta \in \mathring{W}^{1,2}(G)$ of the problems (3.2.6), (3.2.7) and (3.2.8), (3.2.9) are also classical solutions, i.e. $f \in C^4(G) \cap C^3(G \cup \partial G)$, $\theta \in C^2(G) \cap C^1(G \cup \partial G)$. In fact, if $f \in W^{1,2}(G)$, it follows from Equation $(3.2.8)_2$ that $\theta \in W^{2,2}(G)$ and therefore $\theta_x \in W^{1,2}(G)$. From the Equation $(3.2.8)_1$ we have $f \in W^{3,2}(G)$ and so on. The imbedding theorems ensure the continuity to every order of the functions which belong to these Sobolev spaces in G as in $G \cup \partial G$. So, in what follows, we shall refer to the solutions of the above problem as classical solutions. Then, by a suitable multiplication of relations (3.2.8) by f and θ and by integration on G we find that every eigenvalue λ of problem (3.2.8), (3.2.9) is positive, i.e.

$$\lambda = \frac{\int_G \{|\operatorname{grad} f_x|^2 + |\operatorname{grad} f_y|^2\}\, dxdy}{\int_G |\operatorname{grad} \theta|^2\, dxdy.}.$$

(f, θ) above is the eigenvector corresponding to λ. Denoting by $J[f]$ the right-hand side of this equality, we see that Equations (3.2.8) are the Euler equations of the variational problem for the functional $J[f]$ defined in the class of functions infinitely differentiable on $G \cup \partial G$. Equivalently, in our case it is the class of functions $f \in \mathring{W}^{2,2}(G)$, $\theta \in \mathring{W}^{1,2}(G)$, $v \in C^1(G \cup \partial G)$, whence $\lambda_1 = \min J[f]$. Analysis of the variational problems for the functionals

$$J_1[u] = \int_G (|\operatorname{grad} u|^2 + 2\, uv)\, dxdy,$$

$$J_\gamma[f] = \frac{\int_G (|\operatorname{grad} f_x|^2 + |\operatorname{grad} f_y|^2 + \gamma |\operatorname{grad} f|^2)\, dxdy}{\int_G (|\operatorname{grad} \theta|^2 + \gamma \operatorname{Pr} \theta^2)\, dxdy},$$

whose Euler equations are the equations $\Delta u = v$ and (3.2.6) respectively, shows that for $\lambda < \lambda_1$ we have $\gamma < 0$ and for $\lambda > \lambda_1$ we have $\gamma > 0$. λ_1 is the smallest eigenvalue of problem (3.2.8), (3.2.9) and consequently, $\lambda_1 = \mathrm{Ra}_{cr}$. In other words, if $\mathrm{Ra} < \lambda_1$, then the basic flow is linearly stable and for $\lambda_1 < \mathrm{Ra} < \lambda_1 + \delta$ it is unstable.

If now instead of the expressions (3.2.5) we assume that $\bar{f}$ and $\bar{\theta}$ are stationary, then from problem (3.2.1)′–(3.2.3)′ we have

$$(3.2.10)\qquad \begin{cases} \Delta\Delta f = \dfrac{\partial(\Delta f, f)}{\partial(x, y)} + \lambda\theta_x, \\ \lambda\theta = \operatorname{Pr} \dfrac{\partial(\theta, f)}{\partial(x, y)} + f_x, \end{cases} \qquad (x, y) \in G$$

$$(3.2.11)\qquad f = \frac{\partial f}{\partial n} = \theta = 0, \ (x, y) \in \partial G.$$

The existence of a nontrivial solution of this nonlinear system implies that in addition to the basic flow, which corresponds to the solution $(f, \theta) = (0, 0)$ of problem (3.2.10), (3.2.11), there exists also another stationary flow. By a suitable multiplication of the above equations by f and θ and by integration on G, we find that for $\lambda \leqslant \lambda_1$ the unique solution of problem (3.2.10), (3.2.11) is trivial. Therefore, the basic flow is the only possible flow (being also stable) of the fluid in the subcritical and critical domains.

Let us prove by the Leray-Schauder degree method that λ_1 is a branch point for a supercritical solution of the nonlinear problem (3.2.10), (3.2.11). Let $\mathcal{B}$ be the Banach space whose elements are pairs of the form $u = (f, \theta)$, $f \in W^{3,2}(G)$, $\theta \in W^{1,2}(G)$ with the norm

$$\|u\|_{\mathcal{B}} = \|f\|_3 + \|\theta\|_1,$$

where $\|\cdot\|_n$ stands for the norm of the space $W^{n,2}(G)$, and consider the following linear problem corresponding to problem (3.2.10), (3.2.11)

$$\text{(3.2.12)} \quad \begin{cases} \Delta\Delta f' = \dfrac{\partial(\Delta f, f)}{\partial(x, y)} + \lambda\theta_x, \\ \qquad\qquad\qquad\qquad\qquad\qquad (x, y) \in G \\ \Delta\theta' = \Pr \dfrac{\partial(\theta, f)}{\partial(x, y)} + f_x, \end{cases}$$

$$\text{(3.2.13)} \quad f' = \frac{\partial f'}{\partial n} = \theta' = 0, \; (x, y) \in \partial G.$$

For f and θ fixed in $W^{3,2}(G)$ and $W^{1,2}(G)$ respectively this problem has a unique weak solution $f' \in \mathring{W}^{2,2}(G)$, $\theta' \in \mathring{W}^{1,2}(G)$. The existence and uniqueness of the solutions of problems (3.2.10), (3.2.11) and (3.2.12), (3.2.13) in these classes follows from the corresponding theorems for the generalized Dirichlet problem for even-order self-adjoint elliptic maps as particular cases for the maps Δ and $\Delta\Delta$ [35]. In addition, writing

$$g(f, \theta) \equiv \frac{\partial(\Delta f, f)}{\partial(x, y)} + \lambda\theta_x, \quad h(f, \theta) \equiv \Pr \frac{\partial(\theta, f)}{\partial(x, y)} + f_x,$$

we have, in case the boundary is of class C^∞

$$\text{3.2.14)} \quad \|f'\|_r \leqslant c\|g\|_0, \quad \|\theta'\|_s \leqslant c\|h\|_0, \quad (s = 1, 2, \; r = 3, 4)$$

where c is a constant which depends only on the domain G and on the operator Δ and $\Delta\Delta$ respectively.

Equations (3.2.12) define a map $F: \mathcal{B} \to \mathcal{B}$, $F(f, \theta) = (f', \theta')$. The weak solutions of the problem (3.2.10), (3.2.11) are fixed points of the map F and, as in the above, they are also classical solutions. Let us now consider the linear stability problem (3.2.8), (3.2.9), which is the same as that obtained from (3.2.10), (3.2.11) by neglecting the nonlinear terms, and let λ_1 be its least eigenvalue. Then, by the above conditions it follows that for Pr Gr $\leqslant \lambda_1$ the single fixed point of the map F is the point $(0, 0)$. By virtue of (3.2.14) for $u = (f, \theta)$ and $u_n = (f_n, \theta_n)$, with $u, u_n \in \mathfrak{B}$, we have

$$\|F(u_n) - F(u)\|_{\mathfrak{B}} = \|f'_n - f'\|_3 + \|\theta'_n - \theta'\|_1 \leqslant$$
$$\leqslant c\{\|g(f_n, \theta_n) - g(f, \theta)\|_0 + \|h(f_n, \theta_n) - h(f, \theta)\|_0\},$$

and if $\|u_n - u\|_{\mathfrak{B}} \to 0$ as $n \to \infty$, then by the continuity property of the functions g and f it follows that $\|F(u_n) - F(u)\|_{\mathfrak{B}} \to 0$, which means that F is a continuous map.

Take now a bounded sequence of solutions of the problem (3.2.12), (3.2.13), $u_n \in \mathfrak{B}$, $\|u_n\|_{\mathfrak{B}} \leqslant M_1$. For fixed Pr and Gr, we have $\|g(f_n, \theta_n)\|_0 \leqslant \leqslant M_2$, $\|h(f_n, \theta_n)\|_0 \leqslant M_2$ and from inequalities (3.2.14) we find that

$$\|f'_n\|_4 \leqslant c\|g(f_n, \theta_n)\|_0 \leqslant cM_2, \quad \|\theta'_n\|_2 \leqslant c\|h(f_n, \theta_n)\|_0 \leqslant cM_2.$$

Hence, by the corollary of theorem 2.1.4, it follows that from the sequences $\{f'_n\}$ and $\{\theta'_n\}$ of elements of $W^{1,2}(G)$ we can extract two subsequences convergent in $W^{3,2}(G)$ and $W^{1,2}(G)$ respectively.

Therefore F maps a bounded sequence into a compact set; as F is, in addition, continuous, it follows that F is compact.

The Fréchet differential of the map F at the point $u = 0$ is the linear completely continuous map A: $\mathfrak{B} - \mathfrak{B}$, $A(f, \theta) = (f'', \theta'')$, where (f'', θ'') is the weak solution $f'' \in \mathring{W}^{2,2}(G)$, $\theta'' \in \mathring{W}^{1,2}(G)$ of the problem

$$\text{(3.2.15)} \qquad \begin{cases} \Delta\Delta f'' = \lambda\theta_x, \\ \Delta\theta'' = f_x, \end{cases} \quad (x, y) \in G$$

$$\text{(3.2.16)} \qquad f'' = \frac{\partial f''}{\partial n} = \theta'' = 0, \ (x, y) \in \partial G.$$

Similarly, the fixed points of the map A are classical solutions of the problem (3.2.8), (3.2.9). As the fixed points of the Fréchet differential of F are the solutions of the problem governing the neutral linear stability, this implies that the critical points of the linear stability are branch points for the nonlinear stationary problem.

Let us prove that F is uniformly continuous with respect to λ on $\bar{\omega}$. Let ω be a bounded set of solutions of the problem (3.2.10), (3.2.11) such that $\|u\|_{\mathfrak{B}} < M_1$. Consequently, $\|f\|_3 < M_1$, $\|\theta\|_1 < M_1$. Then, for $u \in \bar{\omega}$, we have

$$\|F(u, \lambda) - F(u, \lambda_0)\|_{\mathfrak{B}} = \|f'(\lambda) - f'(\lambda_0)\|_3 \leqslant c\|g(f'(\lambda), \theta') -$$
$$- g(f'(\lambda_0), \theta)\|_0 = c\|(\lambda - \lambda_0)\theta\|_1 = c(\lambda - \lambda_0)\|\theta\|_1 \leqslant c(\lambda - \lambda_0)M_1$$

and choosing $\eta = \varepsilon (cM_1)^{-1}$ from $|\lambda - \lambda_0| < \eta$, it follows that $\|F(u, \lambda) - F(u, \lambda_0)\| < \varepsilon$.

The linear problem

$$u = \mu A(u, \lambda),$$

may be also written as a system of equations

$$\begin{cases} \Delta\Delta f = \mu\lambda\theta_x, \\ \Delta\theta = \mu f_x, \end{cases}$$

under the boundary conditions (3.2.9). For $\lambda = 0$ and for an arbitrary real parameter μ, this problem has only a trivial solution. Therefore, the conditions of Theorem 3.1.1 are satisfied for $\lambda_0 = 0$. On the other hand, writing $A_1(f, \theta) = (B_1(\theta), 0)$, $A_2(f, \theta) = (0, B_2(f))$, where $B_1(\theta)$ and $B_2(f)$ respectively are the solutions of the linear problems

$$\begin{cases} \Delta\Delta f' = \theta_x \\ f' = \dfrac{\partial f'}{\partial n} = 0 \text{ on } \partial G \end{cases} \quad \begin{cases} \Delta\theta' = f_x \\ \theta = 0 \text{ on } \partial G \end{cases}$$

we see that $A(\cdot, \lambda) = A_1 + \lambda A_2$. In this way, Conditions 1^0, 2^0 and 3° of Theorem 3.1.2 are satisfied. Since for $\lambda < \lambda_1$ the problem (3.2.10), (3.2.11) has only the trivial solution, $i[I - F(\cdot, \lambda), 0] = +1$. To prove the existence of secondary solutions, it remains to show that λ_1 is of odd multiplicity. It is easy to prove that the rank of λ_1 is equal to 1, i.e. the solutions of the equations $L^k(u) = 0$, $k = 2, 3, \dots$ are among those of the equation $L(u) = 0$, where $L(u) \equiv u - A(u, \lambda_1)$. Therefore, the multiplicity of λ_1 is equal to the number of the linearly-independent solutions of the equation $L(u) = 0$. By means of a finite-difference scheme, for the circle (approximated by an octagon) and for rectangles having certain ratios of their sides, it is found that $m(\lambda_1) = 1$. Then, by virtue of Theorem 3.1.2, the point λ_1 is a point of supercritical branching. Secondary flows exist for $\lambda < \lambda_1 + \delta$, where $\lambda_1 + \delta < \lambda_2$; and for $\lambda < \lambda_1 + \delta$ the solutions $u(\lambda)$ of the nonlinear problem do not belong to the boundary $\partial\omega$.

3.2.2. Couette flow in the case of a fixed exterior cylinder

We consider the case of Couette flow between rotating cylinders, the exterior one being at rest, under the hypothesis that the flow is axi-symmetric $\left(\dfrac{\partial}{\partial\varphi} = 0\right)$. A study of the branching problem leads to the replacement of the nonlinear problem (3.2.10), (3.2.11) by

$$\begin{cases} \mathfrak{L}\mathfrak{L}f = -\lambda[M(f, v) + av_z], \\ \mathfrak{L}v = -\lambda[N(f, v) + bf_z], \end{cases} \tag{3.2.17}$$

$$f = \frac{\partial f}{\partial r} = v = 0 \text{ for } r = r_1, r_2, \tag{3.2.18}$$

where, in cylindrical co-ordinates r, φ, z, we have

$$\mathfrak{L} \equiv \Delta - \frac{1}{r^2}, \quad a(r) = \frac{2v_0}{r}, \quad b = -\left(\frac{v_0}{r} + (v_0)_r\right),$$

$$v_0(r) = \frac{1}{r_2^2 - 1}\left(\frac{r_2^2}{r} - r\right),$$

$$M(f, v) = -\frac{\partial}{\partial r}(f_z \mathfrak{L} f) + \frac{\partial}{\partial z}\left[\frac{1}{r}(rf)_r \mathfrak{L} f\right] + \frac{1}{r^2}(v^2)_z,$$

$$N(f, v) = -\frac{1}{r}(rv)_r f_z + \frac{1}{r}(rf)_r v_z.$$

$f(r, z)$ is the stream function of the perturbation, $v_0(r)$ represents the Couette basic flow, $v(r)$ denotes the φ-component of the perturbation velocity, r_2 is the radius of the outer cylinder and $r_1 = 1$ is the radius of the inner cylinder. Comparing the two nonlinear problems, we see that the type of the maps is the same, but the linearized problem

$$\text{(3.2.19)} \qquad \begin{cases} \mathfrak{L}\mathfrak{L} f = -\lambda a v_z, \\ \mathfrak{L} v = -\lambda b f_z, \end{cases}$$

$$\text{(3.2.20)} \qquad f = \frac{\partial f}{\partial r} = v = 0 \quad \text{for} \quad r = r_1, \quad r = r_2,$$

obtained by neglecting M and N, unlike the problem (3.2.8), (3.2.9), has λ a factor in the right-hand side of $(3.2.19)_2$. Therefore, in this case the Fréchet differential $F_0'(\cdot, \lambda)$ is of the type λA. Moreover, since the cylinders are unbounded, $-\infty < z < \infty$, we must consider solutions periodic with respect to z

$$f(r, z) = f(r, z + nL); \quad v(r, z) = v(r, z + nL), \quad n = 0, \pm 1, \pm 2, \ldots$$

Instead of G we have a periodicity rectangle $G_L = (r, r_2) \times (0, L)$ and the function spaces $W^{k,2}(G, L)$, $C_0^k(G, L)$, $\mathring{W}^{k,2}(G, L)$ consist of functions periodic with respect to z and having period L. The generalized Dirichlet problems for the equations $-\mathfrak{L} u = g$, $\mathfrak{L}\mathfrak{L} u = g$ will be uniquely soluble in $\mathring{W}^{1,2}(G, L)$ and $\mathring{W}^{2,2}(G, L)$ respectively; the solutions belong to $W^{2m,2}(G, L)$ if the order of the elliptic map $\mathfrak{L}$ is $2m$. Similarly from relation (3.2.14) we have

$$\text{(3.2.21)} \qquad \| u \|_{2m} \leqslant c \| g \|_0,$$

where, this time, c depends only on G_L. The Rellich selection theorem and inequality (3.2.21) allow us to prove the complete continuity and the uniform continuity with respect to λ of the map $F(\cdot, \lambda)$. In order to determine the branching we have to prove that λ_1 has an odd multiplicity. In fact, by representing the eigenvectors of problem (3.2.19), (3.2.20) as Fourier series

$$\text{(3.2.22)} \quad \begin{cases} f(r, z) = \sum_{n=1}^{\infty} f_n(r) \sin n\sigma z + \sum_{n=0}^{\infty} \bar{f}_n(r) \cos n\sigma z, \ \sigma = \dfrac{2\pi}{L} \\ v(r. z) = \sum_{n=1}^{\infty} \bar{v}_n(r) \sin n\sigma z + \sum_{n=0}^{\infty} v_n(r) \cos n\sigma z, \end{cases}$$

for the functions $f_n, v_n, \bar{f}_n, \bar{v}_n$, we derive the following equations

$$\text{(3.2.23)} \quad \begin{cases} [\mathfrak{M} - (n\sigma)^2] f_n(r) = \lambda an\, \sigma\, v_n(r) \\ -[\mathfrak{M} - (n\sigma)^2]\, v_n(r) = \lambda bn\, \sigma f_n(r), \end{cases}$$

with the conditions

$$\text{(3.2.24)} \quad f_n(r) = \frac{\mathrm{d}f_n(r)}{\mathrm{d}r} = v_n(r) = 0, \ \text{for } r = r_1, r = r_2,$$

where $\mathfrak{M} = \dfrac{\mathrm{d}^2}{\mathrm{d}r^2} + \dfrac{1}{r}\dfrac{\mathrm{d}}{\mathrm{d}r} - \dfrac{1}{r^2}$. These admit solutions of the form $f_n(r) \sin n\sigma z$, $v_n(r) \cos n\sigma z$ and $f_n(r) \cos n\sigma z$, $-v_n(r) \sin n\sigma z$. As the number of these solutions is even, in order to apply the topological degree, we restrict $\mathscr{B}$ to $\mathscr{B}^* = \{(f, v) \mid f \in U_3(G, L),\ v \in S_1(G, L)\}$, where $U_3(G, L)$ ($\subset W^{3,2}(G, L)$) consists of functions odd with respect to z and $S_1(G, L)$ ($\subset W^{1,2}(G, L)$) consists of functions even with respect to z. In $\mathscr{B}^*$, problems (3.2.23), (3.2.24) have a unique classical solution (up to the norm). It is shown that F is also compact in $\mathscr{B}^*$. If $\lambda_1(\sigma)$ is the least eigenvalue of these epuations for $n = 1$, it is proved, by reducing the eigenvalue problem (3.2.23), (3.2.24) to an integral equation with positive kernel, that we can choose σ such that for $n > 1$ the problem (3.2.23), (3.2.24) does not have any eigenvalue equal to $\lambda_1(\sigma)$. Therefore, the equation $(I - \lambda A)\, u = 0$ has in B^* (up to the norm) a unique solution for $\lambda = \lambda_1(\sigma)$. This proves that the multiplicity of λ_1 is equal to 1 and then, by Krasnosel'skii theorem, a continuous branch of solutions $u(\lambda)$ of the problem (3.2.17), (3.2.18) appears at $\lambda_1(\sigma)$. Since the solution of the nonlinear problem (3.2.17), (3.2.18) is unique for $\lambda = 0$ and $\lambda = \lambda_1(\sigma)$, by theorems 3.1.2, the branch is supercritical in the neighbourhood of $\lambda_1(\sigma)$. Experimentally it has been observed that the secondary flow in the form of Taylor's toroidal vortices sets in at $\lambda_1 = \min_{\sigma} \lambda_1(\sigma)$ and exists for $\lambda > \lambda_1$.

Generally the existence and uniqueness of this minimum is an open problem. When the ratio between the cylinders radii is equal to 2, Velte [99] (from Ch. 2) made a numerical calculation of λ_1 and the corresponding wave number σ_{cr} and found that these values agree with the experimental data. Hence, for the particular case when $\frac{r_2}{r_1} = 2$, $\omega_2 = 0$, the principle of exchange of stabilities have been numerically proven. For all other cases of the Couette flow, this principle is not yet proved.

For Couette flow with $\omega_2 = 0$ or $\omega_2 \neq 0$, the linearization principle holds. This flow loses its stability at λ_1 (also in the frame of the nonlinear theory), and the nonlinear nonstationary perturbed motion, tends to the Taylor vortices [9] as $t \to \infty$.

3.2.3. Flows between two cylinders rotating in the same direction

In both cases considered above, we have stressed the branching at the point at which linear stability is lost. In the following example, we discuss in detail the eigenvalue problem for the Fréchet differential of F and identify all the branch points, under the implicit hypothesis that all other conditions of Theorem 3.1.2 are fulfilled.

For some flows, by means of the knowledge of the spectrum of F'_0 we determine the spectrum of F, and consequently the intervals of uniqueness of the solution.

Write $a = r_2 r_1^{-1}$, $\omega = \omega_2 \omega_1^{-1}$, take ω_1 and r_1 as characteristic velocity and length respectively, and define the Reynolds number as $\mathrm{Re} = \frac{\omega_1 r_1^2}{\nu}$. Then Couette flow between two cylinders rotating in the same direction, in cylindrical co-ordinates r, φ, z, has the velocity $(0, v_0(r), 0)$ where $\alpha = (1 - a^2\omega)(a^2 - 1)^{-1}$, $\beta = a^2(1 - \omega)(a^2\omega - 1)^{-1}$ and $v_0(r) = -\alpha r - \frac{\alpha\beta}{r}$. We look for axi-symmetric $\left(\frac{\partial}{\partial \varphi} = 0\right)$ flows with the velocity $\left(-\frac{1}{r\,\mathrm{Re}}\frac{\partial}{\partial z}(r^2\psi), -\alpha r - \frac{\alpha\beta}{r} + rv, \frac{1}{r\,\mathrm{Re}}\frac{\partial}{\partial r}(r^2\psi)\right)$, corresponding to the steam function ψ, such that $r\psi = f\,\mathrm{Re}$ (f is the stream function for the case $\omega_2 = 0$). Then, instead of system (3.2.17), we have

$$(3.2.17)' \quad \begin{cases} \Lambda \dfrac{1}{r^3} \Lambda\psi = \lambda \alpha v_0 r^2 v_z - \left[r^2 \dfrac{\partial(r^{-3}\Lambda\psi, r^2\psi)}{\partial(z, r)} + 2r^3 v v_z \right], \\[2ex] \Lambda v = r^3 \psi_z + \dfrac{\partial(r^2 v, r^2\psi)}{\partial(z, r)}, \end{cases}$$

with the boundary conditions

$$(3.2.18)' \quad \psi = \frac{\partial \psi}{\partial r} = v = 0, \quad \text{for } r = 1 \text{ and } r = a,$$

where

$$\lambda = 4\,\mathrm{Re}^2, \quad \Lambda \equiv \frac{\partial}{\partial r} r^3 \frac{\partial}{\partial r} + r^3 \frac{\partial^2}{\partial z^2},$$

$$\frac{\partial(\psi_1, \psi_2)}{\partial(z, r)} \equiv \frac{\partial\psi_1}{\partial z}\frac{\partial\psi_2}{\partial r} - \frac{\partial\psi_1}{\partial r}\frac{\partial\psi_2}{\partial z}.$$

Problem (3.2.17)′, (3.2.18)′ and the linear problem (3.2.19)′, (3.2.20)′, corresponding to the Fréchet differential at the point $\psi = 0$, are put into the generalized form $\psi = \lambda F_1(\psi) + F_2(\psi)$, and $\psi = \lambda A(\psi)$ respectively, where F_1, F_2 and A satisfy [16] the conditions of theorem 3.1.2. Hence, in order to determine the branch points, it remains to find the eigenvalues of odd multiplicity of the linear nonselfadjoint problem

$$\text{(3.2.19)}' \qquad \begin{cases} \Lambda \dfrac{1}{r^3} \Lambda\psi = \lambda\alpha v_0 r^2 v_z, \\ \Lambda v = r^3\, \psi_z, \end{cases}$$

$$\text{(3.2.20)}' \qquad \psi = \frac{\partial\psi}{\partial r} = v = 0 \ \text{ for } r = 1, \quad r = a,$$

Similarly to the case $\omega_2 = 0$, v is an even function periodic with period L in z; ψ is an odd function periodic in z with the same period. From the system (3.2.19)′, it is easy to see that [16]

$$\lambda = \left[\int_1^a \int_0^L \frac{1}{r^3} (\Lambda\psi)^2 \,\mathrm{d}z\mathrm{d}r\right]\left\{-\int_1^a \int_0^L \alpha^2 r^{-1}\left(1 + \frac{\beta}{r^2}\right)\left[\left(\frac{\partial v}{\partial z}\right)^2 + \left(\frac{\partial v}{\partial r}\right)^2\right] \mathrm{d}z\,\mathrm{d}r\right\}^{-1}.$$

It follows that for $a^2\omega > 1$ the eigenvalues are negative. Since the real parameter λ is positive, there exists no branching for motions such that $a^2\omega > 1$. Consequently Couette flow is unique in the class of periodic functions defined above. For this case Synge [75] (from Ch. 1) proved the stability of the Couette flow. Therefore branch points can appear only if the condition $a^2\omega < 1$ is satisfied

$$\text{(3.2.25)} \qquad r_2^2\omega_2 < r_1^2\omega_1.$$

Expanding ψ and v in Fourier series of odd and even functions with respect to z, respectively,

$$\text{(3.2.22)}' \qquad \psi = \sum_{n=1}^{\infty} \psi_n(r) \sin n\sigma z, \quad v = \sum_{n=1}^{\infty} v_n(r) \cos n\sigma z,$$

where σ is the wave number ($\sigma = 2\pi L^{-1}$), we obtain from system (3.2.19)′, (3. 2.20)′

$$(3.2.23)' \qquad \begin{cases} \Lambda_1 \dfrac{1}{r^3} \Lambda_1 \psi_n = \lambda n \sigma \alpha r^3 v_0 v_n, \\ \\ -\Lambda_1 v_n = \sigma n r^3 \psi_n, \end{cases}$$

$$(3.2.24)' \qquad v_n = \psi_n = \frac{\mathrm{d}\psi_n}{\mathrm{d}z} = v_n = 0 \text{ for } r = 1, \quad r = a,$$

where $\Lambda_1 = \dfrac{\mathrm{d}}{\mathrm{d}r} r^3 \dfrac{\mathrm{d}}{\mathrm{d}r} + r^3\sigma^2 n^2$, and $n = 1, 2, \ldots$. Denote by I_1 the Bessel function of imaginary argument and change variables as follows

$$\psi_n = z\varphi, \quad v_n = t_0^2 \sigma z u, \quad t = t_0^{-1} \int_1^r \frac{\mathrm{d}r}{r^3 z^2}, \quad r^6 z^4 = J^{-1}(t),$$

$$(3.2.26) \qquad t_0 = \int_1^a \frac{\mathrm{d}r}{r^3 z^2},$$

where $z = I_1(\sigma r)/r$ is the solution of the equation $\Lambda_1 z = 0$. Then problem (3.2.23)′, (3.2.24)′ takes the form

$$(3.2.23)'' \qquad \begin{cases} \dfrac{\mathrm{d}^2}{\mathrm{d}t^2} J \dfrac{\mathrm{d}^2\varphi}{\mathrm{d}t^2} = \lambda^2 (Jr)^{-1} \alpha v_0 u, \\ \\ -\dfrac{\mathrm{d}^2\varphi}{\mathrm{d}t^2} = J^{-1}\varphi, \end{cases}$$

$$(3.2.24)'' \qquad \varphi = \frac{\mathrm{d}\varphi}{\mathrm{d}t} = u = 0 \text{ for } t = 0 \text{ and } t = 1.$$

By means of Green functions G_φ, G_u, for the operators of system (3.2.23)″, we reduce this problem to the integral equation

$$(3.2.27) \qquad \varphi(t) = \lambda \int_0^1 S(t, s)\varphi(s)\mathrm{d}s,$$

which, for $a^2\omega < 1$ has a positive kernel. In addition, if the following inequalities

$$(3.2.28) \qquad S\begin{pmatrix} t_1 \ldots t_n \\ s_1 \ldots s_n \end{pmatrix} \geqslant 0, \quad \begin{pmatrix} 0 < t_1 < \ldots < t_n < 1, \\ 0 < s_1 < \ldots < s_n < 1, \end{pmatrix} \quad n = 1, 2, \ldots \Bigg),$$

$$(3.2.29) \qquad S\begin{pmatrix} t_1 \ldots t_n \\ t_1 \ldots t_n \end{pmatrix} > 0 \quad (0 < t_1 < \ldots < t_n < 1, \ n = 1, 2, \ldots),$$

hold, then $S(t, s)$ is an oscillation kernel in the sense of Krein [5], [28] and the integral equation (3.2.27) has positive simple eigenvalues

$$0 < \lambda_1(n\sigma) < \ldots < \lambda_s(n\sigma) \to \infty. \tag{3.2.30}$$

The necessary and sufficient condition for these numbers to be simple eigenvalues for problem (3.2.19)', (3.2.20)' also is that for any system of four positive integers, m, n, p, s, the relation $\lambda_p(m\sigma) \neq \neq \lambda_s(n\sigma)$ holds. Necessity is immediate. To prove the sufficiency suppose that there exists at least two members m and p such that

$$\lambda_p(m\sigma) = \lambda_s(n\sigma), \tag{3.2.31}$$

holds. Then problem (3.2.19)', (3.2.20)' should admit two linearly-independent solutions $\psi_m^{(p)}(r)$ sin $m\sigma z$ and $\psi_n^{(s)}(r)$ sin $n\sigma z$ as eigenvectors corresponding to the eigenvalue $\lambda = \lambda_p(m\sigma) = \lambda_s(n\sigma)$. Let us prove, then, that equality (3.2.31) holds at most for one countable set of wave numbers σ. Suppose, without loss of generality, that there exist $m > n$ such that relation (3.2.31) holds. For n, s and m fixed, it follows from inequalities (3.2.30) that there exists at most one number p such that (3.2.31) is valid. For n, s and p fixed, assume that there exists an infinity of numbers m which satisfy equality (3.2.31). Then, for $\lambda = \lambda_s(n\sigma)$, we have an infinity of linearly-independent eigenvectors of problem (3.2.19)', (3.2.20)', corresponding to the eigenvalues $\lambda_p(m\sigma)$. This contradicts the fact that the linear map defined by system (3.2.19)' is completely continuous. Hence, in any case, we can have only a finite number of couples (m, p) such that equality (3.2.31) holds. Let m^* be the greatest among these m. Then $\lambda_s(n\sigma)$ are simple eigenvalues for problem (3.2.19)', (3.2.20)' in the spaces $\mathfrak{B}^*(m^*\sigma)$, since this problem does not admit associate vectors.

From the analyticity of the maps G_φ and G_μ with respect to $n\sigma$, it follows that for fixed p, m, s, n, Equality (3.2.31) holds for at most a countable set of wave numbers. Consequently, for $a^2\omega < 1$, for every σ, up to a countable set, secondary motions periodic with respect to z set in, for $\omega > 0$, at the points $\lambda_s(n)$.

For general flows between rotating cylinders, including Couette flow, an analogous study [16], [15], under the hypothesis

$$\frac{v_0(r)}{r} > 0, \quad -\left(\frac{\mathrm{d}v_0}{\mathrm{d}r} + \frac{v_0}{r}\right) > 0 \text{ for } 1 < r < a, \tag{3.2.25'}$$

leads to the conclusion that secondary flows branch from the basic flow $v_0(r)$ at the points λ_i, and that they exist for λ belonging to the intervals $(\lambda_1, \lambda_2), (\lambda_3, \lambda_4), \ldots,$ where λ_i are the simple eigenvalues of the linear system (3.2.19)'.

By means of the theory of integral equations with oscillation kernel, it may be proved that the linear problem of stability problem for flows with velocity $v_0(r)$, admits eigenvalues, which, if the principle of exchange of stabilities holds, are precisely the λ_i.

The analogy (noticed as early as 1916 [51] (from Ch. 1)) between the mathematical problem of the stability of Couette and Bénard flows, discussed in [54], has recently been extended to the mathematical problem of the branching for these two flows by Kirchgässner and Kielhöfer [26].

3.2.4. Flows in bounded domains

Let S be a smooth surface of rotation, which does not contain points of its rotation axis Oz. Consider the class of flows with symmetry of rotation in the domain bounded by S. If the velocity of these flows depend only on r and z, and if the body forces in cylindrical co-ordinates are $(0, \delta F(r), 0)$ where δ is a parameter, then the Navier-Stokes equations have the following solution

$$\mathbf{v}_0 = (0, \delta v_0(r), 0).$$

Suppose that there exists a new flow with the velocity $\mathbf{v}_0 + \mathbf{v}$, where $\mathbf{v} = \delta(v_r, v_\varphi, v_z)$. Then, if

$$\delta v_0(r) = \lambda r^{-3},$$

where $\lambda = \dfrac{\delta}{\nu}$, it may be proved [18] that the corresponding Stokes map is selfadjoint and that for flows with

$$\delta v_0(r) = \lambda r^{\beta}, \ \beta \neq -1$$

this map is symmetrizable. For flows with velocity

$$(0, \lambda r^{\beta}, 0), \quad \beta \neq -1, \tag{3.2.32}$$

Yudovich [18], putting the hydrodynamic stability problem as a variational principle, proves that the principle of exchange of stabilities holds. From the same variational principle it follows that the flow (3.2.32) is linearly stable for $\lambda < \lambda_1$. In addition, the Fréchet differential F_0' of the Navier-Stokes map has an infinity of positive eigenvalues $0 < \lambda_1 \leqslant \lambda_2 \leqslant \ldots \leqslant \lambda_n \leqslant \ldots$, and $-\lambda_i$ are also eigenvalues. Each point of the spectrum of the map F_0' has the rank equal to 1, so that each positive eigenvalue λ_i of odd multiplicity is, according to Theorem 3.1.2, a branch point.

3.2.5. Kolmogorov's flow

Among the fluid flows for which the stability and branching problems have been entirely solved is the plane Kolmogorov's flow

$$u = \frac{\gamma}{\nu} \sin y, \ v = 0, \quad p = \text{constant}, \quad (\gamma > 0), \tag{3.2.33}$$

which is one of the stationary motions periodic with respect to y and with period 2π of the hydrodynamic system of equations

$$\begin{cases} u_t + u_x u + u_y v = -\rho^{-1} p_x + F_1 + \nu \Delta u, \\ v_t + v_x u + v_y v = -\rho^{-1} p_y + F_2 + \nu \Delta v, \\ u_x + v_y = 0, \end{cases}$$

corresponding to body forces

$$F_1 = \gamma \sin y, \quad F_2 = 0.$$

For this flow the linearization principle holds [36] (from Ch. 2). Hence for stability purposes, it is sufficient to consider the linear problem of stability, which was solved in [60] (from Ch. 3) and [14]. Let us now consider the problem concerning the uniqueness and the branching of the motion (3.2.33), taking into account that the principle of exchange of stabilities holds in this case [60] (from Ch. 2). Since the methods used to prove the branching have been presented in previous sections, we only give the final results.

Kolmogorov's flow loses uniqueness when the linear stability is lost, and new stationary solutions periodic with respect to x, with period $\frac{2\pi}{\alpha_0}$, and with respect to y with period 2π, set in. For $\alpha_0 \geqslant 1$ the motion (3.2.33) is unique and stable. Let m be the integer part of α_0^{-1}. For $0 < \alpha_0 < 1$, the Navier-Stokes map has the form λF $(\lambda = \gamma \nu^{-2})$, the Stokes map is $\lambda F_0'$ and these maps satisfy the hypotheses of Krasnosel'skii's theorem. F_0' has m positive characteristic values λ_i, $0 < \lambda_1 < \ldots < \lambda_m$ and m characteristic values $-\lambda_i$, each of them of multiplicity equal to 1, so that every positive λ_i is a branch point. If $m = 2k$ $(k \in N)$, the intervals $[\lambda_1, \lambda_2]$, $[\lambda_3, \lambda_4]$, ..., $[\lambda_{2k-1}, \lambda_{2k}]$ and then their symmetric intervals with respect to the origin belong to the spectrum of the nonlinear map F. For $m = 2k + 1$, the spectrum of F contains additionally the intervals $(-\infty, -\lambda_{2k+1}]$ and $[\lambda_{2k+1}, \infty)$. For all these intervals, we have $i[I - F, 0] = -1$. Therefore Kolmogorov's flow exists and is unique for $\lambda < \lambda_1$, while at the point λ_1 a continua of motions sets in. These motions exist for $\lambda_1 < \lambda < \lambda_2$, and the point λ_2 is a point of subcritical branching. For $\lambda \in (\lambda_2, \lambda_3)$ the flow (3.2.33) is unique in the considered class, etc.

Observing that the linear eigenvalue problem for a nonstationary perturbation is equivalent to the nonlinear problem for the stationary secondary motion, Yudovich [14] shows the branching of the secondary flows periodic with respect to time, the loss of the stability of these flows being due to the appearance of a motion which is conditionally periodic with respect to time.

The relatively small number of papers dealing with the branching and stability of the solutions of the Navier-Stokes equations by the Leray-Schauder method, allowed us to make a complete survey of the results obtained by means of this method, its generality requiring only little information on the problems considered. Therefore, numerous studies on branching theory use analytic methods allowing the description of the branching and stability phenomena, as will be illustrated in § 3.3., § 3.4 for some particular cases.

§ 3.3. LIAPUNOV-SCHMIDT METHOD

In this section we analyse the solvability problem of nonlinear equations in two cases: a) when the Fréchet differential of the map which defines the equation does not vanish and b) when it does. In the former case the solution is unique; in the latter case, the number of small solutions is given by the branching equation. We shall also indicate the series which define the solutions, first for the integral equations (Section 3.3.1) and then for nonlinear equations in Banach spaces (Section 3.3.2).

3.3.1. The case of integral equations

Consider first the Fredholm linear equation

$$u(s) - \lambda \int_a^b K(s, t)\, u(t)\, dt = f(s). \tag{3.3.1}$$

By Fredholm's alternative, if λ is not a characteristic value of the linear map $A(\cdot) = \int_a^b K(s, t)\,(\cdot)\, dt$, then Equation (3.3.1) has a unique solution given by the expression

$$u(s) = f(s) + \lambda \int_a^b \Gamma(s, t)\, f(t)\, dt, \tag{3.3.2}$$

where $\Gamma(s, t)$ is the resolvent of the kernel K; if $\lambda = \lambda_0$ is a characteristic value of multiplicity n, to which there correspond the orthonormal eigenfunctions $\varphi_1, \ldots, \varphi_n$, then Equation (3.3.1) has solutions if and only if the following orthogonality conditions are satisfied

$$(f(s), \psi_i) = 0, \quad i = 1, \ldots, n, \tag{3.3.3}$$

for the orthonormal system ψ_i $(i = 1, \ldots, n)$ of eigenfunctions of the adjoint map

$$A^*(\cdot) = \int_a^b K(t, s)(\cdot)\, dt, \tag{3.3.4}$$

corresponding to the characteristic value λ_0.

In order to determine the solutions of Equation (3.3.1) in the latter case, we modify the kernel K up to a kernel $K_1(s, t) = K(s, t) - \frac{1}{\lambda_0}\sum_{k=1}^{n} \psi_k(s)\,\overline{\varphi}_k(t)$ such that, by Schmidt's lemma [52], the linear map $\widetilde{A}(\cdot) = \int_a^b K_1(s, t)\,(\cdot)\,dt$ does not have λ_0 as a characteristic value. Consequently the map $I - \lambda_0\widetilde{A}$ possesses a bounded inverse map. If we put

$$\xi_i = \int_a^b \overline{\varphi}_i(t)\,u(t)\,dt, \quad i = 1, 2, \ldots, n, \tag{3.3.5}$$

then (3.3.1) may be written in the equivalent form

$$u(s) - \lambda_0 \int_a^b K_1(s, t)\,u(t)\,dt = f(s) + \frac{1}{\lambda_0}\sum_{k=1}^{n} \xi_k\psi_k(s). \tag{3.3.1$'$}$$

Since λ_0 is not a characteristic value for the map $\widetilde{A}$, this equation will have the solution

$$u(s) = f_1(s) + \lambda_0 \int_a^b \Gamma_1(s, t)\,f_1(t)\,dt, \tag{3.3.2$'$}$$

where $\Gamma_1(s, t)$ stands for the resolvent of the kernel K_1 and f_1 is the right-hand side of Equation (3.3.1)$'$. The undetermined parameters ξ_n in (3.3.2)$'$ may be determined from a system obtained by substituting relation (3.3.2)$'$ in (3.3.5); this system does not have a unique solution.

Consider now the nonlinear integral equation (studied by Schmidt [51])

$$u(s) - \lambda_0 \int_a^b K(s, t)\,u(t)\,dt = -U_{01}\binom{s}{u, v} - \tag{3.3.6}$$
$$- \sum_{m+n\geqslant} U_{mn}\binom{s}{u, v},$$

or, equivalently,

$$u(s) - \lambda_0 \int_a^b K_1(s, t)\,u(t)\,dt = -U_{01}\binom{s}{u, v} - \tag{3.3.6$'$}$$
$$- \sum_{m+n\geqslant 2} U_{mn}\binom{s}{u, v},$$

where $U_{mn}\binom{s}{u, v}$ are finite sums of integrals of the form

$$u^{\alpha_0}(s)\, v^{\beta_0}(s) \int_a^b \ldots \int_a^b C(s, t_1, \ldots, t_\rho)\, u^{\alpha_1}(t_1)\, v^{\beta_1}(t_1) \ldots u^{\alpha_\rho}(t_\rho)\, v^{\beta_\rho}(t_\rho)\, dt_1 \ldots dt_\rho,$$

$\alpha_0, \alpha_1, ..., \alpha_\rho, \beta_0, \beta_1, ..., \beta_\rho$ are integers or zero such $\alpha_0 + \alpha_1 + ... + \alpha_\rho = m$, $\beta_0 + \beta_1 + ... + \beta_\rho = n$; $C(s, t_1, ..., t_\rho)$ is a real or complex continuous function of the arguments $s, t_1, ..., t_\rho \in [a, b]$; $u(s)$ is the unknown function and the function $v(t)$ is the parameter of the problem. Schmidt's equation (3.3.6)′ contains nonlinear terms in the right-hand side. Denoting this unknown side by $f_1(s)$, we obtain an equation to which we can apply the above algorithms for the linear case. Then for u, relation (3.3.2)′ yields the equation

$$u(s) = \sum_{k=1}^{n} \left[\overline{\psi}_k(s) + \lambda_0 \int_a^b \Gamma_1(s, t)\, \overline{\psi}_k(t)\, dt \right] \xi_k - U_{01}\begin{pmatrix} s \\ u, v \end{pmatrix} -$$

$$(3.3.2)'' \qquad - \lambda_0 \int_a^b \Gamma_1(s, t)\, U_{01}\begin{pmatrix} t \\ u, v \end{pmatrix} dt - \sum_{m+n \geqslant 2} \left[U_{mn}\begin{pmatrix} s \\ u, v \end{pmatrix} + \right.$$

$$\left. + \lambda_0 \int_a^b \Gamma_1(s, t)\, U_{mn}\begin{pmatrix} t \\ u, v \end{pmatrix} dt \right].$$

This equation is solved by successive approximations for continuous solutions of the form

$$(3.3.7) \qquad u(s) = \sum_{\gamma_1 + \, ... \, + \gamma_n + \nu \geqslant 1} \xi_1^{\gamma_1} \, ... \, \xi_n^{\gamma_n} V_\nu^{\gamma_1, \, ... \, , \gamma_n} \begin{pmatrix} s \\ v \end{pmatrix},$$

where the integrals V_ν satisfy a recurrent system of infinite order, obtained by substituting expression (3.3.7) in Equation (3.3.2)″. It may be shown that for small $|\xi_1|, ..., |\xi_n|$ and $|v|$, the sum map from the right-hand side of Equation (3.3.6) is a contraction on a sphere of the space of the continuous functions. Therefore, Equation (3.3.2)″ has a unique solution of the form (3.3.7). As in the above, ξ_i are given by the system (which Schmidt called *branching equation*)

$$(3.3.5)' \qquad \xi_k = \sum_{\gamma_1 + \, ... \, + \gamma_n + \nu \geqslant 1} \xi_1^{\gamma_1} \, ... \, \xi_n^{\gamma_n} \int_a^b \psi_k(t)\, V_\nu^{\gamma_1 \, ... \, \gamma_n} \begin{pmatrix} t \\ v \end{pmatrix} dt$$

obtained by substituting expression (3.3.7) in the relations (3.3.5). In system (3.3.5), $\xi_k (k = 1, ..., n)$ does not appear and so, generally, the branching equation does not have a unique solution. Since to each small solution $(\xi_1, ..., \xi_n)$ of this equation there corresponds a single solution of Equation (3.3.6) of the form (3.3.7), it follows that the number of the small solutions of the branching equation is equal to that of the solutions of Equation (3.3.6). As $v \to 0$ all these solutions tend to zero. [4)]

Briefly, the above algorithm—known as the Liapunov-Schmidt method — consists of the following steps: By means of the unknown

---

4) By determining the branching solutions, we also solve the problem of the continuation of a solution of a nonlinear equation as the parameter passes through a point at which the map, linearized with respect to that solution, has an eigenvalue.

parameters ξ_j, the map A is modified up to the linear map E for which λ_0 is not an eigenvalue. Therefore it admits an inverse such that the equation is reduced to the form $x = F(x, y)$, where y is the parameter of the problem. The solution x of this equation is determined by successive approximations as power series in ξ_j with functions of y as coefficients. Introducing x in the expression of ξ_j we obtain the branching equation whose solutions $(\xi_1, \ldots, \xi_n)$ are series of rational powers of y; the number of these small solutions is equal to that of the solutions of the given equation.

By the Liapunov-Schmidt method, the solution of Equation (3.3.6)′ has been reduced to that of the system (3.3.2)″, (3.3.5). The idea of the method consists in writing the nonlinear equation in an equivalent form as a system of two equations, one of them possessing an invertible linear left-hand side. By virtue of the same idea, the Schmidt Equation (3.3.6) could also be solved as follows: we decompose the space of the continuous functions into a direct product of two orthogonal subspaces, one of them being generated by $\varphi_1, \ldots, \varphi_n$. Projecting Equation (3.3.6) on these subspaces we obtain again a system of two equations, equivalent to Equation (3.3.6). One of these equations has an invertible linear left-hand side; the second equation consists of the orthogonality conditions (3.3.3). These two ways of solving an integral equation will be illustrated in the case of Banach spaces, in Theorem 3.3.2 and in Schmidt's lemma.

Expression (3.3.7) is common to all solutions of the equation (3.3.6). That is why it is necessary to specify the particular form of each solution as a power series in the parameter v. As an example, we consider the Hammerstein equation (in which the parameter enters in the same way as in the Navier-Stokes equations)

$$u(x) + \lambda \int_G K(x, y)\, f(u(y), y)\, \mathrm{d}y = 0,$$

where G is a bounded domain belonging to a finite-dimensional Euclidean space and f is an analytic function of the form

$$f(u, x) = \sum_{k=0}^{\infty} b_k(x)\, u^k.$$

We shall study the solutions of the above equation around a fixed eigenvalue μ_0 of the linear map corresponding to the Hammerstein equation; in this way, the parameter $\mu - \mu_0$ plays the role of the function v from Equation (3.3.6). Suppose that $u_0(t)$ is an eigensolution of the nonlinear Hammerstein equation corresponding to the eigenvalue μ_0, and μ_0 is a simple eigenvalue of the linear equation

$$\varphi(x) + \mu_0 \int_G K(x, y)\, f'_u\, (u_0(y), y)\, \varphi(y)\, \mathrm{d}y = 0,$$

to which there corresponds the eigensolution $\varphi(x)$. Putting $\mu = \lambda + \mu_0$, $u = u_0 + v$, the Hammerstein equation becomes

$$v(x) + \mu_0 \int_G K(x, y) f'_u(u_0(y), y)\, v(y)\, \mathrm{d}y + \lambda \int_G K(x, y) [f(u_0(y), y) +$$

$$+ v(y) f'_u(u_0(y), y)]\, \mathrm{d}y + (\lambda + \mu_0) \int_G K(x, y) \left[\frac{v^2(y)}{2!} f''_{u^2}(u_0(y), y) + \right.$$

$$\left. + \frac{v^3(y)}{3!} f'''_{u^3}(u_0(y), y) + \dots \right] \mathrm{d}y = 0.$$

This equation has the form (3.3.6) where $u_0, \lambda_0, v, n, \varphi_1, \varphi_i\ (i \neq 1)$ from the Equation (3.3.6) have been replaced by $v, \mu_0, \lambda, l, \varphi$ and 0, respectively, and $V_\nu\binom{l}{v}$ may be expressed in terms of $b_j(x)$. Then the branching equation for the above equations is

$$\text{(3.3.8)} \qquad \sum_{m=2}^{\infty} L_{m0}\, \xi^m + \sum_{m=0}^{\infty} \xi^m \sum_{l=1}^{\infty} \lambda^l L_{ml} = 0,$$

where L_{ij} are numbers resulting from the integrations on the domain G of some expressions containing u_0, b_k and φ. By solving Equation (3.3.8), it may be shown that ξ_i, and consequently also the small solutions $v(x)$, can be expressed as rational powers of the parameter λ with functions of x as coefficients. In other words, from the solution u_0, at the point μ_0, several branches appear which are analytic with respect to λ, each point of them, corresponding to a number λ, being a solution of the Hammerstein equation. In mechanics the only curves of interest are the real curves, whose number depends of the values taken by L_{ij}. For instance, if $L_{20} = \dots = L_{(n-1),0}$, $L_{n0} \neq 0$, n is even and sgn $L_{n0}L_{01} = -1$, then two real curves branch supercritically and if sgn $L_{n0}L_{01} = +1$, then there are two real subcritical branches. Although the general equations of hydrodynamics are integro-differential, results analogous (§ 3.4) to those obtained for the Hammerstein equation are also found for the Navier-Stokes equations.

We have seen that the branching equation yields the number and the form of small solutions of an integral equation, as a series of rational powers of the parameter of the equation. If these powers are known, then the branching equation is no longer necessary and the series are introduced directly in the initial equation. Taking into account the orthogonality condition, we can deduce the coefficients.

3.3.2. **The case of nonlinear equations in Banach spaces**

We now look for a way of generalizing the Fredholm alternative and the Liapunov-Schmidt method to the case of infinite-dimensional Banach spaces.

Let $\mathcal{B}_1$ and $\mathcal{B}_2$ be two Banach spaces, let $B : \mathcal{B}_1 \to \mathcal{B}_2$ be a linear continuous map, let $B^* : \mathcal{B}_2^* \to \mathcal{B}_1^*$ be the adjoint operator of B defined by the equality

$$\langle x, B^*\psi\rangle = \langle Bx, \psi\rangle, \ (x \in \mathcal{B}_1, \psi \in \mathcal{B}_2^*)\ ^{5)},$$

and let $N(B)$, $N(B^*)$ be the null spaces of the operators B and B^*. Denoting the range of B by $R(B)$, then $N(B^*)$ consists of the linear continuous functionals $\psi \in \mathcal{B}_2^*$ vanishing on $R(B)$. The map B is called *normally resoluble* if every element which determines the vanishing of all the functionals from $N(B^*)$ belongs to $R(B)$, hence the equation

$$Bx = z \tag{3.3.9}$$

has solutions iff z is orthogonal to $N(B^*)$. As a consequence of the Hahn-Banach theorem, it may be easily seen that B is normally resoluble iff $R(B)$ is closed in $\mathcal{B}_2$. A normal resoluble operator is said to be of Fredholm type (or Φ — operator) if $\dim N(B) = \dim N(B^*) = n < \infty$.

Assuming that B is of Fredholm type, denote $N(B) = \mathcal{B}_{1,n}$, $N(B^*) = \mathcal{B}_{2,n}^*$, and $\{\varphi_1, ..., \varphi_n\}$, respectively $\{\psi_1, ..., \psi_n\}$, being two bases of these spaces, let $\{\gamma_1, ..., \gamma_n\}$ and $z_1, ..., z_n$ be their dual bases; therefore, for $1 \leqslant i, j \leqslant n$,

$$\langle \varphi_i, \gamma_j\rangle = \delta_{ij} = \langle z_i, \psi_j\rangle. \tag{3.3.10}$$

Let $\mathcal{B}_{2,n}$ be the linear subspace of $\mathcal{B}_2$ spanned by the vectors $z_1, ..., z_n$, $P_i : \mathcal{B}_i \to \mathcal{B}_{i,n}$ $(i = 1, 2)$ the operators defined by the equalities

$$P_1(x) = \sum_{i=1}^{n} \langle x, \gamma_i\rangle \varphi_i, \ (x \in \mathcal{B}_1)$$

$$P_2(z) = \sum_{i=1}^{n} \langle z, \psi_i\rangle z_i, \ (z \in B_2),$$

respectively and consider the subspace

$$\mathcal{B}_{1,\infty-n} = \{f \in \mathcal{B}_1 \mid \langle f, \gamma_i\rangle = 0, \ (i = 1, ..., n)\},$$

$$\mathcal{B}_{2,\infty-n} = \{g \in \mathcal{B}_2 \mid \langle g, \psi_i\rangle = 0, \ (i = 1, ..., n)\}.$$

From (3.3.10) it follows that $\mathcal{B}_{i,n} \cap \mathcal{B}_{i,\infty-n} = \{0\}$ and, for every $x \in \mathcal{B}_i$, we have $x - P_i(x) \in \mathcal{B}_{i,\infty-n}$. Hence every element $x \in \mathcal{B}_i$ can be uniquely written as a sum of an element $u = P_i(x) \in \mathcal{B}_{i,n}$ and an element $v = x - P_i(x) \in \mathcal{B}_{i,\infty-n}$.

In this way, we get the decomposition in a direct sum of closed linear subspaces

$$\mathcal{B}_i = \mathcal{B}_{i,\infty-n} \oplus \mathcal{B}_{i,n} \ (i = 1, 2).$$

---

[5] If $x \in \mathcal{B}$ and $\psi \in \mathcal{B}^*$ then the value of the functional ψ at the point x is denoted by $\langle x, \psi\rangle$.

We note that since B is normally resoluble and $\psi_1, \ldots, \psi_n$ constitutes a basis of $N(B^*)$, the subspace $\mathcal{B}_{2, \infty-n}$ coincides with $R(B)$. Therefore the restriction of B to $\mathcal{B}_{i, \infty-n}$ defines an operator

$$\widetilde{B} : \mathcal{B}_{1, \infty-n} \to \mathcal{B}_{2, \infty-n}.$$

By construction, we have $R(\widetilde{B}) = \mathcal{B}_{2, \infty-n}$ and $N(\widetilde{B}) = N(B) \cap \mathcal{B}_{1, \infty-n} = \mathcal{B}_{1, n} \cap \mathcal{B}_{1, \infty-n} = \{0\}$. Hence the operator $\widetilde{B}$ is bijective. Since $\widetilde{B}$ is continuous and $\mathcal{B}_{1, \infty-n}$ and $\mathcal{B}_{2, \infty-n}$ are Banach spaces (being closed in $\mathcal{B}_1$ respectively $\mathcal{B}_2$), by Banach's theorem the operator $\widetilde{B}^{-1}$ is also continuous.

In this way, the Fredholm alternative for Φ-operators in Banach spaces can be stated as follows: if $n = 0$ (hence if the operator B is bijective), then Equation (3.3.9) has a unique solution for every $z \in \mathcal{B}_2$; if $n > 0$, then there exist n linearly independent functionals, $\psi_1, \ldots, \psi_n \in \mathcal{B}_2^*$ such that Equation (3.3.9) has solutions iff $\langle z, \psi_i \rangle = 0$, $i = 1, 2, \ldots, n$. If this condition is satisfied, then the equation admits in $\mathcal{B}_{1, \infty-n}$ the unique solution $x = \widetilde{B}^{-1}(z)$ and every other solution from $\mathcal{B}_1$ is of the form

$$x = \widetilde{B}^{-1} z + \sum_{i=1}^{n} c_i \varphi_i,$$

where the c_i denote arbitrary scalars.

Since $\widetilde{B}$ is an isomorphism and both $\mathcal{B}_{1, n}$ and $\mathcal{B}_{2, n}$ are n-dimensional linear spaces, by choosing a function which maps a basis of $\mathcal{B}_{1, n}$ onto a basis of $\mathcal{B}_{2, n}$, we can modify the operator B up to an isomorphism $\hat{B}$ which coincides with B on $\mathcal{B}_{1, \infty-n}$. Thus, we obtain **Schmidt's generalized lemma.** *The operator*

$$\hat{B} = B - \sum_{i=1}^{n} \langle \cdot, \gamma_i \rangle z_i$$

has the bounded inverse operator

$$\hat{B}^{-1} = \widetilde{B}^{-1} - \sum_{i=1}^{n} \langle \cdot, \psi_i \rangle \widetilde{B}^{-1} z_i - \sum_{i=1}^{n} \langle \cdot, \psi_i \rangle \varphi_i,$$

where $\widetilde{B}$ is the restriction of B to $\mathcal{B}_{1, \infty-n}$.

Theorem 3.3.1 [51]. *Let* $\Phi(x, y)$ *be an operator defined on the product of two Banach spaces* $\mathcal{B}, \mathcal{B}_1$ *and taking values in a Banach space* $\mathcal{B}_2$, *continuous in some neighbourhood* ω *of the point* (x_0, y_0) *and such that*

i) $\Phi(x_0, y_0) = 0$;

ii) *in* ω *the Fréchet differential* $\Phi'_y(x, y)$ *exists and is continuous;*

iii) *the linear operator* $B = -\Phi'_y(x_0, y_0)$ *has a bounded inverse. Then for every* x *in a certain neighbourhood* δ *of* $x_0 \in \mathcal{B}$ *there exists a unique solution* $y = f(x)$ *of the equation*

$$\Phi(x, y) = 0, \tag{3.3.11}$$

such that f(x_0) = y_0 *and* f(x) *is continuous in* δ. *If, in addition,* Φ(x, y) *admits in* ω *the power series expansion*

$$\text{(3.3.12)} \qquad \Phi(x, y) = \sum_{r+s\geq 1} C^r_{r+s}\Phi_{rs}(y - y_0)^r (x - x_0)^s, \ ((x, y) \in \omega),$$

where

$$\Phi_{rs} = \frac{1}{(r+s)!} \frac{\partial^{r+s}\Phi(x_0, y_0)}{\partial x^s \partial y^r},$$

then the solution y = f(x) *of* (3.3.11) *also admits a power series expansion in a certain neighbourhood* δ *of* x_0.

Assume now that Conditions i) and ii) of Theorem 3.3.1 are fulfilled together with: iii)′ the Fréchet differential $\Phi'_y(x_0, y_0)$ is a Fredholm operator whose null space is an n-dimensional linear space ($n > 0$). Under these conditions, leaving in the left-hand side the Fréchet differential $\Phi'_g(0, 0)\, g$ only, (3.3.11) may be written as

$$\text{(3.3.11)}' \qquad Bg = L(h, g),$$

where the nonlinear operator L is given by $L(h, g) = \Phi(h, g) - \Phi'_g(0, 0)\, g$, $g = y - y_0$, $h = x - x_0$.

Denote $g = u + v$, where $u = (I - P_1)\, g \in \mathfrak{B}_{1, \infty - n}$ and $v = P_1 g \in \mathfrak{B}_{1, n}$ and project (3.3.11)′ on $\mathfrak{B}_{2, \infty - m}$ and $\mathfrak{B}_{2, m}$. We get equations

$$\text{(3.3.11)}'' \qquad \begin{cases} \widetilde{B}u = (I - P_2)\, L(h, u + v), \\ 0 = P_2 L(h, u + v), \end{cases}$$

or, equivalently,

$$\text{(3.3.11)}''' \qquad \begin{cases} \widetilde{B}u = L(h, u + v), \\ 0 = P_2 L(h, u + v). \end{cases}$$

As $\widetilde{B}$ admits a bounded inverse, Equation $(3,3.11)'''_1$ can be written in a form analogous to (3.3.2)′′

$$u = \widetilde{B}^{-1} L\, (h, u + v)$$

which yields the solution $u = u\,(h, v)$. Introducing $u\,(h, v)$ in Equation $(3.3.11)''_2$, we fiind the branching equation

$$\text{(3.3.13)} \qquad P_2 L(h, u\,(h, v) + v) = 0.$$

We have, therefore, the following extension of the branching theorem [51], [2] to Banach spaces.

Theorem 3.3.2. *If the map* Φ *satisfies the above Conditions i), ii) and iii), then Equation* (3.3.11) *has as many small solutions with* y(x_0) = y_0, *as does the branching equation* (3.3.13). *Here* $\mathrm{v(h)} = \sum_{i=1}^{n} c_i(\mathrm{h})\, \varphi_i$, *and* $c_i(h)$ *are small functionals with* $c_i(0) = 0$, *and* $u(h, v) \in \mathfrak{B}_{1, \infty - n}$ *satisfies the equation*

$$\widetilde{B}u = L(h, u + v). \tag{3.3.14}$$

If Φ *is an analytic operator admitting the expansion* (3.3.12), *then Equation* (3.3.11) *admits solutions of the form*

$$y = y_0 + \sum_{i=1}^{n} \xi_i \varphi_i + \sum_{k+l=1}^{\infty} g_{kl} \left(\sum_{i=1}^{n} \xi_i \varphi_i \right)^k h^l, \tag{3.3.15}$$

their number being given by the branching equation

$$\langle \Phi_{01}, \psi_k \rangle + \left\langle \sum_{r+s\geqslant 2} \Phi_{rs} \left(\sum_{i=1}^{n} \xi_i \varphi_i + \right.\right.$$

$$\left.\left. + \sum_{\nu+l=1}^{\infty} g_{\nu l} \left(\sum_{i=1}^{n} \xi_i \varphi_i \right)^{\nu} h^l \right)^r h^s, \psi_k \right\rangle = 0, \tag{3.3.13$'$}$$

where $\xi_i = c_i(h)$, *and* $g_{kl}\, v^k\, h^l$ *are homogeneous operators, that are solutions of Equation (3.3.14).*

We note that (3.3.13)′ is equivalent to the orthogonality conditions $< L(h, u + v), \psi_k > = 0$, $(h = 1, ..., m)$. These are analogous to (3.3.3), where $L(h, u + v)$ plays the role of the right-hand side of Equation (3.3.6) and $u(s)$ is given by (3.3.7).

We now give another form of the branching equation.

Writing (3.3.11)′ in the equivalent form

$$\hat{B}g = L(h, g) - \sum_{i=1}^{n} \xi_i z_i, \tag{3.3.11$''$}$$

where $\hat{B}$ is invertible, we find an expression analogous to (3.3.2)″,

$$g = \widetilde{B}^{-1} [L(h, g)] + \sum_{i=1}^{n} \xi_i \varphi_i, \tag{3.3.16}$$

to which we can apply Theorem 3.3.1. We obtain

$$g = \sum_{i=1}^{n} \xi_i \varphi_i + G, \tag{3.3.17}$$

where G is a known function of h and $\xi_1, ..., \xi_n$. Then the branching equation is

$$\xi_i = \langle G(h, \xi_1, ..., \xi_n), \gamma_i \rangle + \sum_{j=1}^{n} \xi_j (\varphi_j, \gamma_i). \tag{3.3.8$'$}$$

Taking into account that by the definition of γ_i we have $< \varphi_j, \gamma_i > = \delta_{ij}$, for the particular case $n = 1$, $\xi_1 = \xi$, $h = \lambda h_1$, where h_1 is a unit vector of the space $\mathfrak{B}$, the branching equation becomes

$$\langle G(h, \xi_1, ..., \xi_n), \gamma_i \rangle = 0. \tag{3.3.8$''$}$$

This equation may be also written in the form

$$\sum_{m=2}^{\infty} L_{m0}\, \xi^m + \sum_{m=0}^{\infty} \xi^m \sum_{l=1}^{\infty} \lambda^l L_{ml} = 0, \tag{3.3.8}'''$$

where, this time, L_{kl} are operators defined on $\mathcal{B}$ taking values in $\mathcal{B}_1$ and ξ and λ are complex numbers.

The fact that Φ is analytic was not used in this second way of deducing the branching equation. If Φ is an analytic operator, we may proceed in a simpler way. In fact, since

$$\langle z_i, \psi_j \rangle = \delta_{ij}, \ \langle Bg, \psi_i \rangle = \langle g, B^* \psi_i \rangle = 0$$

and setting

$$\langle g, \gamma_i \rangle = \xi_i,$$

we have, from the definition of g,

$$\langle \hat{B}g, \psi_j \rangle = \langle Bg, \psi_j \rangle - \sum_{i=1}^{n} \langle g, \gamma_i \rangle \langle z_i, \psi_j \rangle = -\xi_j.$$

Hence, multiplying Equation (3.3.11)'' by ψ_i, we obtain

$$-\xi_i = \langle L(h, g), \psi_i \rangle - \xi_i, \ (i = 1, \ldots, m); \tag{3.3.18}$$

and, taking into account the equality

$$\hat{B}g = \Phi_{01} h + \sum_{k+l \geqslant 2} C_{k+l}^{l} \Phi_{kl} g^k h^l - \sum_{i=1}^{n} \xi_i z_i, \tag{3.3.19}$$

we find that

$$0 = \left\langle \Phi_{01} h + \sum_{k+l \geqslant 2} C_{k+l}^{l}\, \Phi_{kl} g^k h^l, \psi_i \right\rangle, \ (i = 1, \ldots, m). \tag{3.3.13}''$$

Since, by the implicit function theorem, the solution of Equation (3.3.11)'' is a function $g\,(h, \xi)$ analytic with respect to h and ξ, it follows that expanding g as a power series with respect to ξ and h, and substituting this series in (3.3.11)'', we can determine the coefficients of the expression. Introducing the determined g in relations (3.3.13)'', we obtain the branching equation.

In applications, the branching equation is reduced to the form (3.3.8) and is solved by a geometric method called the Newton's diagram. The applicability of the Liapunov-Schmidt analytic method is limited by the radii of convergence of the series obtained. Hence, this method gives all the solutions but on small domains of parameter variation.

Some extensions of the Liapunov-Schmidt procedure are used in differential topology under the name of transversality. In the particular case $\mathcal{B}_1$ is a Banach manifold then $\mathcal{B}_{2,\,m}$ is the space on which f is transversal at the point (x_0, y_0) while the bifurcation equation can be written as $\Phi_P(x, y) = 0$ where $\Phi_P = \Phi\,|_{S_P}$ and the smooth manifold S_P is the space of the solutions of the equations $(3.3.11)_1''$ (Appendix 5).

§ 3.4. BRANCHING OF SOLUTIONS OF THE NAVIER-STOKES EQUATIONS BY THE LIAPUNOV-SCHMIDT METHOD

Let $\lambda_1 = \text{Re}_{cr}(\text{Ra}_{cr})$ be the point of loss of the linear stability, which is also the first branch point for the solution of Navier-Stokes equations obtained by the Leray-Schauder method. In all the motions considered, λ_1 is a simple eigenvalue of the linear Stokes map; the principle of the exchange of stabilities holds and the secondary solutions are stationary. Using the Liapunov-Schmidt method, we construct all the branches (whose number is equal to 2) arround λ_1 as power series of $\sqrt{\lambda - \lambda_1}$ in the case of thermal convection (Section 3.4.1), of motion between rotating cylinders (Section 3.4.2) and of motions in bounded domains (Section 3.4.3). For large classes of motions, both methods specified above are used (Section 3.4.4) in order to prove the stability of the supercritical secondary solutions and the instability of the subcritical ones. In the latter case, the branching can take place subcritically, supercritically or both sub- and supercritically.

3.4.1. Convective motions

The existence of the branching, at Ra_{cr}, from the conduction solution, of the convective solutions were proved long ago by means of the method of perturbations with respect to a parameter connected with the Rayleigh number. Recently the branching of the convective motion and its stability are studied by analytic perturbation theory with respect to a parameter connected with the norm of the solution. The Liapunov-Schmidt method was applied to the study of convection beginning in 1954 when Sorokin [48] proved the existence, slightly supercritically, of two branches of convective solutions as a power series of $\sqrt{\text{Ra} - \text{Ra}_{cr}}$. The first rigorous application of the Liapunov-Schmidt method to the problem of thermal convection belongs to Yudovich [19]; subsequently it has also been considered by other authors [6], [26] as well.

Consider the classical problem of free thermal convection, in the usual physical coordinates in the domain Ω, with boundary $\partial\Omega$ [19],

$$(3.4.1)\qquad \begin{cases} \left.\begin{aligned} \nu\Delta\mathbf{v} - \operatorname{grad} p &= (\mathbf{v}\operatorname{grad})\,\mathbf{v} + \beta T\mathbf{g}, \\ \chi\Delta T - \mathbf{v}\operatorname{grad} T &= \lambda v_3, \\ \operatorname{div}\mathbf{v} &= 0, \end{aligned}\right\} \text{in } \Omega \\ \left.\begin{aligned} \mathbf{v} &= 0, \\ T &= 0, \end{aligned}\right\} \text{on } \partial\Omega. \end{cases}$$

The nontrivial solutions of the problem (3.4.1) are secondary flows which set in at the point $\lambda = \lambda_1$ from the solution $\mathbf{v} = T = 0$. Let λ_1 be the least eigenvalue of the linearized problem

$$(3.4.2)\qquad \begin{cases} \left.\begin{aligned} &\nu\Delta\mathbf{v} - \operatorname{grad} p = \beta T\mathbf{g}, \\ &\chi\Delta T = \lambda v_3, \\ &\operatorname{div}\mathbf{v} = 0, \end{aligned}\right\} \text{in } \Omega \\ \left.\begin{aligned} \mathbf{v} &= 0, \\ T &= 0, \end{aligned}\right\} \text{on } \partial\Omega. \end{cases}$$

In order to apply the Liapunov-Schmidt theory to system (3.4.1), we reduce it to a nonlinear equation in a Banach space. Let us denote by H_1 the Hilbert space of solenoidal vector fields in $W^{1,2}$ vanishing on $\partial\Omega$, by H_2 the Hilbert space of the functions from $W^{1,2}$ vanishing on $\partial\Omega$, with the $W^{2,2}$-norm, and let $\mathbf{v} \in H_1$, $f(x) \in L_{\frac{6}{5}}(\Omega)$. If $T' = C(\mathbf{v})f$ is the generalized solution from H_2 of the problem

$$(3.4.3)\qquad \begin{cases} \chi\Delta T' - \mathbf{v}\operatorname{grad} T' = f(x), \text{ in } \Omega \\ T' = 0, \text{ on } \partial\Omega \end{cases}$$

for $f(x) \in L_{\frac{6}{5}}(\Omega)$ and $\mathbf{v} \in H_1(\Omega)$, then the generalized solution of $(3.4.1)_{2,5}$ satisfies the equation

$$(3.4.1)'_{2,5}\qquad T = \lambda C(\mathbf{v})v_3 = \lambda M\mathbf{v}$$

where the nonlinear map $M : H_1 \to H_2$ may be written in the form

$$M\mathbf{v} = \sum_{k=1}^{\infty} M_k\mathbf{v},\ M_1\mathbf{v} = C(\mathbf{o})\,v_3,\ M_k\mathbf{v} = C(\mathbf{o})\,((\mathbf{v}\operatorname{grad})\,M_{k-1}\mathbf{v}),\ (k \geqslant 2).$$

Let $\mathbf{v}' = D\mathbf{f}$ be the generalized solution from H_1 of the problem

$$(3.4.4)\qquad \begin{cases} \left.\begin{aligned} &\nu\Delta\mathbf{v} - \operatorname{grad} p = \mathbf{f}, \\ &\operatorname{div}\mathbf{v} = 0, \end{aligned}\right\} \text{in } \Omega \\ \mathbf{v} = 0, \text{ on } \partial\Omega. \end{cases}$$

The generalized solution of the system $(3.4.1)_{1,3,4}$ satisfies therefore the equation

$$(3.4.1)'_{1,3,4} \qquad \mathbf{v} = \mathrm{D}(\mathbf{v}\ \mathrm{grad})\ \mathbf{v} + \mathrm{D}(\beta\ T\mathbf{g}).$$

Taking into account the relations $(3.4.1)'_{1,3,4}$ and $(3.4.1)'_{2,5}$ the weak form of system (3.4.1) is the following equation in H_1

$$(3.4.1)'' \qquad \mathbf{v} = \lambda A\mathbf{v} + L(\mathbf{v}, \lambda),$$

where

$$A\mathbf{v} = \beta\mathrm{D}(\mathbf{g}M_1\mathbf{v}),\ L(\mathbf{v}, \lambda) = \mathrm{D}(\mathbf{v}\ \mathrm{grad})\ \mathbf{v} + \beta\lambda\sum_{k=2}^{\infty}\mathrm{D}(\mathbf{g}M\ \mathbf{v}).$$

A is a strictly positive completely continuous linear map whose spectrum consists of positive eigenvalues $\lambda_1, \lambda_2, \ldots$ Equation (3.4.1)'' admits the solution $\mathbf{v} = 0$, $T = 0$, for every λ, and this solution is unique [19] for $\lambda \leqslant \lambda_1$. Let us now determine all the solutions which appear at the point $\lambda = \lambda_1$ from $\mathbf{v} = 0$, $T = 0$. Suppose that λ_1 is simple and let us apply Theorem 3.3.2 to an equation more general than (3.4.1)'', i.e.

$$(3.4.5) \qquad g = \lambda Ag + \sum_{k=2}^{\infty} R_k g,$$

in a Banach space, where

$$R_{kg} = R_k(g, g, \ldots, g)$$

is a k-linear map which is analytically dependent on λ

$$R_{kg} = \sum_{l=0}^{\infty} h^l R_{klg},\quad h = \lambda - \lambda_1.$$

As in Theorem 3.3.2, let

$$g = \xi\varphi_1 + u,\quad \xi = (g, \varphi_1)_{H_1},$$

where φ_1 is the eigenfunction of the map A corresponding to the characteristic value λ_1. Equation (3.4.5) corresponds to Equation (3.3.11)' and, consequently,

$$g - \lambda_1 Ag = Bg,\ hAg + \sum_{k=2}^{\infty} R_k g = L(h, g).$$

Let $\widetilde{B}$ be the restriction of B to the subspace orthogonal to the proper subspace of A. Then $\widetilde{B}(\xi\varphi_1) = 0$ and, consequently,

$$(3.4.5)' \qquad \widetilde{B}(u) = L(h, u + \xi\varphi_1),$$

where

$$u = \sum_{i+j=1}^{\infty} g_{ij}\xi^i h^j,\ (u, \varphi_1)_{H_1} = 0,$$

and g_{ij} are determined by equation

$$\sum_{i+j=1}^{\infty} g_{ij}\xi^i h^j - \lambda_1 A \sum_{i+j=1}^{\infty} g_{ij}\xi^i h^j = \frac{h\xi}{\lambda_1} + hA \sum_{i+j=1}^{\infty} g_{ij}\xi^i h^j +$$

$$+\sum_{k=2}^{\infty}\sum_{m=0}^{\infty} h^m R_{km}\left(\xi\varphi_1 + \sum_{i+j=1}^{\infty} g_{ij}\xi^i h^j\right). \tag{3.4.5)''}$$

This equation may be also written as

$$(g_{10} - \lambda_1 A g_{10})\xi + (g_{01} - \lambda_1 A g_{01})\, h + (g_{11} - \lambda_1 A g_{11})\xi h +$$

$$+ (g_{20} - \lambda_1 A g_{20})\xi^2 + (g_{02} - \lambda_1 A g_{02})\, h^2 + (g_{03} - \lambda_1 A g_{03})\, h^3 +$$

$$+ (g_{12} - \lambda_1 A g_{12})\xi h^2 + \ldots = \left[\frac{\varphi_1}{\lambda_1} + A g_{10} + R_{20}(\varphi_1, g_{01}) +\right.$$

$$\left. + R_{20}(g_{01}, g_{10})\right]\xi h + [R_{20}\varphi_1 + R_{20}(\varphi_1, g_{10}) + R_{20}(g_{10}, \varphi_1) +$$

$$+ R_{20}g_{10}]\xi^2 + [A g_{01} + R_{20}(g_{01}, g_{01})]\, h^2 + [A g_{02} + R_{21}g_{01} + \tag{3.4.5)'''}$$

$$+ R_{20}(g_{01}, g_{02}) + R_{30}(g_{02}, g_{01}) + R_{30}g_{01}]\, h^3 + [A g_{11} + R_{20}(\varphi_1, g_{02}) +$$

$$+ R_{20}(g_{02}, \varphi_1) + R_{20}(g_{20}, g_{02}) + R_{20}(g_{02}, g_{10}) + R_{20}(g_{01}, g_{11}) +$$

$$+ R_{20}(g_{11}, g_{01}) + R_{21}(\varphi_1, g_{01}) + R_{21}(g_{01}, \varphi_1) + R_{21}(g_{10}, g_{01}) +$$

$$+ R_{21}(g_{01}, g_{10})]\xi h^2 + \ldots,$$

and the orthogonality condition $(u, \varphi_1)_{H_1} = 0$ becomes

$$\sum_{i+j=1}^{\infty} \xi^i h^i (g_{ij}, \varphi_1)_{H_1} = 0.$$

Denoting by Rg the right-hand side of Equation (3.4.5)''', we obtain a branching equation analogous to (3.3.13)',

$$(Rg, \varphi_1)_{H_1} = 0. \tag{3.4.6}$$

The above procedure of deriving the branching equation, implied in system (3.3.11)''', leads now to elaborate calculus, so that we use the system which corresponds to (3.3.11)'':

$$g - \lambda_1 A g = Rg - (Rg, \varphi_1)_{H_1}\varphi_1, \; (Rg, \varphi_1)_{H_1} = 0. \tag{3.4.7}$$

Let us remark that the branching equation is still (3.4.6), but Equation (3.4.5)''', which yields the g_{ij} 's, has been replaced by the equivalent Equation (3.4.7). From (3.4.7) it is found that $g_{10} = g_{01} = g_{11} = g_{02} = g_{12} = g_{03} = 0$ and

$$g_{20} - \lambda_1 A g_{20} = P_0 R_{20} \varphi_1,$$

$$g_{30} - \lambda_1 A g_{30} = P_0\{R_{20}(\varphi_1, g_{20}) + R_{30}\, \varphi_1\},$$

$$g_{21} - \lambda_1 A g_{21} = P_0\{A g_{20} + R_{21}\varphi_1\},$$

where $P_0 g = g - (g, \varphi_1)\varphi_1$, $g \in \mathfrak{B}$. Therefore,

$$g = g_{20}\xi^2 + g_{30}\xi^3 + g_{21}\xi^2 h + \ldots, \tag{3.4.8}$$

so that the branching equation becomes

$$\frac{h}{\lambda_1}\xi + \xi^2(R_{20}\, \varphi_1, \varphi_1) + \xi^3[(R_{20}(\varphi_1, g_{20}) + R_{20}(g_{20}, \varphi_1), \varphi_1) +$$

$$+ (R_{30}\, \varphi_1, \varphi_1)] + \xi^2 h(R_{21}\, \varphi_1, \varphi_1) + \ldots = 0. \tag{3.4.6)'$$

From Equation (3.4.5)'', it is easy to obtain that $(R_{20}\, \varphi_1, \varphi_1) = 0$ so that, writing $-c_3$ for the coefficient of ξ^3, the branching equation takes the form

$$\frac{\xi h}{\lambda_1} - \xi^3 c_3^3 + h\xi^2(R_{21}\varphi_1, \varphi_1) + \ldots = 0, \tag{3.4.6)''$$

which has the following small solutions

$$\xi_1 = 0, \;\; \xi_{2,3} = \pm \sqrt{\frac{h}{\lambda_1 c_3}}\, \varphi_1 + \mathrm{O}(h).$$

They are real if $c_3 > 0$, that is if

$$(R_{20}(\varphi_1, g_{20}) + R_{20}(g_{20}, \varphi_1), \varphi_0) + (R_{30}\varphi_1, \varphi_1) < 0. \tag{3.4.9}$$

Now, coming back to the convection problem (3.4.1)'', we may state that around λ_1, slightly supercritically, as the unicity and the stability of the solution $\mathbf{v} = 0$, $T = 0$, are lost, just two convective solutions appear which are power series of $\sqrt{\lambda - \lambda_1}$. This result is conditioned on the positivity of c_3 and by the simplicity of λ_1. For an arbitrary bounded domain Ω, it can be shown [19] that $c_3 > 0$; if Ω is a vertical circular or rectangular cylinder of large height or if the convection is two- or three-dimensional and takes place in horizontal channels, then it can be shown numerically that λ_1 is a simple eigenvalue. In the last two cases, the domain Ω being unbounded, the motion is assumed periodic sucht hat the hydrodynamics problem is reduced to a problem for finite domains (periodicity cells). On the other hand, λ_1 is a priori a multiple eigenvalue and it becomes simple only if the solution is an even function: therefore we seek the solution in the corresponding subsets; in Sections 3.2.2 and 3.2.3 we used an analogous procedure. Finally, a theorem due to Yudovich [19] — which can be applied to convection and to the motion between rotating cylinders — shows that there is no other solution besides the periodic even solutions [6)].

A similar study concerning the branching at the other eigenvalues λ_i $(i > 1)$ will be performed in Section 3.4.4. Unlike the secondary solutions corresponding to λ_1, the branching solutions corresponding to λ_i $(i > 1)$ are unstable (Section (3.4.5).

3.4.2. **Couette motion**

In Section 3.2.3. we proved that in the case of motion between coaxial cylinders rotating in the same direction all the eigenvalues $\lambda_{\bar{i}}$ of the Stokes map are simple and therefore they are branch points of new space periodic motions for all the periods up to a countable set. By Yudovich's theorem, there exist no other solutions besides the periodic ones (Section 3.2.3). In what follows, we shall show by the Liapunov-Schmidt method that up to translation from the Couette basic motion there appears at Re_{cr} a single branch of solutions corresponding to the Taylor vortices observed in experiments. Let us consider, hence [24], [43] (from Ch. 2), the motion between two cylinders of radii r_1, r_2 $(r_1 < r_2)$, which are rotating in the same direction with constant angular velocities ω_1 and ω_2 respectively; let $r_2 - r_1$, $r_1\omega_1$ and $\rho r_1^2\omega_1^2$ be the characteristic length, velocity and pressure and let the Reynolds number $\lambda = \mathrm{Re} = (r_2 - r_1)\, r_1\omega_1\nu^{-1}$. For every λ there exists a solution corresponding to Couette motion with velocity $(0, v_0(r), 0)$, $v_0(r) = -\alpha r - \alpha\beta r^{-1}$, where α and β were defined in Section 1.1.3; other axi-symmetric periodic stationary motions appear at λ_i when relation (3.2.25) which expresses Rayleigh's criterion, holds. Let us look for these motions as sums $\mathbf{v}(r, z) = (0, v_0(r), 0) + \mathbf{v}^{(1)}(r, z)$, where $v_1^{(1)} = u_1$, $v_3^{(1)} = u_3$, $v_2^{(1)} = u_2\sqrt{|\alpha|\dfrac{r_1}{r_2 - r_1}}$. The secondary motions are then solutions of the following problem

$$
(3.4.10)\quad \begin{cases} \mathbf{Du} = -\lambda\nabla p + \lambda A_1(v_0)\,\mathbf{u} + \lambda B_1(\mathbf{u})\,\mathbf{u} + \lambda\mathbf{u}\nabla\mathbf{u}, \text{ in } \Omega \\ \nabla\mathbf{u} = 0, \\ \mathbf{u} = 0 \text{ for } r = \dfrac{r_1}{r_2 - r_1} \text{ and } r = \dfrac{r_2}{r_2 - r_1}, \end{cases}
$$

---

6) To prove this theorem, it is shown that when the eigenvalue λ_1 is multiple and when its multiplicity is a result of the invariance of the problem under a certain group of transformations, then under certain conditions, all solutions can be obtained by means of the transformations of this group acting on a single solution which has the form $g = \xi\varphi_1 + u$. Further this solution is constructed by Liapunov-Schmidt method.

where $\mathbf{u}(r, z + 2\pi\sigma^{-1}) = \mathbf{u}(r, z)$, σ is the wave number, Ω is the strip $\left(\frac{r_1}{r_2 - r_1}, \frac{r_2}{r_2 - r_1}\right) \times (-\infty, +\infty)$ and

$$\mathrm{D} = \begin{pmatrix} \Delta - r^{-2} & 0 & 0 \\ 0 & \Delta - r^2 & 0 \\ 0 & 0 & \Delta \end{pmatrix},$$

$$A_1(v_0) = \begin{pmatrix} 0 & -2\,[|\alpha|(r_2 - r_1)\,r_1^{-1}]^{\frac{1}{2}}\,v_0 r^{-1} & 0 \\ -2\,[|\alpha|(r_2 - r_1)\,r_1^{-1}]^{\frac{1}{2}} & 0 & 0 \\ 0 & 0 & 0 \end{pmatrix}$$

$$B_1(\mathbf{u}) = \begin{pmatrix} 0 & -\alpha r_1 u_2 r^{-1}(r_2 - r_1)^{-1} & 0 \\ u_2 r^{-1} & 0 & 0 \\ 0 & 0 & 0 \end{pmatrix},$$

$$\Delta = \frac{1}{r}\frac{\partial}{\partial r}\left(r\frac{\partial}{\partial r}\right)\frac{\partial^2}{\partial z^2}, \quad \nabla = \left(\frac{\partial}{\partial r}, 0, \frac{\partial}{\partial z}\right).$$

The weak solutions of the problem (3.4.10) satisfy equation

(3.4.11) $$\mathbf{u} = \lambda A\mathbf{u} + \lambda T\mathbf{u}, \quad \mathbf{u} \in H_\sigma,$$

where the Hilbert space H_σ is the completion in the norm $\|\cdot\|$, corresponding to the scalar product $((\mathbf{u}, \mathbf{v})) = \int_{\Omega_\sigma}\left\{\nabla\mathbf{u}\cdot\nabla\mathbf{v} + \frac{1}{r^2}(u_1 v_1 + u_2 v_2)\right\} r\,\mathrm{d}r\,\mathrm{d}z$, of the set of real solenoidal vector functions $\mathbf{u}(r, z)$ with compact support in Ω, and Ω_σ is the periodicity domain

$$\left(\frac{r_1}{r_2 - r_1}, \frac{r_2}{r_2 - r_1}\right) \times (z_0, z_0 + 2\pi\sigma^{-1}).$$

By some *a priori* inequalities, similar to those satisfied in the energy method (Section 2.6.1), it can be proved that the functionals $(\mathbf{u}\nabla\mathbf{u}, \cdot)$ and $(B_1(\mathbf{u})\,\mathbf{u}, \cdot)$ (where $(\cdot, \cdot)$ is the scalar product in $L_2(\Omega_\sigma)$) are linear and continuous on H_σ for every fixed $\mathbf{u}$ in H_σ; therefore we can define the nonlinear map $T\colon H_\sigma \to H_\sigma$ by the relation

(3.4.12) $$-((T\mathbf{u}, \mathbf{v})) = (\mathbf{u}\nabla\mathbf{u}, \mathbf{v}) + (B_1(\mathbf{u})\,\mathbf{u}, \mathbf{v}).$$

We define also the linear map A by the equality

$$(A\mathbf{u}, \mathbf{v}) = -(A_1(v_0)\,\mathbf{u}, \mathbf{v}), \quad \mathbf{v} \in H_\sigma;$$

it is easy to see that A is a completely continous map.

In this way, the classical equations of hydrodynamics can be expressed by a nonlinear equation in H_σ; to this equation we now apply the Liapunov-Schmidt method. Let $\sigma > 0$ be the wave number so that $\lambda_1(\sigma)$ is a simple characteristic value of the map A corresponding to the eigenfunction φ_1, and denote ba ψ_1 the eigenfunction of the adjoint map A^* corresponding to the characteristic value λ_1 such that $((\psi_1, \varphi_1)) = 1$. Equation (3.4.11) may be put in the form (3.3.11)′, where

$$Bg = g - \lambda_1 Ag, \quad L(h, g) = hAg + \lambda Tg, \quad \mathbf{u} = \mathbf{g}, \quad h = \lambda - \lambda_1.$$

Writing $\hat{A} = A - \lambda_1^{-1} ((\psi_1, \cdot))\, \varphi_1$ and $R(\lambda_1) = I - \lambda_1 \hat{A})^{-1}$ such that $\hat{B}^{-1} = R(\lambda_1)$, by Schmidt's lemma it follows that the map $R(\lambda_1)$ exists and is bounded. Putting $\widetilde{A} = R(\lambda_1)\, \hat{A}$, $\widetilde{T} = R(\lambda_1)\, T$, $\xi = \frac{\lambda}{\lambda_1} ((g, \psi_1))$ and taking into account the fact that $R(\lambda_1)\, \varphi_1 = \varphi_1$, we may write Equation (3.4.11) in the form

$$\hat{B}g = Ag - \frac{1}{\lambda} \xi \varphi_1,$$

or, equivalently,

$$\hat{B}g = h\hat{A}g + (\lambda_1 + h)\, Tg + \xi\varphi_1. \tag{3.4.13}$$

Multiplying Equation (3.4.13) by $\hat{B}^{-1} = R(\lambda_1)$, we obtain the nonlinear equation

$$g = \xi\varphi_1 + h\hat{A}g + (\lambda_1 + h)\, \widetilde{T}g, \tag{3.4.13$'$}$$

which plays the role of relation (3.3.16). Equation (3.4.13)′ is solved iteratively; for small values of $|\, \xi \,|$ and $|\, h \,|$ the obtained solution is a series of ξ and h similar to expression (3.3.17). Multiplying now Equation (3.4.11) by ψ_1, we obtain the equality

$$\begin{aligned} 0 &= h\xi(\lambda_1 + h)^{-1} + (\lambda_1 + h)\, ((\psi_1, Tg)) = \\ &= h\xi(\lambda_1 + h)^{-1} - (\lambda_1 + h)\, (\psi_1, \mathbf{u} \cdot \nabla \mathbf{u} + B_1(\mathbf{u})\, \mathbf{u}), \end{aligned} \tag{3.4.14}$$

which, after the introduction of the above series, yields a branching equation, corresponding to (3.3.8), of the form

$$0 = h\xi(\lambda_1 + h)^{-1} + \sum_{k=2}^{\infty} a_k \xi^k + \sum_{\substack{k=1\\ l=2}}^{\infty} a_{k.l}\, \xi^k h^l, \tag{3.4.14$'$}$$

where $a_{2n} = 0$, $a_{2n,\, m} = 0$ for $n, m = 1, 2, \ldots$. The branching equation can also be written as

$$0 = \frac{h\xi}{\lambda_1} + a_3 \xi^3 + \ldots, \tag{3.4.14$''$}$$

where $a_3 = \lambda_1^2((\psi_1, T(T\varphi_1 + \varphi_1) - T(T\varphi_1) - T\varphi_1))$, so that its solutions are

$$\xi_1 = 0,$$
$$\xi_2 = \sqrt{\frac{-h}{a_3\lambda_1}} + \ldots \text{ higher order terms}, \ \xi_3 = -\xi_2.$$

There correspond to these solutions the trivial solution and two other solutions

$$\begin{cases} u_2 = u(r, z, \xi_2, h), \\ u_3 = u(r, z, -\xi_2, h). \end{cases}$$

It can be shown that u_2 and u_3 are equal up to a $\frac{\pi}{\sigma}$ translation in the z — direction.

Unlike convection, in the case of the motion between rotating cylinders, the condition $a_3 > 0$ was proved only numerically [99] (from Ch. 2). Hence, at $\lambda_1(\sigma)$, up to a translation along the cylinders axis, a single branch of z-periodic stationary solutions bifurcates supercritically from the basic Couette motion, these solutions being power series of $\sqrt{\mathrm{Re} - \mathrm{Re}_{cr}}$. The effective derivation of these series can be found in [43] (from Ch. 2) and [37].

The case of a fixed cylinder can be obtained [42] (from Ch. 2).

3.4.3. **Motions in bounded domains**

The Liapunov-Schmidt method was initiated to solve a problem of hydrodynamics [33]; but to a large extent, its subsequent development is due to the neccessity of solving some integro-differential or partial differential nonlinear equations, which appear in various branches of fluid mechanics [51], [32], [20]. The first application of this method to the study concerning the branching of the solutions of the Navier-Stokes equations is devoted to motions in bounded domains and belongs to Odqvist, who in 1930, obtained [36] a sufficient condition for branching; his proof is sketched below.

Let us consider the stationary Navier-Stokes equations in the integro-differential form given by Odqvist and let us look for solutions $\mathbf{u}_1(x)$, $p_1(x)$ as the sum $\mathbf{u}(x) + \mathbf{v}(x)$, $p(x) + q(x)$ of a known solution $\mathbf{u}(x), p(x)$ and an unknown perturbation $\mathbf{v}(x)$, $q(x)$ which satisfies the system

$$(3.4.15) \quad \begin{cases} v_i(x) = -\rho \displaystyle\int_\Omega G_{ij}(x, y) \left[v_k \frac{\partial u_j}{\partial y_k} + u_k \frac{\partial v_j}{\partial y_k} + v_k \frac{\partial v_j}{\partial y_k} \right] \mathrm{d}y, \\ \dfrac{\partial v_i}{\partial x_l}(x) = -\rho \displaystyle\int_\Omega \frac{\partial G_{ij}}{\partial x_l} \left[v_k \frac{\partial u_j}{\partial y_k} + u_k \frac{\partial v_j}{\partial y_k} + v_k \frac{\partial v_j}{\partial y_k} \right] \mathrm{d}y, \\ q(x) = -\rho \displaystyle\int_\Omega g_j(x, y) \left[v_k \frac{\partial u_j}{\partial y_k} + u_k \frac{\partial v_j}{\partial y_k} + v_k \frac{\partial v_j}{\partial y_k} \right] \mathrm{d}y. \end{cases}$$

$\mathbf{v} = (v_1, v_2, v_3)$; $\mathbf{u} = (u_1, u_2, u_3)$, G_{ij}, g_j are elements of Green's hydrodynamic tensor and Ω is a bounded two- or three-dimensional domain. $\mathbf{v}$ must satisfy, in addition, the adherence condition at the boundary of Ω. Neglecting the nonlinear terms with respect to the perturbation, we obtain the linear system

$$(3.4.16)\quad \begin{cases} v_i(x) = -\rho\displaystyle\int_\Omega G_{ij}(x, y)\left[v_k\frac{\partial u_j}{\partial y_k} + u_k\frac{\partial v_j}{\partial y_k}\right]\mathrm{d}y, \\ \dfrac{\partial v_i}{\partial x_l}(x) = -\rho\displaystyle\int_\Omega \frac{\partial G_{ij}}{\partial x_l}\left[v_k\frac{\partial u_j}{\partial y_k} + u_k\frac{\partial v_j}{\partial y_k}\right]\mathrm{d}y, \\ q(x) = -\rho\displaystyle\int_\Omega g_j(x, y)\left[v_k\frac{\partial u_j}{\partial y_k} + u_k\frac{\partial v_j}{\partial y_k}\right]\mathrm{d}y. \end{cases}$$

According to Liapunov-Schmidt theory, the branching takes place if system (3.4.16) has real eigenvalues μ (where μ is the coeficient of the dynamic viscosity which enters the expression of the Green tensor) and real corresponding eigensolutions. Therefore, sufficient conditions for non-branching are obtained by requiring this system to have no real eigenvalue. Under suitable hypotheses on the smoothness of the integrands, we find that system (3.4.16) is equivalent to the linear differential Navier-Stokes system

$$(3.4.16)'\quad \mu\Delta v_i = \frac{\partial q}{\partial x_i} + \rho\left(v_k\frac{\partial u_i}{\partial x_k} + u_k\frac{\partial v_i}{\partial x_k}\right), \quad \frac{\partial v_i}{\partial x_i} = 0.$$

By Green's formula from system (3.4.16)$'$ we deduce the relation

$$(3.4.16)''\quad \int_\Omega\left\{\mu\left(\frac{\partial v_i}{\partial x_k} + \frac{\partial v_k}{\partial x_i}\right)^2 + \rho\left(\frac{\partial u_i}{\partial x_k} + \frac{\partial u_k}{\partial x_i}\right)v_i v_k\right\}\mathrm{d}x = 0.$$

Proposition *(Odqvist). A sufficient condition for the stationary solutions of the Navier-Stokes equations to possess no branching is that the integrand in formula* (3.4.16)$''$ *should be a positively defined quadratic form.*

In particular, from the rigid motion for which the equality $\dfrac{\partial u_i}{\partial x_k} + \dfrac{\partial u_k}{\partial x_i} = 0$ holds, there is no branching.

By studying the form of the solutions of the Navier-Stokes equations in bounded domains around Re_{cr}, Sorokin [49] effectively built secondary solutions as power series of $\mathrm{Re} - \mathrm{Re}_{cr}$ or $\sqrt{\mathrm{Re} - \mathrm{Re}_{cr}}$; he also showed branching from the secondary solution for some Re', $\mathrm{Re}' < \mathrm{Re}_{cr}$ and thus verified Landau's cellular instability hypothesis [30]. Using the Liapunov-Schmidt method for particular flows, Yudovich [18] found again Sorokin's results, and proved that the first coefficient of the series is real and nonvanishing. The treatment is similar to that exposed in Sections 3.4.2 and 3.4.3. The proof of the fact that this coefficient is real and does not vanish for general motions is still an open problem; it amounts proving that some integral in a certain scalar product expression is strictly positive [29].

3.4.4. The stability of branching solutions

In the above discussion we have assumed that the principle of the exchange of stabilities holds, which implies the fact that at the point λ_1 where the linear stability is lost, the stability spectrum has a simple eigenvalue at the origin ($\sigma = 0$), the rest of the eigenvalues lying in the left halfplane. Then, for some fluid flows the Stokes linear map had λ_1 as a simple eigenvalue, so that by Leray-Schauder degree, λ_1 is a branch point, and by the Liapunov-Schmidt method it was found that two continuous branches of stationary solutions branch supercritically. Since at λ_1 the solution has lost its physical sense, we have to find which secondary flows are stable; these flows will correspond to actual motions for $\lambda > \lambda_1$.

In general, the stability of secondary solutions is studied by analytic perturbation methods. But for certain maps, and among them for Navier-Stokes map, this study has been accomplished in [44], [45] by Leray-Schauder degree theory using some results concerning the number and the form of the branches yielded by the Liapunov-Schmidt method. These results generalize the corresponding finite-dimensional results (Section 3.1.1) and will be briefly exposed below.

Let the stationary equation in the Banach space $\mathfrak{B}$

$$(L_1 - \lambda L_2)\, x + F(x, \lambda) = 0, \tag{3.4.17}$$

admit the solution $x = 0$ for every value of the real parameter λ. L_1 and L_2 are linear maps and F is a nonlinear map in $\mathfrak{B}$. Let λ_1 be the least eigenvalue of the Fréchet differential at the point $\tilde{x}$ of the form (3.4.17),

$$(L_1 - \lambda L_2) + F'_x(\tilde{x}, \lambda) \tag{3.4.18}$$

and assume that $F'_x(0, \lambda) = 0$. Then for $\tilde{x} = 0$, λ_1 is an eigenvalue of the map $L_1 - \lambda L_2$. Suppose that the solution $\tilde{x} = 0$ is stable for $0 \leqslant \lambda < \lambda_1$ and unstable for $\lambda > \lambda_1$; L_1 has a compact inverse map L_1^{-1}; $L_1^{-1} L_2 = A$ and $L_1^{-1} F = \mathfrak{F}$ are completely continuous maps and the equation in weak form

$$(I - \lambda A)\, x + \mathfrak{F}(x, \lambda) = 0, \tag{3.4.17$'$}$$

corresponding to Equation (3.4.17), admits regular solutions, such that (3.4.17) is equivalent to the Equation (3.4.17)$'$. Suppose in addition that $\mathfrak{F}$ has a continuous linear Fréchet differential in a neighbourhood of the origin $\tilde{x} = 0$ in the Banach space $\mathfrak{B}$; $F'_x(\tilde{x}, \lambda)$

depends continuously on λ and on $\widetilde{x}$; $F'_x(0, \lambda) = 0$; F'_λ exists and $\mathfrak{F}$ is uniformly continuous with respect to λ. All the maps are assumed to be real. By the principle of exchange of stabilities, it follows that $L_1 - \lambda L_2$ has the simple eigenvalue $\sigma_1 = 0$, the rest of the stability spectrum lying at the left-side of the vertical line $\sigma_r = \eta$ in the complex plane (σ_r, σ_i). Suppose that for small variations of λ, the eigenvalues of the map (3.4.18) vary continuously with respect to λ and apart from the eigenvalue near the origin, these eigenvalues remain at the left-side of $\sigma_r = \frac{\eta}{2}\cdot$ Letting $h = \lambda - \lambda_1$ and $x = g$, Equation (3.4.17)′ becomes

$$(I - \lambda_1 A)\, g - hAg + \mathfrak{F}(g, \lambda) = 0. \tag{3.4.17''}$$

Decomposing the above equation by Schmidt's procedure, we obtain the system

$$\begin{cases} g = h(I - \lambda_1 A)^{-1} g - (I - \lambda_1 A)^{-1}\, \mathfrak{F}(g, \lambda), \\ f(h, \xi) = -\dfrac{h\xi}{\lambda_1} + \langle \mathfrak{F}(g, \lambda), \psi_0\rangle = 0, \end{cases} \tag{3.4.19}$$

where $\xi = (g, \psi_0)$, φ_0 is the eigenfunction of the map A corresponding to the characteristic value λ_1 and ψ_0 is the eigenfunction of the adjoint map A^* corresponding to the characteristic value λ_1. By an iterative solution of Equation $(3.4.19)_1$, we obtain g as a series of ξ and h; then we introduce this series in Equation $(3.4.19)_2$ and find the branching equation. Assume that $\mathfrak{F}(\xi g, \lambda) = \xi^2 F_1(g, \lambda, \xi)$ (this condition is satisfied by the nonlinear term (**u** grad) **u** from the Navier-Stokes equations) and $\mathfrak{F}_1$ is Fréchet differentiable at (g, λ, ξ). Putting $g = \xi g_1$ in Equation $(3.4.19)_2$ we obtain the branching equation

$$f(h, \xi) = \xi f_1(h, \xi) = 0, \tag{3.4.20}$$

where

$$f_1(h, \xi) = -\frac{h}{\lambda_1} + \xi \langle F_1(g, \lambda, \xi), \psi_0\rangle. \tag{3.4.21}$$

By differentiation with respect to h, we find that

$$\left.\frac{\partial f_1}{\partial h}\right|_{h,\, \xi = 0} = -\frac{1}{\lambda_1}\cdot \tag{3.4.22}$$

Expression (3.4.22) shows that the implicit function theorem can be applied and therefore Equation (3.4.20) has a single solution $h = h(\xi)$ for small ξ[7]. The curve $h(\xi)$ corresponds to the branch

[7] Notice that the number of branches is given by that of the solutions $\xi(h)$ of Equation (3.4.20).

which appears from the solution $g = 0$ at the point λ_1. Let

$$I - \lambda A + \mathfrak{F}'_x(\tilde{x}, \lambda) \tag{3.4.23}$$

be the Fréchet differential of the map from the left-hand side of Equation (3.4.17)′ and let us call $(0, \lambda_1)$ a regular branch point if the map (3.4.23) is invertible for every λ sufficiently near λ_1 and $\lambda \neq \lambda_1$. In connection with the regular branch point [44] we have the following.

Lemma 3.4.1. *The point* $(0, \lambda_1)$ *is a regular branch point iff-* $h'(\xi) \neq 0$ *for small* ξ, $\xi \neq 0$. *In particular, the map* (3.4.23) *is inver tible only if* $h'(\xi) \neq 0$.

Lemma 3.4.2. *If the nonlinear map* $\mathfrak{F}$ *is analytic with respect to* x *and* λ, *then a sufficient condition for* $(0, \lambda_1)$ *to be a regular branch point is that the branching should not be vertical, i.e. for* $\lambda = \lambda_1$ *no family of solutions of Equation* (3.4.17)′ *depending continuously on one parameter should exist.*

In order to apply topological degree theory, we must find a number $\delta > 0$ such that for $\lambda - \lambda_1 < \delta$ in the neighbourhood of the origin $0 \in \mathfrak{B}$, there exists a domain $\Omega = \{x \in \mathfrak{B} \mid \|x\| < \varepsilon\}$ containing only the solutions of Equation (3.4.17)′ branching from λ_1, and such that on $\partial\Omega$ no solution of this equation should exist. In our case, it is sufficient to show that we do not have vertical branching, i.e. there exists a number $\varepsilon > 0$ such that the Equation (3.4.17)′ should have no solution with $\|x\| \leqslant \varepsilon$ for $\lambda = \lambda_1$. The proof runs by *reductio ad absurdum*, using the complete continuity of A and $\mathfrak{F}$. Therefore, $d(I - \lambda A, \Omega)$ exists and is constant for $\lambda_1 - \delta \leqslant \lambda \leqslant \lambda_1 + \delta$.

As in the finite-dimensional case (Section 3.1.1), it is shown that if λ_1 is a regular branch point, then in a small neighbourhood of $0 \in \mathfrak{B}$ and for λ near λ_1, besides the trivial solution, Equation (3.4.17) has either two supercritical solutions with the index $+1$ and none subcritical, or a supercritical solution of index $+1$ and another subcritical solution of index -1 or two subcritical solutions of index -1. The fact that their number is just 2 comes out from the Liapunov-Schmidt method. In our case, the solutions with a negative index are unstable, whereas those of positive index are stable, as follows from the following.

Theorem (*Sattinger*). *Let* $(0, \lambda_1)$ *be a regular branch point of Equation* (3.4.17)′ *and let* $\mathfrak{F}$ *be twice continuously differentiable with* $\mathfrak{F}(\xi x; \lambda) = \xi^2 \mathfrak{F}_1(x, \lambda, \xi)$ *where the map* $\mathfrak{F}_1$ *is Fréchet differentiable with respect to* x, λ *and* ξ. *Then the supercritical secondary solutions are stable and the subcritical ones are unstable.*

Proof. Let $\tilde{x}$ be a supercritical secondary solution with the index $+1$ and let the maps

$$A_3(s, \lambda) = A_1 - s\lambda_1 A_2 - s[hA_2 - F'_x(\tilde{x}, \lambda)],$$
$$A_4(s, \lambda) = I - s[\lambda A - \mathfrak{F}'_x(\tilde{x}, \lambda)],$$

which can be obtained by applying to the map (3.4.18) and (3.4.23) certains homotopies. For $\lambda = \lambda_1$, $s < 1$ and s very close to 1, by the assumption on the spectrum of the map (3.4.18), it follows that A_4 has the eigenvalue $\sigma(s, \lambda_1)$ on the left-side of the origin and very close to it, the rest of the stability spectrum lying on the left-side of the line $\sigma_r = \eta$. λ_1 is a simple characteristic value of $A - \frac{1}{\lambda}\mathfrak{F}'_x(\cdot, \lambda)$; hence, the map $\lambda A - \mathfrak{F}'_x(\tilde{x}, \lambda)$ has a simple eigenvalue ζ in the neighbourhood of $\zeta = 1$. As $(0, \lambda_1)$ is a regular branch point, it follows that for small $|\lambda - \lambda_1|$, we have $\zeta \neq 1$ if $\lambda \neq \lambda_1$ and from the properties of the topological degree we have that $\zeta < 1$ iff $\mathrm{i}[-\lambda A + \mathfrak{F}(\tilde{x}, \lambda), \tilde{x}] = +1$. That is why as λ slightly surpasses λ_1, then $\lambda_1(s, \lambda)$ varies continuously with respect to λ without touching the origin for small values of λ. By the above considerations, the map $\lambda A - \mathfrak{F}'_x(\tilde{x}, \lambda)$ has a simple eigenvalue $\zeta < 1$. Hence, as $s \to 1$, the eigenvalue of A_4 does not come to the origin and as A_3 and A_4 have the same eigenvalues, it follows also that for A_3 the eigenvalue does not come to the origin as $s \to 1$. Since by a homotopy the eigenvalue remains real, it follows that the map (3.4.18) has a negative eigenvalue, and hence $\tilde{x}$ is stable. The above reasoning can be done also for solutions with negative index; it is found that they are unstable as long as on the branch $\xi = \xi(h)$ we have $\frac{\partial f}{\partial \xi} \neq 0$.Thus the theorem is proved.

In the case of motion between rotating cylinders and for the convection, the hypotheses of the above theorem are satisfied. Hence the supercriterial solutions, which branch at the point where the linear stability is lost, are stable.

Sattinger's theorem has been generalized [45] to the case of n curves branching from $(0, \lambda_1)$ and for S-shaped branches. It has been applied to the derivation of the stability of some subcritical convective motions. The stability of these motions has been studied by Joseph [21] using the analytic perturbation methods.

The hypothesis that λ_1 is a simple eigenvalue was essential for the above considerations. The examination, in certain degeneracy hypotheses, by topological degree methods of the stability of the secondary solutions branching at λ_1 of multiplicity 2, shows [34] that there is a possibility for the supercritical secondary solutions to be unstable and the subcritical ones — stable, or that there exists no stable secondary solution at all. In the general case, i.e. when those degeneracies can appear and λ_1 are of arbitrary multiplicity, no results are available concerning the stability of the secondary solutions, deduced by topological arguments. We refer the reader interested in branching for multiple λ_1, to [34], [25], [42].

§ 3.5. HOPF BIFURCATION BY THE JOSEPH-SATTINGER METHOD

Suppose that the principle of exchange of stabilities does not hold but at $\lambda = \lambda_0 = \frac{1}{\mathrm{Re}_{cr}}$ the stability spectrum contains two complex conjugate simple eigenvalues with a vanishing real part, the other eigenvalues lying in the left-half plane. Then, at λ_0 two branches of time-periodic secondary solutions bifurcate from the basic solution of the Navier-Stokes equations. They are both either subcritically unstable or supercritically stable; moreover, for every fixed λ, the branching solution is unique up to phase shifts. These results have been obtained by means of the Joseph-Sattinger method and they concern general fluid flows [22], Poiseuille flow in pipes [30] (from Ch. 1) and Poiseuille flow in annular ducts [23].

The branching of (nonconstant) periodic solutions from stationary solutions is called *Hopf bifurcation*. For the case of ordinary differential equations the existence of this branching is given by the *Hopf bifurcation theorem*. This theorem has been extended to the case of the Navier-Stokes equations, among others, by Sattinger and Joseph [22].

The Joseph-Sattinger method generalizes to the infinite-dimensional case Hopf's method for ordinary differential equations, by combining the Poincaré-Lindstedt perturbation method with the ideas of the Liapunov-Schmidt method in order to reduce the bifurcation problem to an implicit functions theorem. After the presentation of Hopf theorem in Section 3.5.1 we shall analyse the Joseph-Sattinger method and in Section 3.5.2 — the linear stability of the secondary solutions, using the Floquet method.

3.5.1. **Deduction of secondary solutions**

Consider first the case of evolution equations in finite dimensional spaces, namely $\frac{dx}{dt} = f(x, \nu)$ where $x \in R^n$, $\nu \in I \subset R$. A solution $x_*(\nu)$ of $f(x, \nu) = 0$ is called a *stationary point* of this system. The number ν_c is called a *critical value* of ν if $\omega(\nu_c) = 0$ where $\omega(\nu)$ is the imaginary part of an eigenvalue of the Jacobian matrix $(D_x f(x_*(\nu), \nu)_{\substack{i=1,2,\ldots,n \\ j=1,2,\ldots,n}}$. Letting $X = x - x_*$ and $\mu = \nu - \nu_c$ the above equations becomes $(*)\ \frac{dX}{dt} = F(X, \mu)$. This equation has $X = 0$ as stationary point and $\mu = 0$ is the critical value. Then the following analytic version of Hopf's theorem holds.

Theorem *(Hopf)*. Suppose 1) $F(0, \mu) = 0$ for μ in an open interval containing 0 and $0 \in R^n$ is an isolated stationary point of F; 2) F is analytic in X and μ in a neighbourhood of (0, 0) in $R^n \times R^1$; 3) $A(\mu) = D_X F(0, \mu)$ has a pair of complex conjugate eigenvalues λ and $\bar{\lambda}$ such that $\lambda(\mu) = \alpha(\mu) + i\omega(\mu)$, where $\omega(0) = \omega_0 > 0$, $\alpha(0) = 0$, $\alpha'(0) = 0$; 4) the remaining $n-2$ eigenvalues of $A(0)$ have strictly negative real parts. Then the system (∗) has a family of periodic solutions: there is an $\varepsilon_H > 0$ and an analytic function $\mu^H(\varepsilon) = \sum_2^\infty \mu_i^H \varepsilon^i$, $(0 < \varepsilon < \varepsilon_H)$ such that for each $\varepsilon \in (0, \varepsilon_H)$ there exists a periodic solution $P_\varepsilon(t)$ occurring for $\mu = \mu^H(\varepsilon)$. If $\mu^H(\varepsilon)$ is not identically zero, the first nonvanishing coefficient μ_i^H has an even subscript, and there is an $\varepsilon_1 \in (0, \varepsilon_H]$ such that $\mu^H(\varepsilon)$ is either strictly positive or strictly negative for $\varepsilon \in (0, \varepsilon_1)$. For each $L > 2\pi/\omega_0$ there is a neighbourhood $\mathcal{N}$ of $X = 0$ and an open interval I containing 0 such that for any $\mu \in I$ the only nonconstant periodic solutions of (∗) with periods less than L which lie in $\mathcal{N}$ are members of the family $P(t)$ for values of ε satisfying $\mu^H(\varepsilon) = \mu$, $\varepsilon \in (0, \varepsilon_H)$. The period $T^H(\varepsilon)$ of $P_\varepsilon(t)$ is an analytic function $T^H(\varepsilon) = \frac{2\pi}{\omega_0}\left[1 + \sum_2^\infty \tau_i^H \varepsilon^i\right]$, $(0 < \varepsilon < \varepsilon_H)$. Exactly two of the Floquet exponents of $P_\varepsilon(t)$ approach 0 as $\varepsilon \downarrow 0$. One is 0 for $\varepsilon \in (0, \varepsilon_H)$, and the other is an analytic function $\beta^H(\varepsilon) = \sum_2^\infty \beta_i^H \varepsilon^i$, $(0 < \varepsilon < \varepsilon_H)$. The periodic solution $P_\varepsilon(t)$ is orbitally asymptotically stable with asymptotic phase if $\beta^H(\varepsilon) < 0$ but is unstable if $\beta^H(\varepsilon) > 0$.

So, Hopf's theorem states: if two complex conjugate eigenvalues of the linearized problem cross the imaginary axis at the bifurcation point while the rest of the spectrum remains in the left half complex plane, then a single branch of periodic solutions emerges. It is either subcritical or supercritical according to whether the sign of the first nonvanishing coefficient μ_i^H (which has even i) is negative or positive. The stability or instability is given by the sign of the Floquet exponent. The same assertions hold for the Navier-Stokes equations as will be show in the following.

Let $\tilde{\mathbf{u}}(\mathbf{x}, \mathrm{Re})$ be a steady solution of the Navier-Stokes equations in the bounded domain $\Omega \subset R^3$ and consider hydrodynamical problem linearized around this solution

(3.5.1) $$\frac{\partial \mathbf{u}}{\partial t} + A\mathbf{u} + \operatorname{grad} p = 0,$$

(3.5.2) $$\operatorname{div} \mathbf{u} = 0,$$

(3.5.3) $$\mathbf{u}|_{\partial\Omega} = 0,$$

where $A\mathbf{u} = -\dfrac{1}{\mathrm{Re}}\Delta\mathbf{u} + (\tilde{\mathbf{u}}\ \mathrm{grad})\ \mathbf{u} + (\mathbf{u}\ \mathrm{grad})\ \tilde{\mathbf{u}}$. Suppose that the map A has a pair of simple eigenvalues, i.e. the problem (3.5.1)—(3.5.3) possesses at $\mathrm{Re} = \mathrm{Re}_{cr}$ only two time-periodic solutions

$$\mathbf{u}_1 = \mathrm{e}^{-\gamma t}, \quad \mathbf{u}_2 = \overline{\mathbf{u}}_1,$$

where $\gamma = \xi + \mathrm{i}\eta$ and $\eta(\mathrm{Re}_{cr}) = 0$. These solutions satisfy the problem consisting of the equation

$$-\gamma\varphi + A\varphi + \mathrm{grad}\ p = 0 \tag{3.5.4}$$

and Conditions (3.5.2) and (3.5.3) (i.e. the linear stability problem). We shall show that at $\mathrm{Re} = \mathrm{Re}_{cr}$, periodic solutions $\mathbf{v}(\mathbf{x}, t, \mathrm{Re})$ with the period $2\pi\omega^{-1}(\mathrm{Re})$ branch from the solution $\tilde{\mathbf{u}}$; hence, at $\mathrm{Re} = \mathrm{Re}_{cr}$, we have $\xi(\mathrm{Re}_{cr}) \equiv \omega_0$, and $\mathbf{v}$ is a solution of the problem

$$\frac{\partial\mathbf{v}}{\partial t} + A\mathbf{v} + (\mathbf{v}\ \mathrm{grad})\mathbf{v} + \mathrm{grad}\ p_1 = 0, \tag{3.5.5}$$

$$\mathrm{div}\ \mathbf{v} = 0, \tag{3.5.6}$$

$$\mathbf{v}|_{\partial\Omega} = 0, \tag{3.5.7}$$

$$\mathbf{v}(\mathbf{x}, t, \mathrm{Re}) = \mathbf{v}(\mathbf{x}, t + 2\pi\omega^{-1}, \mathrm{Re}). \tag{3.5.8}$$

Let P_2 be the space of time periodic vector functions with the period $2\pi\omega^{-1}$, considered to be a Hilbert space with respect to the norm generated by the scalar product $(\mathbf{a}, \mathbf{b})_{P_2} \equiv \dfrac{\omega}{2\pi}\displaystyle\int_0^{\frac{2\pi}{\omega}} (\mathbf{a}, \mathbf{b})\,\mathrm{d}t$, $\mathbf{a}, \mathbf{b} \in P_2$, where $(\mathbf{a}, \mathbf{b}) \equiv \displaystyle\int_\Omega \mathbf{a}\overline{\mathbf{b}}\,\mathrm{d}\mathbf{x}$, and let $\mathbf{u}_i^*(\mathrm{Re}_{cr})$ be the eigenfunctions of the adjoint map A^* defined by the relation

$$(A\mathbf{a}, \mathbf{b}) = (\mathbf{a}, A^*\mathbf{b}).$$

By definition, let ε be the Poincaré-Lindstedt parameter

$$\varepsilon \equiv (\mathbf{v}, \mathbf{u}_1^*(\mathrm{Re}_{cr}))_{P_2}. \tag{3.5.9}$$

We use the notation

$$t = s\omega^{-1}, \quad \mathbf{v} = \varepsilon\hat{\mathbf{u}}(\mathbf{x}, s, \varepsilon), \quad p_1 = \varepsilon\hat{p}(\mathbf{x}, s, \varepsilon), \tag{3.5.10}$$
$$\omega = \omega(\varepsilon), \quad \mathrm{Re} = \lambda^{-1}(\varepsilon),$$

where, as $\varepsilon \to 0$, we have $\mathrm{Re} \to \mathrm{Re}_{cr}$, $\lambda \to \lambda_0$, $\hat{\mathbf{u}} \to \mathbf{u}_0(\mathbf{x}, s)$, $\hat{p} \to p_0(\mathbf{x}, s)$, $\omega \to \omega_0$. Then Equation (3.5.5) becomes

$$\omega\frac{\partial\hat{\mathbf{u}}}{\partial s} + A\hat{\mathbf{u}} + \varepsilon(\hat{\mathbf{u}}\ \mathrm{grad})\ \hat{\mathbf{u}} + \mathrm{grad}\ \hat{p} = 0. \tag{3.5.11}$$

As $\varepsilon \to 0$, we obtain the equation

$$\omega_0 \frac{\partial \mathbf{u}_0}{\partial s} + A_0 \mathbf{u}_0 + \operatorname{grad} p_0 = 0, \tag{3.5.12}$$

(where $A_0 = A(\lambda_0) = -\lambda_0 \Delta \cdot + (\widetilde{\mathbf{u}}_0 \operatorname{grad}) \cdot + (\cdot \operatorname{grad}) \widetilde{\mathbf{u}}_0$, $\lambda_0 = \mathrm{Re}_{cr}^{-1}$), which has the real solution

$$\mathbf{u}_0 = 2\mathscr{Re}\{\mathrm{e}^{-\gamma s}\overline{\zeta}\}.$$

As $\varepsilon \to 0$, the secondary nonstationary solutions tend to the null solution and the perturbed motion to $\widetilde{\mathbf{u}}$.

Taking into account the above notation and putting $\mu = \lambda_0 - \lambda$, the problem (3.5.5)—(3.5.8) becomes

$$\omega \frac{\partial \hat{\mathbf{u}}}{\partial s} + A_0 \hat{\mathbf{u}} + \mu \Delta \hat{\mathbf{u}} + \varepsilon (\hat{\mathbf{u}} \operatorname{grad}) \hat{\mathbf{u}} + \operatorname{grad} \hat{p} = 0,$$

$$\operatorname{div} \hat{\mathbf{u}} = 0, \quad \hat{\mathbf{u}}|_{\partial\Omega} = 0,$$

$$\hat{\mathbf{u}}(\mathbf{x}, s, \varepsilon) = \hat{\mathbf{u}}(\mathbf{x}, s + 2\pi, \varepsilon), \quad (\hat{\mathbf{u}}(\varepsilon), \mathbf{u}_1^*(\mathrm{Re}_{cr}))_{P_2} = 1.$$

Expanding the solution of this problem into a Taylor series

$$\begin{Bmatrix} \hat{\mathbf{u}}(\mathbf{x}, t, \varepsilon) \\ p(\mathbf{x}, t, \varepsilon) \\ \omega(\varepsilon) \\ \mu(\varepsilon) \end{Bmatrix} = \sum_{l=0}^{\infty} \varepsilon^l \begin{Bmatrix} \mathbf{u}_l(\mathbf{x}, t) \\ \hat{p}_l(\mathbf{x}, t) \\ \omega_l \\ \mu_l \end{Bmatrix},$$

where $\mu_0 = 0$, $\mu_l = -\lambda_l$ and $\hat{\mathbf{u}}_l = \dfrac{1}{l!} \dfrac{\partial^l \hat{\mathbf{u}}(\mathbf{x}, t, 0)}{\partial \varepsilon^l}$, we find that $\mathbf{u}_0$ satisfies Equation (3.5.12), the solenoidality conditions, and vanishes at the boundary. Moreover $\mathbf{u}_0(s) = \mathbf{u}_0(s + 2\pi)$, $(\mathbf{u}_0, \mathbf{u}_1^*)_{P_2} = 1$, and $\mathbf{u}_m (m > 0)$ are solutions of the problem

$$\omega_0 \frac{\partial \mathbf{u}_m}{\partial s} + A_0 \mathbf{u}_m + \omega_m \frac{\partial \mathbf{u}_0}{\partial s} - \lambda_m \Delta \mathbf{u}_0 + \mathbf{F}_m + \operatorname{grad} p_m = 0,$$

$$\operatorname{div} \mathbf{u}_m = 0, \tag{3.5.13}$$

$$\mathbf{u}_m|_{\partial\Omega} = 0, \quad \mathbf{u}_m(s) = \mathbf{u}_m(s + 2\pi), \quad (\mathbf{u}_m, \mathbf{u}_1^*)_{P_2} = 0,$$

where

$$\mathbf{F}_m = \sum_{\substack{l+j=m-1 \\ l, j \geqslant 0}} (\mathbf{u}_l \cdot \operatorname{grad}) \mathbf{u}_j + \sum_{\substack{l+j=m \\ l,j \geqslant 0}} \left(\omega_l \frac{\partial \mathbf{u}_j}{\partial s} - \lambda_l \Delta \mathbf{u}_j \right).$$

We notice that the role of the first eigenvalue λ_1 of the preceding sections is taken here by λ_0, which is the first approximation in the above perturbation scheme. We consider $\tilde{\mathbf{u}}$ as independent of Re (i.e. $\tilde{\mathbf{u}}_0 = \tilde{\mathbf{u}}$); the general case may be treated in an analogous manner [22]. To system (3.5.13) we must add the orthogonality conditions

$$(3.5.14) \qquad (\lambda_m \Delta \mathbf{u}_0 - \mathbf{F}_m - \operatorname{grad} p_m, \quad \mathbf{u}_1^*(\mathrm{Re})_{cr}))_{P_2} = 0, \quad (\mathrm{i} = 1, 2),$$

which represent the necessary and sufficient condition for the uniqueness of the solution.

In the above we decomposed the solution $\hat{\mathbf{u}}$ into the sum

$$\hat{\mathbf{u}} = \hat{\mathbf{u}}_0 + \mathbf{w},$$

where $(\mathbf{u}_0, \mathbf{u}_1^*(\mathrm{Re}_{cr}))_{P_2} = 1$ and

$$(3.5.15) \qquad (\mathbf{w}, \mathbf{u}_1^*) \ \mathrm{Re}_{cr}))_{P_2} = 0;$$

such that the equation in $\mathbf{u}$ splits into an equation in $\mathbf{w}$ and an orthogonality condition. In other words, if $A : \mathscr{B}_1 \to \mathscr{B}_2$ is the map considered, we decompose both $\mathscr{B}_1$ and $\mathscr{B}_2$ in direct sums of orthogonal subspaces, using two projection maps defined by means of the eigenvectors of A. We obtain in this way a system analogous to $(3.3.11)'''$; the equation analogous to $(3.3.11)_1'''$ is that from which we obtain system (3.5.13) by seeking a power series expression for $\mathbf{w}$. Conditions (3.5.14) are analogous to Equation $(3.3.11)_2'''$.

Let us now deduce the branching equation. First note that $\mathbf{F}_1 = (\hat{\mathbf{u}}_0 \operatorname{grad}) \hat{\mathbf{u}}_0$, whence $(\mathbf{F}_1, \mathbf{u}_1^*(\mathrm{Re}_{cr}))_{P_2} = 0$. Therefore,

$$(3.5.16) \qquad \omega_1 = \lambda_1 = 0.$$

On the other hand, we have

$$\mathbf{F}_2 = (\mathbf{u}_0 \operatorname{grad}) \mathbf{u}_1 + (\mathbf{u}_1 \operatorname{grad}) \mathbf{u}_0, \quad \omega_2 = \mathscr{I}m(\mathbf{F}_2, \mathbf{u}_1^*(\mathrm{Re}_{cr}))_{P_2}$$

and

$$(3.5.17) \qquad \lambda_2 = \frac{\mathscr{R}e\{(\mathbf{F}_2, \mathbf{u}_1^*(\mathrm{Re}_{cr}))_{P_2}\}}{\xi'}.$$

Therefore, the perturbation method can be applied if $\xi_i' \neq 0$, where $\xi' = \dfrac{d\xi}{d\lambda}$. It is easy to be proved [22] that for a wide class of operators (including the Navier-Stokes operator), we have $\dfrac{d\xi}{d\lambda} > 0$ at $\xi = 0$ and $\lambda = \lambda_0$, hence λ_2 can be calculated. Since $\lambda_1 = 0$, the branching equation

$$(3.5.18) \qquad \lambda(\varepsilon) = \frac{1}{\mathrm{Re}(\varepsilon)} = \frac{1}{\mathrm{Re}_{cr}} + \lambda_1(\varepsilon) + \lambda_2(\varepsilon^2) + 0(\varepsilon^2)$$

admits two solutions $\varepsilon_1(\lambda)$ and $\varepsilon_2(\lambda)$. Therefore, branching takes place at λ_0. Since the solution of the system (3.5.13) exists for those λ which appear at ε_1, and ε_2 and are given by (3.5.18), it follows that the branching is either subcritical (when $\lambda_2 > 0$) or supercritical (when $\lambda_2 < 0$).

For plane Couette-Poiseuille flow, the effective numerical calculation of λ_2 is carried out in [41], the branching of this motion being found to be subcritical, which coincides with the experimental results.

3.5.2. Stability of secondary solutions

Let us consider the time-periodic solution found in the previous section, $\mathbf{u} = \tilde{\mathbf{u}} + \hat{\mathbf{v}}$. Superpose on the motion corresponding to $\tilde{\mathbf{u}}$ an infinitesimal perturbation $\mathbf{u}$, which satisfies the linearized problem

$$\omega \frac{\partial \mathbf{u}}{\partial s} + \mathbf{A}(\lambda)\,\mathbf{u} + (\hat{\mathbf{v}} \text{ grad})\,\mathbf{u} + (\mathbf{u} \text{ grad})\,\hat{\mathbf{v}} + \text{grad}\, p = 0, \tag{3.5.19}$$

$$\text{div}\,\mathbf{u} = 0, \quad \mathbf{u}|_{\partial\Omega} = 0.$$

Let us look for the solution of this problem in the form

$$\mathbf{u} = e^{-\sigma s}\mathbf{\Gamma}(\mathbf{x}, s, \varepsilon), \quad p' = e^{-\sigma s} p(\mathbf{x}, s, \varepsilon), \tag{3.5.20}$$

where the complex number $\sigma(\lambda)$ is called the Floquet exponent and $\mathbf{\Gamma}$ and p are periodic with respect to s, having the period 2π. From (3.5.19) and (3.5.20) we obtain that

$$-\sigma\omega(\varepsilon)\,\mathbf{\Gamma} + \omega \frac{\mathrm{d}\mathbf{\Gamma}}{\mathrm{d}s} + A\mathbf{\Gamma} + (\hat{\mathbf{v}} \text{ grad})\,\mathbf{\Gamma} + (\mathbf{\Gamma} \text{ grad})\,\hat{\mathbf{v}} + {} \tag{3.5.21}$$

$$+ \text{grad}\, p = 0,$$

hence the stability of the solution $\tilde{\mathbf{u}} + \hat{\mathbf{v}}$ to infinitesimal perturbations is given by the sign of the real parts of the eigenvalues σ of Equation (3.5.21).

For $\varepsilon = 0$ ($\lambda = \lambda_0$) we have $\hat{\mathbf{v}} = 0$, $\omega = \omega_0$, $A = A_0$ and Equation (3.5.21) becomes

$$-\sigma\mathbf{\Gamma}\omega_0 + \omega_0 \frac{\partial \mathbf{\Gamma}}{\partial s} + A_0\mathbf{\Gamma} + \text{grad}\, p = 0, \tag{3.5.22}$$

which admits solutions of the form $\mathbf{\Gamma}(\mathbf{x}, s, 0) = e^{iks}\mathbf{\Phi}(\mathbf{x})$, where k is an integer. Then $\mathbf{\Phi}$ is the eigensolution corresponding to the eigenvalue σ of the problem

$$-\sigma\mathbf{\Phi} + (\mathrm{i}\, k\omega_0)\,\mathbf{\Phi} + A_0\mathbf{\Phi} + \text{grad}\, p = 0, \tag{3.5.23}$$

$$\text{div}\,\mathbf{\Phi} = 0, \quad \mathbf{\Phi}|_{\partial\Omega} = 0.$$

In other words $-\sigma + ik\omega_0$ is a an eigenvalue γ of A_0. Hence, for $\varepsilon = 0$, we have

$$\sigma = -\gamma + \mathrm{i}\, k\omega_0. \tag{3.5.24}$$

Since, by hypothesis, all but two of the eigenvalues of A_0 have negative real parts, the same is true for the Floquet exponents. For the two Floquet coefficients at the origin ($\sigma = 0$), the corresponding solutions of Equation (3.5.22) are

$$\mathbf{\Gamma}_1 = \mathrm{e}^{-\mathrm{i}s}\xi(\mathbf{x}) \quad \text{and} \quad \mathbf{\Gamma}_2 = \mathrm{e}^{\mathrm{i}s}\xi(\mathbf{x}).$$

Let us examine now the Floquet exponents for $\varepsilon \neq 0$.

One of the Floquet coefficients, $\sigma = 0$, for which the system (3.5.19) admits the solution $\mathbf{u} = \dfrac{\mathrm{d}\hat{\mathbf{u}}}{\mathrm{d}s}$, with $\hat{\mathbf{u}}$ defined in Section 3.5.1, remains at the origin even for $\varepsilon \neq 0$. To study the other Floquet exponent for which $\sigma = 0$ at $\varepsilon = 0$, we look for solutions of Equation (3.5.21) in the form

$$\mathbf{\Gamma}(\mathbf{x}, s, \varepsilon) = a(\varepsilon)\, \frac{\partial \hat{\mathbf{u}}}{\partial s} + \boldsymbol{\gamma}\,(\mathbf{x}, s, \varepsilon), \tag{3.5.25}$$

where $\boldsymbol{\gamma}(\mathbf{x}, s, \varepsilon) = \mathbf{u}_0(\mathbf{x}, s) + \varepsilon\boldsymbol{\gamma}_1(\mathbf{x}, s)$. Therefore, $\boldsymbol{\gamma}$ satisfies the equation

$$\omega \frac{\partial \boldsymbol{\gamma}}{\partial s} - \sigma\omega\boldsymbol{\gamma} + A\boldsymbol{\gamma} + \varepsilon\,[(\hat{\mathbf{u}}\ \mathrm{grad})\,\boldsymbol{\gamma} + (\boldsymbol{\gamma}\ \mathrm{grad})\,\mathbf{u}] -$$

$$- \sigma\omega a\, \frac{\partial \hat{\mathbf{u}}}{\partial s} + \mathrm{grad}\, p = 0,$$

and to this equation we apply a perturbation scheme

$$\begin{Bmatrix} \boldsymbol{\gamma}(\mathbf{x}, s, \varepsilon) \\ p(\mathbf{x}, s, \varepsilon) \\ \sigma(\varepsilon) \\ a(\varepsilon) \end{Bmatrix} = \sum_{l=0}^{\infty} \varepsilon^l \begin{Bmatrix} \boldsymbol{\gamma}_l(\mathbf{x}, s) \\ p_l(\mathbf{x}, s) \\ \sigma_l \\ a_l \end{Bmatrix},$$

where, as in the Section 3.5.1, $\boldsymbol{\gamma}$ and p are periodic with respect to s, having the period 2π, and satisfing the conditions

$$(\boldsymbol{\gamma}_0, \mathbf{u}_1^*(\mathrm{Re}_{cr})) = 1, \quad (\boldsymbol{\gamma}_l, \mathbf{u}_1^*(\mathrm{Re}_{cr})) = 0, \quad (l > 0).$$

Writing explicitly the problems satisfied by σ_l and $\boldsymbol{\gamma}_l$ and comparing problems for $\mathbf{u}_i$, we find $\boldsymbol{\gamma}_0 = \mathbf{u}_0$, $\boldsymbol{\gamma}_1 = 2\mathbf{u}_1$, $\sigma_0 = 0$ and

$$\sigma_2 = -\frac{2\lambda_2 \xi'}{\omega_0}. \tag{3.5.26}$$

Formula (3.5.26) shows that the subcritical solutions (for which $\lambda_2 < 0$, see Section 2.5.1) are stable (since they have $\sigma = \sigma_2\varepsilon^3 + O(\varepsilon^3) > 0$) and the subcritical solutions are unstable.

Among the studies on the branching and stability of the periodic solutions we quote also [11], [12], [13].

§ 3.6. GENERATION OF TURBULENCE BY INSTABILITY AND LOCAL BRANCHING

The Navier-Stokes equations for some domains of motion admit a basic solution for every value of the physical parameter $\lambda = \mathrm{Re}$ (or Ra), but this solution is unique only for some intervals of variation of λ, and the motion observed in nature corresponds to the basic solution only for small λ. The branching and stability theory allow us to choose, from the set of solutions corresponding to a given λ, that one which has a physical reality.

A complete study of the ramified solutions includes that of the secondary solutions, of the solutions branching from these last ones and so on; in the literature there are only a few papers [22], [21], [23] in which secondary solutions are studied. On the other hand, all considered motions have λ_1 as a simple eigenvalue. The existing studies concern mainly the steady basic flows which are described by exact solutions of the Navier-Stokes equations. Finally we recall that branching theory does not provide an answer to the full uniqueness problem; the fact that no other solutions exist apart from the branching ones has been proved only for convective motions in bounded domains [19].

The loss of linear stability at $\lambda = \lambda_1(\mathrm{Re}_{cr}$ or $\mathrm{Ra}_{cr})$ may occur in one of the following ways: a) by continuous amplification of the small perturbations, eventually the motion becoming turbulent; b) by the appearance of new steady motions sub-, super — or sub — and supercritical ones; c) by the appearance of new time — periodic sub — or supercritical motions. In their turn, the last two cases may lead to the following situations:

b$_1$) For some motions (for instance, Couette flow), at λ_1 a branch of stable, steady, secondary flows appear, which, for each λ, correspond to a real motion. It is possible that for some $\lambda' > \lambda_1$, the secondary flow branch gives rise to another branch of stable steady solutions which exist for $\lambda > \lambda'$ and so on, so that transition to turbulence takes place by repeated branchings.

b$_2$) The equilibrium of the fluid may become unstable by the appearance of two branches of supercritical, linearly stable, steady secondary solutions. The actual motion corresponds to that solution which is nonlinearly stable and hence, has a larger domain of attraction. This is, for instance, the case with convection.

b_3) At λ_1, from the basic solution a supercritically branch of linearly stable, steady solutions and another subcritical branch of unstable, steady solutions appears. In the supercritical regime, a repeated branching takes place; therefore, if the basic flow is not the actual one, in the subcritical regime a branching of secondary nonlinearly stable solutions occurs which correspond to the actual flow.

c_1) At λ_1, from the basic flow a supercritical branch of time-periodic, linearly stable solutions appears, which describe the actual motion (Hopf bifurcation).

c_2) At λ_1, from the basic flow (for instance the Poiseuille flow), there appears a subcritical branch of time-periodic, linearly unstable solutions, which for some λ co-exist with the linear stable basic flow and with another stable motion, branched from a time-periodic secondary solution. In this case, the actual flow is described by the linearly stable solution having the largest zone of attraction.

In nature, the branching process does not remain the same from the first stages of transition ($\lambda = \lambda_1$) to the appearance of turbulence; a stationary motion may be replaced by time-periodic flows, these last ones, in turn, by other almost-periodic solutions etc., such that the turbulence is described by functions admitting Fourier transformations.

It follows that the supercritical branching process may be described by the *Landau-Hopf conjecture*[8)] [30], [8] regarding the passing to the turbulence: for $0 \leqslant \lambda < \lambda_1$ the fluid flow is unique and stable; at λ_1 by the loss of stability appear new stable steady solutions. The process continues with a repeated branching accompanied by loss of stability of the former solutions and the appearance of some other time-periodic, steady solutions and also by the appearance of time-almost periodic solutions branching from the periodic solutions. Number $N(\lambda)$ of branching solutions is an increasing function of λ which remains constant for λ belonging to the intervals between the eigenvalues of the Stokes map, it has a jump when λ crosses such a value and it becomes infinite as $\lambda \to \infty$.

The branching solutions may be extended by continuity for every λ, but among all the solutions existing at a given λ, only those are stable which newly bifurcated from the last eigenvalue smaller than λ. As λ is increased and therefore as we approach the turbulent regime, the concept of stability no longer refers to a single solution but to the entire set of solutions existing at a given λ, so that, according to the Landau-Hopf conjuncture, turbulence is reached when there exists a stable set of solutions possessing certain common statistical properties and forming together the so-called steady turbulent solution.

---

8) For the mathematical justification of the equation proposed by Landau see [26] and [33] from Ch. 2.

For the case of two-dimensional flows, Foiaş and Prodi [20] (from Ch. 2) have proved that the set of all stationary solutions of the hydrodynamic mathematical problem is homeomorphic with a bounded set of Euclidean space of finite dimension $n(\mathrm{Re})$, where $n(\mathrm{Re})$ increase as Re is increased and as $t \to \infty$, the convergence of the nonsteady solutions follows from the convergence of its projection on a finite-dimensional space [29]. These results give to the Landau-Hopf conjecture an increased degree of plausibility. Nevertheless neither they, nor the branching theory (at its today stage) prove it completely. A further discussion in a larger context on Landau-Hopf conjecture and Foiaş and Prodi results may be found in Ch. 4.

In the subcritical regime, the instability of the secondary time-periodic solutions is explained by the "snap through" instability. Let λ' $(< \lambda_1)$ be the global stability limit, such that for $0 \leqslant \lambda < \lambda'$, all the perturbations are damped out. For every given λ, $\lambda' < \lambda < \lambda_1$ three motions exist; the basic one, a time-periodic Tollmien-Schlichting wave (branched at λ_1 from the basic flow) and the "stable turbulent flow" branched at λ' from a Tollmien-Schlichting wave. Among these flows, the first and the third ones are linearly stable, the actual flow being the turbulent flow. Hence, at a given λ, no finite perturbation can be kept in the domain of attraction of the basic flow. It passes through the Tollmien-Schlichting wave, and enters the zone of attraction of the turbulent solution and for $\lambda \to \infty$, it attains a steady limit state, contained in the stable turbulent solution [30] (from Ch. 1).

REFERENCES

[1] Babskii, V. G., *On the appearance of the steady convection in a heated fluid layer lying in a selfgravitational field*, 3rd Scientific Conference of young mathematicians from Ukraina, Naukova Dumka, Kiev, 1967, 184–189. (In Ukrainian).

[2] Bartle, R. G., *Singular points of functional equations*, "Trans. Amer. Math. Soc.", **75**, 2, 1953, 366–384.

[3] Bénard, H., *Les tourbillons cellulaires dans une nappe liquide*, Rev. Générale Sci. Pure Appl.", **11**, 1900, 1261–1328.

[4] Fife, P. C., D. D. Joseph, *Existence of convective solutions of the generalized Bénard problem which are analytic in their norm*, "Arch. Rational Mech. Anal.", **33**, 2, 1969, 116–138.

[5] Gantmacher, F. R., M. G. Krein, *Oscillation matrices and small vibrations of mechanical systems*, Gostehizdat, 1950. (In Russian).

[6] Görtler, H., K. Kirchgässner, P. Sorger, *Branching solutions of the Bénard problem*, "Problems of Hydrodynamics and Continuum Mechanics", Nauka Moscow, 1969.

[7] Heinz, E., *An elementary analytic theory of the degree of mapping in n-dimensional space*, "J. Math. Mech.", **8**, 1959, 231–247.

[8] Hopf, E., *A mathematical example displaying features of turbulence*, "Comm. Appl. Math.", **1**, 4, 1948, 303–322.

[9] Iooss, G., *Théorie non linéaire de la stabilité des écoulements laminaires, dans le cas de l'échange des stabilités,* "Arch. Rational Mech. Anal.", **40,** 3, 1971, 166—208.

[10] Iooss, G., *Sur la stabilité de la solution périodique secondaire intervenant dans certains problèmes d'évolution,* "C.R. Acad. Sc. Paris", Série A, **273,** 1971, 912—915.

[11] Iooss, G., *Bifurcation des solutions périodiques de certains problèmes d'évolution,* "C. R. Acad. Sc. Paris", Série A, **273,** 1971, 624—627.

[12] Iooss, G., *Stabilité de la solution périodique secondaire intervenant dans certaines problèmes d'évolution,* "C. R. Acad. Sc. Paris", Série A, **274,** 1972, 108—111.

[13] Ivanilov, Yu. I., G. N. Yakovlev, *On bifurcation of the fluid flow between rotating cylinders,* Prikl. Mat. Meh., **30,** 4, 768—773 (1966). (In Russian).

[14] Yudovich, V. I., *Example of generation of the steady or periodic secondary flow by loss of stability of laminar flow of the viscous incompressible fluid,* Prikl. Mat. Meh., **29,** 3, 453—467 (1965). (In Russian).

[15] Yudovich, V. I., *On bifurcation of rotational fluid flows,* Dokl. Acad. Nauk SSSR, **169,** 2, 306—309 (1966). (In Russian)

[16] Yudovich, V. I., *Secondary flow and fluid instability between rotating cylinders,* Prikl. Mat. Meh., **30,** 4, 688—698 (1966). (In Russian).

[17] Yudovich, V. I., *On the origin of convection,* Prikl. Mat. Meh., **30,** 6, 1000—1005 (1966). (In Russian).

[18] Yudovich, V. I., *An example of loss of stability and of generation of secondary flow for fluid in bounded containers,* Prikl. Mat. Meh., **74,** (**116**), 4, 565—579 (1967). (In Russian).

[19] Yudovich, V. I., *Free convection and branching,* Prikl. Mat. Meh., **31,** 1, 101—111 (1967). (In Russian).

[20] Jacob C., *Sur la détermination des fonctions harmoniques conjuguées par certains conditions aux limites, Application à l'hydrodynamique.* Thèse, 1935.

[21] Joseph, D. D., *Stability of convection in containers of arbitrary shape,* „J. Fluid Mech.1, **47,** 2, 1971, 257—282.

[22] Joseph, D. D., D. H. Sattinger, *Bifurcating time periodic solutions and their stability,* "Arch. Rational Mech. Anal.", **45,** 1972, 79—109.

[23] Joseph, D. D., T. S. Chen, *Friction factors in the theory of bifurcating Poiseuile flow through annular ducts,* "J. Fluid Mech.", **66,** 1, 1974, 189—207.

[24] Kirchgässner, K., *Verzweigungslösungen einer stationären hydrodynamischen Randwertproblems,* Habilitation, Freiburg, 1966.

[25] Kirchgässner, K., *Multiple eigenvalue bifurcation for holomorphic mappings,* Contributions to nonlinear functional analysis, ed. Zarantonello, Academic Press, 1971, 69—99.

[26] Kirchgässner, K., H. Kielhöfer, *Stability and bifurcation in fluid dynamics,* "Rocky Mountain J. Math.", **3,** 2, 1973, 275—318.

[27] Krasnosel'skii, M. A., *Topological methods in the theory of nonlinear integral equations,* Gosud. Izd. Teh.-Teoret. Lit., Moscow, 1956. (In Russian).

[28] Krei, M. G., *On non-symmetric oscillation Green functions for ordinary differential operators,* Dokl. Akad. Nauk SSSR, **25,** 8, 643—648 (1939). (In Russian).

[29] Ladyzenskaya, O. A., *The survey of the results and urgent problems connected with the Navier-Stokes equations (On the hydrodynamic stability),* X^{th} Symposinm of Fluid Dynamics, Rynia (Poland), 1971.

[30] Landau, L., *On the problem of turbulence,* "C.R. Acad. Sc. USSR", **44,** 1944, 311—314.

[31] Leray, J., J. Schauder, *Topologie et équations fonctionnelles,* "Annales, Sc. de l'Ecole Normale Supérieure", Ser. 3, **51,** 1934, 45—78.

[32] Lichtenstein, L., *Vorlesungen über einige Klassen nichtlinearer Integralgleichungen und Integro-Differentialgleichungen nebst Anwendungen,* Berlin, 1931.

[33] Liapunov, A. M., *Sur les figures d'équilibre peu différentes des ellipsoïdes d'une masse liquide homogène douée d'un mouvement de rotation,* P. I.,„Zap. Akad. Nauk., S. Peterburg", 1, (1906).

[34] Mc Leod, J. B., D. H. Sattinger, *Loss of stability and bifurcations at a double eigenvalue,* "J. Funct. Anal.", **14,** 1, 1973, 62—84.

[35] Nirenberg, L., *Remarks on strongly elliptic partial differential equations,* "Comm. Pure Appl. Math.", **8,** 1955, 648—674.

[36] Odqvist, F. K. G., *Über die Randwertaufgaben der Hydrodynamik zäher Flüssigkeiten*, "Math., Zeitschr"., **32**, 1930, 329—375.

[37] Ovchinnikova, S. N., V. I. Yudovich, *Calculation of the secondary stationary flow between rotating cylinders*, Prrikl. Mat. Meh., **32**, 5, 1968, 858—868.

[38] Pascali, D., *Nonlinear mappings*, Bucharest, Ed. Acad. R.S.R., 1974. (In Romanian).

[39] Rabinovitz, P. H., *Existence and nonuniqueness of rectangular solutions of the Bénard problem*, "Arch. Rational Mech. Anal." **29**, 1, 1968, 32—57.

[40] Rabinovitz, P. H., *Some aspects of nonlinear eigenvalue problems*, "Rocky Mountain J. Math.", **3**, 2, 1973, 161—202.

[41] Reynolds, W. C., M. C. Potter, *Finite-amplitude instability of parallel shear flows*, "J. Fluid Mech.", **27**, 1967, 465—492.

[42] Sather, D., *Branching of solutions of nonlinear equations*, "Rocky Mountain J. Math.", **3**, 2, 1973, 203—250.

[43] Sattinger, D. H., *Bifurcation of periodic solutions of the Navier-Stokes equation*, "Arch. Rational Mech. Anal.", **41**, 1, 1971, 66—80.

[44] Sattinger, D. H., *Stability of bifurcating solutions by Leray-Schauder degree*, "Arch. Rational Mech. Anal.", **43**, 2, 1971, 154—166.

[45] Sattinger, D. H., *Stability of solutions of nonlinear equations*, "J. Math.Anal. Appl.", **39**, 1, 1972, 1—12.

[46] Sattinger, D. H., *Topics in stability and bifurcation theory*, Lecture Notes in Mathematics, nr. 309, Springer, Berlin — Heidelberg, N.Y., 1973.

[47] Schwartz, J. T., *Nonlinear functional analysis*, Lecture Notes, N.Y.U., (1963—1964).

[48] Sorokin, V. S., *On stationary fluid flows heated from below*, Prikl. Mat. Meh., **18**, 2, 197—204 (1954). (In Russian).

[49] Sorokin, V. S., *Nonlinear phenomena in bounded flows around critical Reynolds numbers*, Prikl. Mat. Meh., **25**, 2, 248—258 (1961). (Iu Russian).

[50] Uhovskii, M. R., V. I. Yudovich, *On equations of stationary convection*, Prikl. Mat. Meh., **27**, 2, 295—300 (1963). (In Russian).

[51] Vainberg, M. M., V. A. Trenoghin, *Liapunov and Schmidt methods in the theory of nonlinear equations and their further development*, Uspehi, Mat. Nauk, **17**, 2 (104), 13—75 (1962). (In Russian).

[52] Vainberg, M. M., V. A. Trenoghin, *The theory of the branching of the solutions of nonlinear equations*, Nauka, Moscow, 1969. (In Russian).

[53] Velte, W., *Stabilitätsverhalten und Verzweigung stationärer Lösungen der Navier-Stokesschen Gleichungen*, "Arch. Rational Mech. Anal.", **16**, 2, 1964, 97—125.

[54] Veronis, G., *The analogy between rotating and stratified fluids*, "Annual Review of Fluid Mechanics", **2**, 1970, 37—67.

Chapter 4

NATURE OF TURBULENCE

Among the theoreticians of the Navier-Stokes equations there is an increasing belief that the last stage of transition and the onset of turbulence may be explained by phenomenological theories based on these equations. All these lines concern only the temporal (and not spatial) irregularities of the solutions and are framed in the dynamical systems theory.

In this chapter we present three of these phenomenological theories: analyticity properties of the Leray turbulent solutions (§ 4.1); the Landau- Hopf conjecture on the (infinitely many times) repeated branching of the solutions of the Navier-Stokes equations (§ 4.2); the Ruelle-Takens theory, according to which after a finite number of branchings the solutions of the Navier-Stokes equations get trapped into a strange attractor (§ 4.3). The stochastic properties of the Lorenz strange attractor (corresponding to a truncation of the Navier-Stokes equations) are presented in Section 4.3.2. Section 4.4. presents the generic finiteness of the set of stationary solutions of the boundary value problem for the Navier-Stokes equations. A new trend in bifurcation theory mainly developed by Sattinger concerns the connection between bifurcation and symmetry breaking and pattern formation; Section 4.5 gives some ideas in this respect. In § 4.6 the above theories are commented upon and their connectiones with statistical theories and approximate approaches often used in turbulence are discussed.

§ 4.1. **LERAY MODEL**

The first phenomenological model intended to describe the onset of turbulence by loss of analyticity was given by Leray in 1933 (Section 2.2.3). It was the first time that the velocity of the fluid flows was modelled by an element of $L^2(\Omega)$. This velocity ceased to satisfy the Navier-Stokes equations in the usual sense. The sufficient smoothness of this generalized velocity would turn this velocity into a classical solution of the Navier-Stokes equations which is the case for small times. So, turbulence appears if for finite times the classical Navier-Stokes equations break down.

The studies initiated by Leray in hydrodynamics concern bounded domains of motion or the exterior of many bounded domains. The analyticity of the turbulent Leray solutions is still an open problem. It seems that for flows in domains with noncompact boundary there may exist generalized solutions which are not classical [18].

For finite initial energy ($\|\mathbf{v}_0\| < \infty$) Leray stated: *There exists a finite or countable sequence* $\mathrm{J}_0, \mathrm{J}_1, \ldots$ *such that* $\mathrm{J}_\mathrm{q} \subset \mathrm{R}^+ = \{t \in \mathrm{R} \mid \mathrm{t} \geqslant 0\}$, $\mathrm{J}_0 = \{\mathrm{t} \mid \mathrm{t} > \mathrm{a}\}$ *for some* a, J_q *is an open interval for* $\mathrm{q} > 0$, *the* J_q *are disjoint, the Lebesgue measure of* $\mathrm{R}^+ \setminus \bigcup_{\mathrm{q} \geqslant 0} \mathrm{J}_\mathrm{q}$ *is zero,* $\mathbf{v}$ *can be modified on a set of Lebesgue measure zero so that its restriction to each* $\mathrm{R}^3 \times \mathrm{J}_\mathrm{q}$ *becomes smooth and* $\sum_{\mathrm{q}>0} (\textit{length } (\mathrm{J}_\mathrm{q}))^{\frac{1}{2}}$ *is finite.*

These results concerning the regularity of the turbulent solutions have been continuously refined [9]; in 1976 Scheffer proved the following [37].

Theorem 4.1.1. *Let* $\mathbf{v}$ *be a turbulent solution with finite initial kinetic energy such that the initial conditions are smooth. Let* $\mathrm{T} > 0$ *be given and set* $\mathrm{A} = \{\mathbf{x} \in \mathrm{R}^3 \mid$ *the restriction of* $\mathbf{v}$ *to* $\{\mathbf{x}\} \times ([0, \mathrm{T}] \cap (\bigcup_{\mathrm{q} \geqslant 0} \mathrm{J}_\mathrm{q}))$ *is bounded*$\}$. *Then the Hausdorff dimension of* $\mathrm{R}^3 \setminus \mathrm{A}$ *is at most* 5/2.

Theorem 4.1.2. *The* 1/2 *dimensional Hausdorff measure of* $\mathrm{R}^+ \setminus \bigcup_{\mathrm{q} \geqslant 0} \mathrm{J}_\mathrm{q}$ *is zero.*

By a method based on the complexification of the functional spaces Foias and Temam [16] proved the analyticity of $\mathbf{v}$ as a function of t with values in $(H^2(\Omega))^n$ ($n = 2, 3$) on a sufficiently small interval $(t_0, t_0 + \varepsilon)$ whenever $\mathbf{v}_0 \in (H^1(\Omega))$. It follows that for $n = 2$, $\mathbf{v}(\mathbf{x}, t)$ is a continuous function of $(\mathbf{x}, t)$ for all $\mathbf{x} \in \Omega$ and $t > 0$; while for $n = 3$ this is true only for $t \in (0, \infty) \setminus \sigma$ where the dimension of σ is smaller or equal to 1/2, as is stated in theorem 4.1.2.

We mention a very important result for studying the Leray solution: the Foias and Temam decomposition theorem concerning the splitting of $L^2(\Omega)$ into four orthogonal subspaces. This splitting was obtained by a further analysis of the projector P (Section 2.2.3) early studied by Krein which is an extension of the operator curl. This finer splitting allowed for the connection between existence results for weak solutions of Ladyzhenskaya and Lions with those for turbulent Leray solutions. A connection with Hodge theory is also possible [15].

§ 4.2. THE LANDAU-HOPF CONJECTURE

In this section we present the motivation of the Landau-Hopf conjecture according to which turbulence is described by infinitely many branchings of solutions of the Navier-Stokes equations. Since to calculate every subsequent branching it is necessary to know at least numerically the former branching, the Landau-Hopf conjecture has not been proved. Due to the difficulties encountered in computing the branching solutions nobody was able to prove more than a second branching (except for few cases).

The two fundamental ideas on the actual turbulence go back to E. Hopf. His first idea is that of repeated branching of the solutions of the Navier-Stokes equations [8] (from Ch. 3), proved by himself for the Burgers-like equations, and according to which the origin of the turbulence is explained by means of quasi-periodic solutions. This idea generated a strong current into fluid dynamics.

The second main idea of Hopf [19] is that turbulence (more exactly fully developed turbulence rather than its first stages) is given by the evolution of some probability measures described by the famous Hopf equation. This idea was suggested by the lack of a uniqueness theorem for solutions in the nonstationary three-dimensional case for given initial conditions. It seemed naturally to put the problem in some another, more physical (say) way, i.e. to carry out a statistical study of the evolution of given probability measures on the set of the initial conditions. The new trend in turbulence based on statistical solutions of Hopf's equation was developed by C. Foiaş in the now well-known paper [8], Vishik M. I. [43], Arsenev A. A. [2], Bensoussan A., R. Temam [3], Ladyzhenskaya O. A., A. M. Vershik [22], Vishik M. I., A. V. Foursikov [44].

The Hopf repeated branching for a Burgers-type model. The model considered by Hopf (in his notations) is

$$\frac{\partial u}{\partial t} = -z \circ z^* - u \circ 1 + \mu \frac{\partial^2 u}{\partial x^2}, \tag{4.2.1}$$

$$\frac{\partial z}{\partial t} = z \circ u^* + z \circ F^* + \mu \frac{\partial^2 u}{\partial x^2}, \tag{4.2.2}$$

where $\mu > 0$ is a parameter (i.e. the coefficient of the kinematic viscosity), $F(x) = a(x) + \mathrm{i}\, b(x)$ is an arbitrary given complex-valued space function, * denotes the complex conjugate, $\circ$ stands for the convolution product $f \circ g = \dfrac{1}{2\pi}\displaystyle\int_0^{2\pi} f(x+y)\, g(y)\, \mathrm{d}y$, the unknowns $u(x, t)$, $z(x, t)$ are complex-valued functions, all the functions from (4.2.1), (4.2.2) are periodic in x with period 2π, the domain of motion is a one-dimensional circular line and the space variable is an angular variable $x \pmod{2\pi}$. In addition $F(-x) = F(x)$, u is real and u

and z are even functions of x for every $t \geqslant 0$. To (4.2.1), (4.2.2) the following boundary conditions at $x = 0$ and $x = \pi$ are added

$$\frac{\partial u}{\partial x} = \frac{\partial z}{\partial x} = 0. \tag{4.2.3}$$

The set of solutions of (4.2.1)—(4.2.3) changes its properties as μ decreases from ∞ to 0 i.e. on the measure the viscous fluid tends to become an ideal fluid and the laminar motion becomes turbulent. These properties concerning the number and the stability of the solutions (u, z) are supposed by Hopf to hold also for the solutions of the Navier-Stokes equations. Whether this conjecture is true or not remains an open problem. We give now a short sketch of the Hopf conjecture on repeated bifurcations. For large μ there exists a single stationary solution stable against the perturbations of the initial conditions. Let μ_0 be the value at which this steady solution loses its stability. For μ belonging to (μ_1, μ_0) the problem (4.2.1)—(4.2.2) has some solutions which constitute a manifold $M(\mu)$ in the phase space invariant under the phase flow. Due to the influence of the viscosity the dimension $N(\mu)$ of $M(\mu)$ is supposed to be finite. As μ_1 is crossed $N(\mu)$ has a jump and the solutions of $M(\mu)$ lose their stability although they can be analytically continued through μ_1. For $\mu = \mu_1$ another manifold of solutions $M_1(\mu)$ branches from the stationary solution and it is stable for $\mu_2 < \mu < \mu_1$. The nature of the solutions of M_1 (i.e. their stationarity or periodicity with respect to time) depends on the way the eigenvalues (which depend on μ) of the Fréchet derivative of the operator from (4.2.1), (4.2.2) crosses the imaginary axis as μ crosses μ_1. This branching goes on infinitely many times; new stable manifolds of solutions branching from the previous manifold, which becomes unstable as μ crosses a critical value μ_k, appear and $N(\mu) \to \infty$ as $\mu \to 0$. In the first stages of this process the solutions from $M(\mu)$ are stationary, then periodic with respect to time; those are followed, as μ decreases, by solutions $u(\varphi_1, \varphi_2, \mu)$ periodic of φ_1, φ_2, where $\varphi_1 = a_1 t$, $\varphi_2 = a_2 t$ and $a_1 = a_1(\mu)$, $a_2 = a_2(\mu)$ are not rationally related, therefore u are almost periodic with respect to t. In this way the solutions from $M(\mu)$ belong to some torus whose dimension increases as μ decreases, therefore u depends on more and more functions φ_i. By Hopf, the turbulent solutions are quasi-periodic functions of time.

Prandtl's photographs of the flow around a sphere for various μ (J. R. Aeronaut. Soc., 1927) [37] exhibit the main features of the Hopf scheme. The same holds for Conette and Bénard flows. In the first stages Hopf's scheme agrees perfectly with the experiments; nevertheless nobody has proved that the vortices behind the sphere are branching solutions. There is the unproved conjecture that the Kármán street consists also of branching solutions.

We mention that in [8] (from Ch. 3) Hopf showed that the turbulent flows are essentially viscous ones.

§ 4.3. THE RUELLE-TAKENS THEORY

Necessarily in turbulence the solutions are discontinuous but this is not sufficient to define turbulent flow [25]. Another necessary characteristic of the turbulent solutions is their chaoticness. A theory of the onset of turbulence based on strange attractors is used to show that at high Reynolds numbers the Navier-Stokes equations carry sufficient information and for large time t they have such intricate solutions that, after a finite number of branchings, the motion becomes chaotic. This is known as the Ruelle-Takens theory [31], [30], [28], [40].

The Navier-Stokes equations are sufficiently complicated to present turbulent (i.e. chaotic) behaviour. In fact, it was already proved numerically that many other very simple dynamical systems have turbulent behaviour. Ideas underlying the Ruelle-Takens theory are now used to describe turbulent behaviour also for other dynamical systems for celestial mechanics, physics, chemistry, biology, ecology, sociology, weather prediction etc. These systems are studied by synergetics.

4.3.1. The case of the Navier-Stokes equations

The Ruelle-Takens theory is closely related to the Hopf repeated branching but instead of the quasi-periodic solutions the turbulence is thought of as a more complicate motion (strange attractor) whose main feature is chaos. In fact, Ruelle-Takens' first paper follows the stages described by Hopf, but, unlike Hopf having in view some particular results for dynamical dissipative systems (to which the Navier-Stokes equations belong), they state that the quasi-periodic solutions are not attractive. Therefore they are not stable and, consequently, cannot describe the turbulence. Instead, as the Reynolds number is increased, sets of more complicate nonstationary attracting solutions appear. Among them a "strange attractor" seems to be the one which describes the emergence of turbulence at $t \to \infty$. The Ruelle-Takens theory has been verified for the Lorenz model and for particular cases of the Bénard and Couette flows. Recently, especially due to Ya. Sinai's works, the mechanism of the appearance of the chaotic motion from the Lorenz strange attractor has been clarified [38], [4], [5].

As usual, by the phase space H of a dynamical system corresponding to the problem (4.2.1)—(4.2.3) we mean the space of the solenoidal vectors $(u(x), z(x))$. Denote by $M(\mu) \subset H$ the manifold of solutions of this problem i.e. $M(\mu)$ consists of the points $(u(x), z(x))$ where $u(x)$, $z(x)$ are the values for a fixed t of the nonstationary solution $(u(t, x), z(t, x))$ of (4.2.1)—(4.2.3). Let $(u_0(x), z_0(x))$ be the initial data for (4.2.1)—(4.2.3) and let $(u(t, x), z(t, x))$ stand for the corresponding nonstationary solution such that $u(0, x) = u_0(x)$, $z(0, x) = z_0(x)$. The curve of H consisting of the points $(u(t, x), z(t, x)) \in M$ as t vary is called a *phase flow* or a *trajectory*. Then a stationary fluid motion will be a point of M; a periodic motion will be a closed orbit of M; an almost (i.e. quasi)-periodic motion will be a trajectory situated on a torus in M and so on. Correspondingly a stable steady fluid motion $(\bar{u}, \bar{z})$ will be an attractive point of H i.e. each trajectory which starts near $(\bar{u}, \bar{z})$ is ultimately trapped by $(\bar{u}, \bar{z})$ as $t \to \infty$. To a stable periodic (with respect to time) fluid motion corresponds in H an attractive closed orbit etc.

We now discuss some results concerning the topological and statistical properties of the dynamical system associated with the initial value problem for the Navier-Stokes equations in order to point out the progress made by Ruelle and Takens in extending the Hopf study. Suppose that the Navier-Stokes equations have been put in the form of the dynamical system

$$\frac{\mathrm{d}v}{\mathrm{d}t} = X_\mu(v) \tag{4.3.1}$$

where $v \in H$ is the velocity, H is an appropiate Banach space, X_μ is a vector field over H. Based on some (in general unproved) assumptions on X_μ Ruelle and Takens [31] have developed a new theory of turbulence.

A stationary solution v_0 of (4.3.1) is called a *fixed point* of X_μ (it has the property that $X_\mu(v_0) = 0$). Denote $v(t) = D_{X_{\mu}, T}(v_0)$ the dependence of the integral curve $v(\cdot)$ on its initial value v_0. If H is finite dimensional, a point $v \in H$ is nonwandering if for every neighbourhood U of x and every $T > 0$ there exist $t > T$ such that $D_{X_{\mu}, t}(U) \cap \neq \varnothing$. A closed subset A of *nonwandering* points is an *attractor* if it has a neighbourhood U such that $\bigcap_{t>0} D_{X_{\mu}, t}(U) = A$. The attractors are manifolds. A more general attracting set which is not a manifold is a *"strange attractor"*. As $t \to \infty$ an integral curve $v(t)$ of X_μ may be attracted by a fixed point or a closed orbit, or a strange attractor. (The last case takes place when there is a *sensitive dependence on the initial conditions.*) It is assumed that the attractor is a generic set (i.e. its complement is a closed rare set of H). On the other hand, unlike Hamiltonian systems which are conservative, viscous fluids are dissipative systems. For some finite dimensional dissipative systems it has been shown that quasi-periodic motions on a torus constitute a rare set. Therefore, by the above assumption, the set of quasi-periodic motions of the fluids is, perhaps, not an attractor and, consequently, these motions cannot describe turbulence. This reasoning was the starting point in the Ruelle-Takens theory. Thus, by Ruelle and Takens, a fluid system is turbulent when this motion is described by an integral curve of the vector field X_μ which tends to a set A which is neither empty nor a fixed point nor a closed orbit. In this definition they disregarded nongeneric possibilities. This definition as well as many of the unproved assumptions, abounding in this theory, requires an attentive analysis [31]. That is why this theory remains only promising for the general Navier-Stokes equations; for particular cases it has been proved.

Inside the strange attractors the trajectories wander in a chaotic way and are highly sensitive to initial conditions. This last property means that for bounded perturbations of the initial condition the perturbed solution is an increasing function of time and an irregular behaviour of these solutions could occur. Thus the Ruelle-Takens theory introduces chaos as a main characteristic of turbulence. The description of turbulence by a chaotic process implies the fact that the prediction of the future states is limited by the accuracy with which the initial conditions can be determined whereas in the random process this prediction is limited by the range of the correlation of the random process. It follows that an important problem in the Ruelle-Takens theory, which may be posed for some other dynamical systems as well, is the appearance of stochasticity from these attractors; in the case of the Lorenz strange attractor this has been solved recently [5], [39]. In this respect we note the papers of R. Bowen on topological entropy and subshifts which connects the topological characteristics of the dynamical systems with various aspects of their statistical behaviour. Together with Ya. G. Sinai and D. Ruelle he developed the theory of the metric properties of the basic sets for Axiom A systems [26].

An original dynamics to generate turbulence was conceived by St. Savulescu. According to his theory the turbulent motions travel between two energy levels. It is possible that this mechanism is the one which takes place inside the Lorenz strange attractor [33].

The sensitive dependence on initial conditions depends on a probability measure S invariant under time evolution; many such ergodic measures exist (all of them are singular with respect to the Lebesgue measure), but how to determine that one for which the notion of sensitive dependence on initial conditions has a natural meaning is an open question. Some answer can be given for a large class of maps (Axiom A diffeomorphisms) when the measure ρ is characterized by the fact that, on unstable manifolds it is absolutely continuous with respect to Lebesgue measure. For the Hénon attractor (corresponding to a two-dimensional evolution system) this characterization seems to hold, although it is not an Axiom A attractor. On the basis of some results on Axiom A attractors concerning the stability against stochastic perturbations one can argue that the sensitive dependence on initial condition amplifies even the small thermal fluctuations to macroscopic level in a relatively short time [32]. Unfortunately no definite result can be stated in this respect.

The sensitive dependence on initial condition implies that the Fréchet derivative $X'_v: T_vH \to T_{X(v)}H$ of X from (4.3.1) has very large norms for large t; this may be the case if in some direction X_v is very large whereas in some others it has small values. Therefore in some directions v may go further very fast, in others—it comes closer. Correspondingly X'_v (as a function of t) stretches in some directions and contracts in others. This combination is known as *hyperbolicity* or, by Smale [30], as Axiom A. If the asymptotic stretching occurs on the attractor then it is called a strange attractor. For Axiom A dynamical systems a nonstrange attractor is an attractive periodic orbit or an attractive fixed point; this justifies the above definitions for the Navier-Stokes case. Nevertheless we must emphasize that the Navier-Stokes system does not satisfy Axiom A; this fact produces discrepancies between the consequences drawn from the Ruelle and Takens theory of turbulence and the characteristics of the observed natural phenomenon. For topological reasons a flow in two dimensions cannot be turbulent; for fluids this reads: the quasi-periodic motions on a two-dimensional torus T^2 are not turbulent. Ruelle and Takens proved that the fluid turbulence may appear by small perturbations of quasi-periodic flow on a T^4. A stronger result replacing T^4 by T^3 is quoted in [30] where the result of Chenciner and Iooss on bifurcation of a torus T^2 into a torus T^3 is discussed. The particular case (when the equations are in R^3) of this last result has been worked in [21]. Another improvement of the results from [30] were given in [25]. The original theory of Ruelle and Takens as well as its improvements differ from the Hopf theory fundamentally. In the Hopf conjecture after n bifurcations the attracting set would be a n-dimensional torus T^n and $n \to \infty$ as the Reynolds number tends to the critical value, i.e. the origin of the turbulence implies an infinite number of bifurcations. By Ruelle and Takens on the contrary, it would be possible to have a strange attractor belonging to a torus T^4 after four bifurcations. To obtain their result Ruelle and Takens used the extension of the Hopf's bifurcation theorem to diffeomorphisms and the center manifold theorem which reduces the case of infinite dimensional manifolds to the finite dimensional manifolds.

We want also to point out that the chaos ecountered in the Ruelle-Takens theory is with respect to time only [29].

4.3.2. **The Lorenz model**

Consider a fluid layer heated from below i.e. the Bénard problem where the Boussinesq approximation is taken into account. The motion of this configuration is governed by the so-called Rayleigh-Bénard system consisting of the Navier-Stokes equations and an equation for the heat. Expanding the unknowns into Fourier series and retaining only the first three modes the Lorenz equations

$$\begin{cases} \dot{x} = -\sigma x + \sigma y, \\ \dot{y} = rx - y - xz, \\ \dot{z} = -bx + xy, \end{cases} \tag{4.3.2}$$

are obtained. r stands for the Rayleigh number; $\sigma = 10$ and $b = 8/3$. The numerical analysis shows that in the phase-space the trajectories corresponding to this simple ordinary differential system of equations which truncates the Navier-Stokes system become more and more intricate as r is increased. Eventually they wander chaotically between the components of a strange (Lorenz) attractor and the solutions of the Lorenz system are highly sensitive to the initial condition. In this way, according to Ruelle and Takens, the corresponding Lorenz system has a turbulent behaviour. Recent numerical analysis make it evident that the solutions of very many other simple dynamical systems behave chaotically [17]. The strange Lorenz attractor locally appears as an infinite sheeted surface. In fact it consists of the product of a Cantor set and a piece of a two-dimensional manifold. A similar structure is presented by the strange attractor corresponding to a system studied by Hénon and Pomeau [40].

The structure of a strange attractor is vizualized by intersecting it with surfaces S transverse to the flow and investigating the corresponding return map. Let $P(x)$ be the intersection of the integral curve through x with S; P is called the Poincaré map. If π_0 is a torus on which are located the integral curves then $\bigcap_{n>0} P^n\pi_0$ is an attractor.

The conduction state of the fluid layer at rest corresponds to the trivial solution (0, 0, 0) of the Lorenz system. Let us now linearize this system about (0, 0, 0); analysing the eigenvalues of the Fréchet derivative F_0 at (0, 0, 0) corresponding to the linearized system the following assertions can be made. For $-2.025 < r < 1$ the only steady solution of the Lorenz system (4.3.2) is $\mathbf{0} = (0, 0, 0)$ and it is (globally) attracting (i.e. $\mathbf{0}$ is a global sink) because all the eigenvalues of F_0 are real and negative. As r is increased two new stable steady solutions (convective motions) (depending on the parameter r) $C = (\sqrt{b(r-1)}, \sqrt{b(r-1)}, r-1)$, $C' = (-\sqrt{b(r-1)}, -\sqrt{b(r-1)}, r-1)$ bifurcate from the former trivial solution which becomes unstable (i.e. it is attracting for points of a plane through (0, 0, 0) (a two-dimensional stable manifold of $\mathbf{0}$) and repelling for points of a line (unstable manifold)). A further increase in r produces a Fréchet derivative of the Lorenz operator at C and C' having two complex conjugate and a real eigenvalue. At $r \approx 13.926$ C and C' will lose their stability and a Hopf bifurcation takes place;

stable and unstable manifolds of the origin become identical (the origin being now a homoclinic point) and thus two closed unstable orbits arise. As $r > 13.926$, C and C' are attracting fixed points and the unstable manifold of the origin is attracted to the opposite point. There is an invariant set (consisting of infinitely many periodic orbits) and points near it eventually leave it in a chaotic way to one of the attracting points C and C'. This "preturbulent" stage is similar to Smale's horseshoe. At $r > 24.06$ a Lorenz strange attractor bifurcates. The trajectories wander between the periodic orbits of the strange attractor in a chaotic way. For greater r the unstable closed orbits shrink to the fixed unstable points C and C' such that for $r \approx 24.74$ a standard Lorenz attractor sets in. As $r \geqslant 50$ the unstable manifold of C and C' developes a fold and stable large amplitude closed orbits bifurcate (possibly). The Lorenz attractor was extensively discussed in [28], [40]. We mention now three properties of the solutions of (4.3.2) which can be rigorously proved: a) there is a bounded region $B \subset \mathbb{R}^3$ such that every solution of (4.3.2) eventually becomes trapped in B; b) The time evolution given by (4.3.2) contracts volumes in $\mathbb{R}^3$ at a constant rate; c) The system (4.3.2) is invariant under the symmetry $(x, y, z) \to (-x, -y, z)$. The frontier of the Lorenz strange attractor has been localized in [1].

§ 4.4. GENERIC FINITENESS OF THE SET OF THE SOLUTIONS OF THE NAVIER-STOKES EQUATIONS

The branching analysis is just part of the study of the structure of the set of solutions of a problem. This study also includes the analysis of the finiteness of this set, which is closely connected with the branching analysis. In the following we shall survey the existing (few) results concerning the generic finiteness of the set of stationary solutions of the boundary-value problem for the Navier-Stokes equations (shortly reffered to as finiteness of $S(f, v_1, \nu)$). There are two topological methods (based on transversality) used in proving these results. The first one concernes the ∞-dimensional version of Sard's theorem which was given in 1965 by Smale. (Obviously as the solutions of the Navier-Stokes equations belong to infinite-dimensional Banach spaces theorems of differential topology of infinite dimensional manifolds are to be used.) The other technique was used as early as 1967 by C. Foias and G. Prodi in the theory of the Navier-Stokes equations. It consists in projecting the problem onto finite dimensional vector spaces spanned by eigenvectors of $\tilde{\Delta}^{-1}$ (Section 2.2.2) and then to applying the Sard theorem (i.e. in the finite-dimensional case) to the projected equation. (Appendix 5).

Roughly speaking, a property of a solution of a problem in a Banach space B is generic with respect to a datum $\mathbf{f}$ of that problem if it holds for almost all $\mathbf{f} \in B$, or, more precisely, if it holds up to a rare closed subset of B (i.e. except for a set of the first category), therefore for $\mathbf{f}$ belonging to an open dense subset of B. We recall that the general setting of the problem governing stationary flows of viscous incompressible fluids contains three data: the Reynolds number Re (in the dimensionless form of the Navier-Stokes equations or the coefficient of the kinematic viscosity ν, if these equations are written in physical coordinates), the body forces $\mathbf{f}$ and the boundary data $\mathbf{v}_1$. It is with respect to some of these data that the generic properties of the turbulent solutions of the Navier-Stokes equations are studied. The analysis of the finiteness of $S(\mathbf{f}, \mathbf{v}_1, \nu)$ started with C. Foias and R. Temam's paper [12] (in which the second method was used) followed by [11], [13] where the bounded domain Ω of motion is 2- or 3-dimensional, the genericity is with respect to the body forces $\mathbf{f}$ and the two above mentioned techniques are used. Using the first method, the genericity with respect to $(\mathbf{f}, \mathbf{v}_1)$ and $\mathbf{f}$ for the same case is considered in [14]; in addition the boundary $\partial\Omega$ consists of a finite number of connected components, Ω being locally located on one side of $\partial\Omega$. Refinements of the second method permitted a more complete investigation of $S(\mathbf{f}, \mathbf{v}_1, \nu)$ in [15]; additional results concerning the generic finiteness of S with respect to $\mathbf{f}$ and ν for $\mathbf{v}_1$ fixed are deduced. In [15] it is proved that the number of bifurcation points for $\nu > \nu_0 > 0$ is finite. On the basis of some results on real analytic sets due to F. Bruhat and H. Whiteny, information is obtained about $S = \bigcup_{\nu>0} S(\mathbf{f}, \mathbf{v}_1, \nu)$ and, in particular about the generic finiteness of primary and secondary bifurcations for $\nu_0 > 0$ and about the bifurcations as $\nu \to 0$. The theory developed in [15] is also adapted to unbounded domains of motion and results on the finiteness of $S(\mathbf{f}, \mathbf{v}_1, \nu)$ in the case of Couette flows between infinite concentric cylinders and Bénard convection between (infinite) horizontal plates are deduced. Finally we mention the generic finiteness of S for the case of the periodic (with respect to time) solutions of the nonstationary Navier-Stokes equations [41], [16]. The generic finiteness of S with respect to the domain Ω was done by J. Marsden and A. Tromba.

Foias' analysis of the generic finiteness of $S(\mathbf{f}, \mathbf{v}_1, \nu)$ belongs to a more general investigation of the invariants of the Navier-Stokes equations put in the form of an evolution equation

$$\frac{\mathrm{d}\mathbf{v}}{\mathrm{d}t} = F(\mathbf{v}) \tag{4.4.1}$$

and an initial condition

$$\mathbf{v}(0) = \mathbf{v}_0. \tag{4.4.2}$$

S is the simplest invariant manifold of this equation i.e. if $\mathbf{v}_0 \in S$ then $\mathbf{v}(t) \in S$ for every t, where $\mathbf{v}$ is a solution of (4.4.1), (4.4.2). The study of invariants is intended to explain the appearance of turbulence. We mention that Foias' theory of turbulence, based on his statistical solutions of the Navier-Stokes equations, was extended in many respects (e.g. to random $\mathbf{f}$ and $\mathbf{v}_0$); use was made of theorems of measurable section [22]. For other remarks concerning the connection of the generic finiteness of S with turbulence we quote [10].

Let us now apply the first of the two above mentioned methods to the following boundary-value problem for the stationary Navier-Stokes equations $F\mathbf{v} = A\mathbf{v} + B(\mathbf{v}) = \mathbf{f}$ where $\mathbf{v} \in N^1(\Omega)$ is the turbulent solution, $\mathbf{f} \in N$ is the body force. In Section 2.2.3 we wrote the integral relation satisfied by $\mathbf{v}$ in the nonstationary case for $\mathbf{f} = 0$. Here $A: N^1 \to N^{1\prime}$, is the linear continuous operator defined by $(A\mathbf{u}, \mathbf{v}) = ((\mathbf{u}, \mathbf{v}))$ for every $\mathbf{u}, \mathbf{v} \in N^1$ and $B: N^1 \times N^1 \to N^{1\prime}$ is a bilinear continuous operator defined by $(B(\mathbf{u}, \mathbf{v}), w) = \sum_{i,j=1}^{n} \int_\Omega u_i \partial \mathbf{v}_j / \partial x_i w_j \, dx$, ($n = \dim \Omega = 2$ or 3) for $\mathbf{u}, \mathbf{v}, \mathbf{w} \in N^1(\Omega)$. In Section 2.2.2 instead of A a we wrote $\widetilde{\Delta}$. The nonlinear mapping $F: \mathfrak{D}(A) \to N$ is a Fredholm mapping and its Fréchet derivative at $\mathbf{u}$ applied to $\mathbf{v}$ is $F'_{\mathbf{u}}\mathbf{v} = \nu A\mathbf{v} + B(\mathbf{u}, \mathbf{v}) + B(\mathbf{v}, \mathbf{u})$ for every $\mathbf{u}, \mathbf{v} \in \mathfrak{D}(A)$, where $\mathfrak{D}(A) = A^{-1} N$ is endowed with the norm $|A\mathbf{u}|$ which is equivalent to the norm induced by $H^2(\Omega)$. (In other words, $\mathfrak{D}(A)$ is the energy space of A.) In this problem we supposed implicitely that $\mathbf{v}|_{\partial\Omega} \equiv \mathbf{v}_1 = 0$. It was proved that $S(\mathbf{f}, \mathbf{0}, \nu)$ is compact in H^2, N^1 and N; this property enables one to prove the generic finiteness of S. In fact assume that $S(\mathbf{f}, \mathbf{0}, \nu)$ is not finite. Then there exists a sequence of mutually distinct elements $\mathbf{v}_j \in S(\mathbf{f}, \mathbf{0}, \nu)$, converging in $\mathfrak{D}(A)$ to $\mathbf{v}' \in S(\mathbf{f}, \mathbf{0}, \nu)$ (due to the compactness of S). Therefore by passing to the limit in the relation $\nu A(\mathbf{v}_j - \mathbf{v}') + B(\mathbf{v}_j, \mathbf{v}_j) - B(\mathbf{v}', \mathbf{v}') = 0$ we find a nonzero $\mathbf{v}'' = \lim_{j \to \infty} \frac{\mathbf{v}_j - \mathbf{v}'}{|\mathbf{v}_j - \mathbf{v}'|}$ such that $F'_{\mathbf{v}}\mathbf{v}'' = \nu A\mathbf{v}'' + B(\mathbf{v}', \mathbf{v}'') + B(\mathbf{v}'', \mathbf{v}') = 0$. Hence $F'_{\mathbf{v}}$ is not injective and, by a remark of Appendix 5 $\mathbf{v}'$ is not a regular point of F; consequently $\mathbf{v}'$ is a singular point of F and, correspondingly, $F(\mathbf{v}') = \mathbf{f}$ is a singular value of F. On the other side F belongs at least to C^1 and index $F = 0$, therefore, by the Sard-Smale theorem the set of all singular values $\mathbf{f}$ of F belongs to a closed rare subset of N. The set $S(\mathbf{f}, \mathbf{0}, \nu)$, corresponding to those $\mathbf{f}$ which do not belong to this subset, is finite or equivalently, S is generically finite with respect to $\mathbf{f}$.

The second method is essentially based on the use of the finite dimensional spaces V_m instead of the space N^1 as follows. Denote by $\mathbf{w}_i, \lambda_i, i = 1, 2, \ldots$ the eigenfunctions and eigenvalues respectively of the selfadjoint operator $\widetilde{\Delta}^{-1}$, compact in N; and let V_m be the space spanned by $\mathbf{w}_1, \mathbf{w}_2, \ldots, \mathbf{w}_m$. It can be proved that for sufficiently large m (which depend on ν and the L^2-norm of $\mathbf{f}$) the solution $\mathbf{v}$ of $F(\mathbf{v}) = \mathbf{f}$ has a projection $Q_m\mathbf{v}$ on V_m which depends only on $P_m\mathbf{v}$. For this $\mathbf{f}$ this equation is equivalent to $\Phi_m(\xi) = \mathbf{f}$, $\xi \in 0_m$, $\xi = P_m\mathbf{v}$, 0_m is an open subset of V_m, $\Phi_m\colon 0_m \to V_m$ is analytic and $P_m\mathbf{v} \to \mathbf{v}$ is one to one and analytic. The classical Sard theorem applied to Φ_m allows one to obtain finiteness of S for generic values of $\mathbf{f}$ belonging to a subset of P_mN and then to N.

A further use of Sard-Smale theorem for Fredholm mappings of nonvanishing index allows one to deduce other details concerning the generic finiteness of S.

Finally we mention the general properties and the generic finiteness of S given in [14] and [15] for $\mathbf{f} \in H^{-1}(\Omega)$, $\mathbf{v}_1 \in H^{1/2}(\partial\Omega)$ and $\int_{\partial\Omega} \mathbf{v}_1\mathbf{n}\,d\sigma = 0$, $\partial\Omega = \partial\Omega_1 \cup \partial\Omega_2 \cup \ldots \cup 0\,\partial\Omega_s$;

1) $S(\mathbf{f}, \mathbf{v}_1, \nu)$ is nonempty;

2) $S(\mathbf{f}, \mathbf{v}, \nu)$ is closed and bounded in $H^1(\Omega)$, compact in $L^2(\Omega)$;

3) $S(\mathbf{f}, \mathbf{v}_1, \nu)$ is reduced to one point for $\nu > \nu_1$ (see [14] for more precise value for ν_1);

4) For $\mathbf{f} \in L^2(\Omega)$ and $\mathbf{v}_1 \in H^{3/2}(\partial\Omega)$ then $S(\mathbf{f}, \mathbf{v}_1, \nu) \subset H^2(\Omega)$;

5) $S(\mathbf{f}, \mathbf{v}_1, \nu)$ is a compact subset of $H^2(\Omega)$;

6) $S(\mathbf{f}, \mathbf{v}_1, \nu)$ is homeomorphic to a compact set of $\mathbf{R}^m$ for sufficiently large m;

7) There exists a dense open set $0 \subset H(\Omega) \in H^{3/2}(\partial\Omega)$ such that for every $\nu > 0$ and every pair $(\mathbf{f}, \mathbf{v}_1) \in 0$ the set $S(\mathbf{f}, \mathbf{v}_1, \nu)$ is finite. For every connected component 0_α of 0, the number of points in $S(\mathbf{f}, \mathbf{v}_1, \nu)$ for $\{\mathbf{f}, \mathbf{v}_1\} \in 0_\alpha$ is constant and every solution is a C^∞ function of the pair $\{\mathbf{f}, \mathbf{v}_1\}$;

8) For every $\nu > 0$ and $\mathbf{v}_1 \in H^{3/2}(\partial\Omega)$, there exists a dense open set $0' \subset H$ such that $S(\mathbf{f}, \mathbf{v}_1, \nu)$ is finite for every $\mathbf{f} \in 0'$. For every connected component 0_α of 0 the number of points in $S(\mathbf{f}, \mathbf{v}_1, \nu)$ for $\mathbf{f} \in 0_\alpha$ is odd and constant and every solution is a C^∞ function of $\mathbf{f}$; this last property is derived by applying the topological degree theory of I. D. Elworthy and A. J. Tromba;

9) For every $\nu > 0$ there exists an open dense set $G \subset N$ such that for every $\mathbf{f} \in G$ and for $\mathbf{v}_1$ belonging to every open dense subset of $H^{3/2}(\partial\Omega)$ $S(\mathbf{f}, \mathbf{v}_1, \nu)$ is finite.

10) Let $\mathbf{v}_1$ be fixed in $H^{3/2}(\partial\Omega)$ [42]. Then there exists an open dense set $G(\mathbf{v}_1)$ of N such that for every $\mathbf{f} \in G(\mathbf{v}_1)$ $S(\mathbf{f}, \mathbf{v}_1, \nu)$ is finite for all $\nu > 0$ which do not belong to a closed rare set of $\mathbb{R}^+$;

11) For given $\mathbf{f}$ in N, $\mathbf{v}_1$ given in $\dot{H}^{3/2}(\partial\Omega)$ and m large enough P_m is injective and P_mS is a compact real C-analytic set;

12) For every $\mathbf{f} \in G(\mathbf{v}_1)$ there G is an open dense set of N, the continuum of the solutions $D(\mathbf{f}, \mathbf{v}_1) = \bigcup_{\nu>0} S(\mathbf{f}, \mathbf{v}_1, \nu)$ has the following form

a) it contains isolated points and isolated C-analytic manifolds located over some values of ν whose number is finite at the right of every number $\nu_0 > 0$;

b) it contains one or more analytic manifolds of dimension 1 whose projection on the ν-axis goes from 0 to $+\infty$. The set of singular points of this continuum (and consequently the set of bifurcation points ν) is finite to the right of every number $\nu_0 > 0$.

§ 4.5. PATTERN FORMATION; SYMMETRY BREAKING INSTABILITY

This section is intended to put forward a new field of research in fluid dynamics: the study of stability and branching of the solutions of Navier-Stokes equations from the point of view of group representation theory.

The Bénard cells and Taylor vortices are patterns branching from the former stable equilibrium or Couette motion as the physical parameters cross critical values Ra_{cr} or Re_{cr}. The primary equilibria are invariant under some group of symmetry transformations while the branching patterns are invariant only to transformations of a subgroup of the former one. So far, Ra_{cr} and Re_{cr} were characterized as points where stability is lost and branching takes place. In virtue of the above assertions Ra_{cr} and Re_{cr} received a new characterization: they are points where the instability occurs due to symmetry breaking. This new point of view, framed in the language of group representation theory, was developed mainly by D. H. Sattinger. As a large list of references ir now available [35], [36] we confine ourselves to an indication of the bridge between the algebra and linear functional analysis involved in this theory.

A representation of a group G on a vector space V is a homomorphism $g \to T(g)$ of G into the group of invertible transformations on V $T(g_1g_2) = T(g_1)\,T(g_2)$. For instance if V is the space of continuous functions on R^3 and G is the group of rigid motions then a representation of G is $(T_gu)(x) = U(g^{-1}x)$, where $gx = 0x + a$, 0 being an orthogonal matrix. In the Bénard case $u = (u_1, u_2, u_3, \theta, p)$ where u_i are the components of the velocity, θ is the temperature and p is the pressure.

§ 4.6. CONCLUDING REMARKS; OPEN PROBLEMS

The nature of turbulence still remains the most difficult problem in fluid mechanics. A primary method of investigation of this field used by engineers continues to be that based on the Reynolds equations to which various kinds of closure relations connecting correlations have to be added. Since these relations are extra informations on the physics of fluids it follows that in these approaches the mathematical model governing the laminar flows and turbulence are no longer the same. We mention also Savulescu's theory based on an integrodifferential form of the equations of motion [34] allowing a more rational derivation of a closure relation.

Opposite to this situation, there is a belief among mathematicians that turbulence is well described by the single Navier-Stokes equations whose solutions for large time t have bad properties: they are no larger regular in t, no longer simple, no longer unique, no longer stable etc. Correspondingly there are different ways of dealing with turbulence from the mathematical point of view. The first problem to be faced by a mathematician willing to deal with turbulence is the lack of a single precise definition of what is meant by turbulent motion. As a consequence every theoretician begins his rigorous study by stating first a definition of what he means by turbulent flows.

In §§ 4.1—4.3 three typical definitions were given. We say typical in the following sense. Leray type models are those in which generalized solutions are considered and consequently global existence theorems can be proved; the mechanical characteristics of the motion are no longer classical functions but elements of $L^2(\Omega)$. In the other two models turbulence involves an infinite number of branchings or it is defined as a chaotic motion. A particular model corresponds to a particular choice of the function space to which the solution belongs.

All three typical definitions and their corresponding theories concern the preturbulent stage [16]. They describe the motion which is no more laminar but which is not yet completely turbulent.

There is no doubt that phenomenological theories are best suited to laminar nonsteady fluid flows. On the other hand, a phenomenological description of turbulence by means of instantaneous quantities is irrelevant instead, its description in terms of averaged quantities and statistical correlations seems adequate. The experimentalists are unanimous in their view that turbulent flows are extremely complicated nonstationary, highly irregular (temporally and spatially) and chaotic. As far as the last stages of the transition and the first stage of turbulence are concerned the situation is worse. The instantaneous values of the fluctuations are no longer small compared with the basic or averaged amounts. It may happen that their ratio varies widely [18]. The phenomenological description of transition in almost all the particular flows of interest is alread known in detail. Nevertheless there is not yet a unique mathematical characterization of the point λ_0 (of the parameter space) of the onset of turbulence. Stability theory yields the opint λ_s where stability is lost; the generalized models furnish the points λ_g where the solutions cease to be regular; the Ruelle-Takens theories show at which dimensional number λ_c the solutions become chaotic and so on. None among these points is the point of onset of turbulence. We have only that λ_s, λ_g, λ_c are smaller than λ_0 if the bifurcations are supercritical. All the characterizations provided by these theories are necessary for the onset of turbulence but not sufficient. Maybe the only sufficient characterization of the origin is the point where the instantaneous description becomes irrelevant (Savulescu). In this respect Savulescu proposes an integro-differential formulation for the equations of the shear layer flows. The form of these equations seems adequate to mimic a new diffusion process generated by $\frac{d^2u}{d^2y}$ like in the random evolution. This is due to the fact that the large number of mathematical fluctuations occurring in this formulation can be interpreted as particle interactions. All these fluctuations (i.e. the terms of the integro-differential equations) are of the same order of magnitude and none of them can be neglected. In the Ruelle-Takens theory $\lambda_c \to \lambda_0$ when the stochasticity can be derived from the strange attractor.

The frequent breakdowns ask for a Leray model; the intricate form of the phase-space trajectories is similar to that predicted by Ruelle-Takens theory; and the existence of many higher harmonics corresponds to the Landau-Hopf conjecture. Nevertheless turbulence does not depend on the geometry of the fluid configuration while transition differs very much from one domain of motion to another. Thus in the first stages of transition on the flat plate the Leray model works. In convection and rotating flows the transition corresponds to the Landau-Hopf picture throughout the domain of motion. The last stages of transition for every fluid flow are better characterized by chaos. So, none of the three theories in this chapter is able to describe the onset of turbulence; all of them participate in the description of this complex phenomenon with different weights. These theories may be considered as simple elements whose suitable combinations give rise to various complex models agreeing with real fluid flows. In the light of these theories the mechanisms generating turbulence are: the discontinuity, the branching due to the loss of stability, and the chaos of the solutions of the Navier-Stokes equations.

None of these theories has been rigorously proved for the full Navier-Stokes equations. On the other hand these theories have been verified on models approximating or truncating the Navier-Stokes equations (Burgers-like equations, Lorenz equation) or on particular flow s(e.g. Couette flow between concentric cylinders and Bénard convection). For a discussion of these theories into the general context of the turbulence we quote [29].

All these theories give a qualitative insight into the phenomenon of the origin of turbulence. The qualitative agreement between hydrodynamic stability (and to an extent branching) theory and experiments for the first stages of transition of particular fluid flows may be found in [30] (from Ch. 1). Experiments on fluid layers heated from below give evidence of the Rayleigh number Ra_1 where the quilibrum of the fluid layer is lost and steady convection branches from it, $Ra_2 > Ra_1$ where steady convection branches into unsteady convection, and Ra_3 around which the fluid motion becomes chaotic [7], [6].

Let us finally remark that early studies in branching theory concern mainly the onset of steady secondary solutions. Since the emergence of secondary periodic motions (known as Hopf bifurcation) occurs widely in nature and technics and consequently in the origin of turbulence of synergetic systems and impressively large literature is now available. We mention only that the Hopf bifurcation theorem for ordinary differential equations has been generalized to the case of Navier-Stokes partial differential equations by Iooss [20], Yudovich [45] and Joseph and Sattinger [22] (from Ch. 3) by means of three different methods. Other recent extensions and applications of the Hopf bifurcation theorem in fluid dynamics may be found in [17], [24].

All the problems discussed in this chapter are similar to those encountered in the study of turbulence of many other dynamical systems from physics, chemistry, biology, etc. This study is framed into the so-called global analysis theory. Fluid dynamics provides this theory with a very rich source of examples which waits for classification and rigorous investigation.

REFERENCES

[1] Afraimovich, V. S., V. V. Bikov, L. A. Shilnikov, *The origin and structure of the Lorenz attractor*, DAN, **234** (1977) 336–339.

[2] Arsenev, A. A., *Construction for turbulent measures for Navier-Stokes equations*, Soviet Math. Dokl., **16** (1975), 1422–1424. (DAN, **225**, (1975) 18–20)

[3] Bensoussan, A., R. Temam, *Equations stochastiques du type Navier-Stokes*, J. Funct. Analysis, **13**, 2 (1973) 195–222.

[4] Bul, E. B., Ya. G. Sinai, *On some structurally stable mechanism of onset of invariant hyperbolic sets*, Nauka, Moskow, 1978, 104–112 (Russian).

[5] Bunimovich, L. A., Ya. G. Sinai, *Stochasticity of the attractor from Lorenz model*, Nonlinear waves, Nauka, Moskow, 1979, 212–226 (Russian).

[6] Busse, F. H., J. A. Whitehead, *Instabilities of convection rolls in a high Prandtl number fluid*, J. Fluid Mech., **47**, 2 (1971) 305–320.

[7] Dubois, M., P. Bergé, *Velocity field in the Rayleigh-Bénard instability: transitions to turbulence*, Synergetics vol. 3, A. Pacault, C. Vidal (eds.), Springer, Berlin 1979, 85–93.

[8] Foias, C., *Statistical study of Navier-Stokes equations*, Rend. Sem. Mat. Padova, I: **48** (1973) 219–349; II: **49** (1973) 9–123.

[9] Foias, C., *Quelques resultats sur les fluides en rotation*, Séminaire Goulaouic-Lions-Schwartz, 1974–1975, exposé n° VI, VI. 1–VI. 8.

[10] Foias, C., *A survey on the functional dynamical system generated by the Navier-Stokes equations*, Lecture Notes in Mathematics **771**, Springer, Berlin, 1980, 196—202.

[11] Foias, C., R. Temam, *Sur certaines propriétés génériques des équations de Navier-Stokes*, C. R. Acad. Sci. Paris, Série A., **280** (1975) 563—565.

[12] Foias, C., R. Temam, *On the stationary statistical solutions of the Navier-Stokes equations and turbulence*, Publications Mathématiques d'Orsay, n⁰ 120—75—28, 1975.

[13] Foias, C., R. Temam, *A generic property of the set of stationary solutions of Navier-Stokes equations*, Lecture Notes in Mathematics **565**, Springer, Berlin, 1976, 24—28.

[14] Foias, C., R. Temam, *Structure of the set of stationary solutions of the Navier-Stokes equations*, Comm. Pure Appl. Math., **30** (1977) 149—164.

[15] Foias, C., R. Temam, *Remarques sur les équations de Navier-Stokes stationnaires et les phénomènes successifs de bifurcation*, Annali Scuola Norm. Sup. di Pisa, Série IV, **5**, 1 (1978) 29—63.

[16] Foias, C., R. Temam, *Some analytic and geometric properties of the solutions of the evolution Navier-Stokes equations*, J. Math. Pures et Appl., **58** (1979) 339—368.

[17] Hassard, B. D., N. D. Kazarinoff, Y-H. Wan, *Theory and applications of Hopf bifurcation*, Cambridge University Press, 1981.

[18] Heywood, J. G., *Auxiliary flux and pressure conditions for Navier-Stokes problems*, Lecture Notes in Mathematics **771**, Springer, Berlin, 1980, 223—234.

[19] Hopf, E., *Statistical hydromechanics and functional calculus*, J. Rational Mech. Analysis, **1** (1952) 87—123.

[20] Iooss, G., *Existence et stabilité de la solution périodique secondaire intervenant dans les problèmes d'évolution du type Navier-Stokes*, Arch. Rational Mech. Analysis, **47** (1972) 301—329.

[21] Iooss, G., *Bifurcation of invariant tori in* R^3, *Bifurcation and nonlinear eigenvalue problems*, Lecture Notes in Mathematics, **782**, Springer-Verlag, Berlin, 1980, 192—200.

[22] Landyzhenskaya, O. A., A. M. Vershik, *Sur l'évolution des mesures déterminées par les équations de Navier-Stokes et la résolution du problème de Cauchy pour l'équation statique de E. Hopf*, Annali della Scuola Norm. Sup. di Pisa, Série IV, 2 (1977) 209—230.

[23] Lakshmikantham, V (ed.), *Nonlinear systems and applications*, Acad. Press, 1977, 199—210.

[24] Marsden, J., M. McCracken, *The Hopf bifurcation and its applications*, Applied Math. Sci., vol. 19, Springer-Verlag, New York, 1976.

[25] Newhouse, S., D. Ruelle, F. Takens, *Occurrence of strange Axiom A attractors near quasi-periodic flows on* T^m, $m \geqslant 3$, Commun. Math. Phys. **64** (1978) 35—40.

[26] Nitecki, Z. (ed.), *Approximation methods for Navier-Stokes problems*, Lecture Notes in Mathematics **771**, Springer-Verlag, Berlin, 1980.

[27] Poston, T., I. Stewart, *Catastrophe theory and its applications*, Surveys and references works in mathematics, Pitman, London, 1978.

[28] Ratiu, T., P. Bernard, (eds.), *Turbulence*, Lecture Notes in Mathematics **615**, Springer-Berlin, 1977.

[29] Rose, H. A., P. H. Sulem, *Fully developed turbulence and statistical mechanics*, Le Journal de Physique, **39**, 5 (1978) 441—484.

[30] Ruelle, D., *Dynamical system with turbulent behaviour*, Lecture Notes in Physics **80**, Springer, Berlin, 1978, 341—360.

[31] Ruelle, D., F. Takens, *On the nature of turbulence*, Comm. Math. Phys., **20**, (1971) 167—192; **23** (1971) 343—344.

[32] Ruelle, D., *Microscopic fluctuations and turbulence*, Phys. Lett., **72 A** (1979) 81—82.

[33] Savulescu, St. N., *A stochastic model of fluctuations in boundary layer flows*, Lecture Notes in Physics, **76**, Springer, Berlin, 1978, 289.

[34] Savulescu, St. N., A. Georgescu, H. Dumitrescu M. Bucur *Cercetări matematice în teoria modernă a stratului limită*, Ed. Acad. R. S. România, 1981.

[35] Sattinger, D. H., *Group theoretic methods in bifurcation theory*, Lecture Notes in Mathematics **762**, Springer Verlag, Berlin, 1979.

[36] Sattinger, D. H., *Bifurcation and symmetry breaking in applied mathematics*, Bull. (New Series) of the Amer. Math. Soc., **3**, 2 (1980) 779–819.

[37] Scheffer, V., *Turbulence and Hausdorff dimension*, Lecture Notes in Mathematics **565**, Springer, Berlin, 1976.

[38] Sinai, Ya. G. *Ergodic theory as a useful mathematical theory*, Recent advances in statistical mechanics, Proc. of the Brasov Intern. School, August-September, 1979, 109–147.

[39] Sinai, Ya. G., *Stochasticity of dynamical systems*, Nonlinear waves, Nauka, Moscow, 1979, 192–211 (Russian).

[40] Temam, R. (ed.), *Turbulence and Navier-Stokes equations*, Lecture Notes in Mathematics **565**, Springer, Berlin, 1976.

[41] Temam, R. *Une propriété générique de l'ensemble des solutions stationnaires ou périodiques des équations de Navier-Stokes*, Actes du Colloque Franco-Japonais, Tokyo, sept. 1976.

[42] Temam, R., *Navier-Stokes equations, theory and numerical analysis*, North Holland, Amsterdam, 1977.

[43] Vishik, M. I., *Cauchy problem for the Hopf equation corresponding to quasilinear parabolic equations*, Soviet Mat. Dokl., **16** (1975) 1126–1130 (DAN, **224** (1975) 23–26).

[44] Vishik, M. I., A. V. Foursikov, *Solutions statistiques homogènes des systèmes différentiels paraboliques et du systéme de Navier-Stokes*, Annali della Scuola Norm. Sup. di Pisa, Série IV, 3 (1977) 531–576.

[45] Yudovich, V. I., *The onset of auto-oscillations in a fluid*, Prikl. Mat. Mekh., **35**, (1971) 638–655.

Chapter 5

THE INFLUENCE OF THE PRESENCE OF A POROUS MEDIUM ON HYDRODYNAMIC STABILITY

After deriving the eigenvalue problems in § 5.1 which describe the classical linear Rayleigh-Taylor and Kelvin-Helmholtz instabilities in the presence of porous media, various particular equilibria and motions characterized by linear velocity profiles are analysed in § 5.2 and § 5.3. In this last case stability curves in the (η_*, G_3^2) plane are deduced; G_3 is the dimensionless number introduced by Şt. I. Gheorghitza. The Rayleigh-Taylor instability is also analysed for the case of a cylindrical domain of motion.

§ 5.1. THE MATHEMATICAL PROBLEM

In the previous chapters the density was assumed to be constant; suppose now that it depends on the height, the simplest case being that of two horizontal layers of fluids in static equilibrium having different but constant densities inside each layer. The configuration in which the density of the lower fluid is greater than that of the upper one is stable, the inverse one — unstable, this second case giving rise to a *gravitational convective motion*. This type of *instability* is called the *Rayleigh-Taylor* instability and is studied in meteorology, the Earth's atmosphere being a fluid with stratified density.

Another type of instability is that of the flow of two horizontal fluid layers whose velocities are constant and parallel to their frontier; for small differences between velocities, the configuration is stable and for velocities surpassing some critical value — unstable. An instability of this type is called the *Kelvin-Helmholtz instability* and appears, for instance, at the surface of the seas and oceans or at the frontier between water and petroleum. A detailed study of these instabilities may be found in [5] (from Ch. 1).

The presence of the porous medium is extremely important for many problems of practical interest, whence the necessity of studying the Rayleigh-Taylor or Kelvin-Helmholtz instability in the presence of these media. In the classical linear frame this theory has been initiated [4] in 1967 and subsequently developed [5]—[7] by

Şt. I. Gheorghitza; it has been extended to nonuniform flows [9] and to domains of motions bounded in two directions [10]. The most general case which may be treated by Gheorghitza's theory is that in which the velocity, density, and filtration coefficient are dependent on height and the free fluid is viscous; hitherto only the cases where two of these conditions are fulfilled have been treated.

Remark that the theory considered treats the instability of free fluid motions bounded by porous media (by free fluid we mean a fluid in the absence of the porous medium); but there exist instability theories for motions in porous media, as for instance in thermal convection [11], [1], [2],[8].

The mathematical problem of the linear stability of stratified free flows in the presence of porous media consists of the equations of the perturbation in the porous medium and in the free fluid, the continuity conditions for the pressure and normal velocity at the frontier of two free fluids and at the surface bounding the porous medium.

Let us consider as horizontal layer a porous medium, characterized by dimensionless quantities $m > 1$ (porosity), and K (filtration coefficient), unbounded along the x and y directions. Writing a star for the quantities which characterize the linear fluid motion in this layer, the equation of motion is

$$\frac{\rho}{m}\frac{\partial \mathbf{u}_*}{\partial t} = -\operatorname{grad}(p_* + \rho g z) - \rho g K^{-1}\mathbf{u}_*,$$

where g is the gravitational acceleration. Let $\mathbf{V}_*, p_*$ be a steady fluid motion in this layer and let $\mathbf{u}_* = \mathbf{V}_* + \mathbf{v}_*$, $q_* = P_* + p_*$ be the perturbed motion whose perturbations $\mathbf{v}_*, p_*$ satisfy the equation

$$\frac{\rho}{m}\frac{\partial \mathbf{v}_*}{\partial t} = -\operatorname{grad} p_* - \rho k K^{-1}\mathbf{v}_*.$$

Assume that the density is constant in each layer, hence, $\operatorname{div}\mathbf{v}_* = 0$. Take the perturbations as normal modes $\mathbf{v}_*(x, y, z, t) = \mathbf{v}_{0*}(z)\exp[\mathrm{i}(k_x x + k_y y) + nt]$ and denote $k^2 = k_x^2 + k_y^2$, $D = \dfrac{\mathrm{d}}{\mathrm{d}z}$, $\mathbf{v}_{0*} = (u_*, v_*, w_*)$.

We obtain

$$(\mathrm{D}^2 - k^2)\,p_* = 0, \tag{5.1.1}$$

$$u_* = \frac{-\mathrm{i}\,k_* p_*}{\rho(nm^{-1} + kK^{-1})},\quad v_* = \frac{-\mathrm{i}\,k_y p_*}{\rho(nm^{-1} + gK^{-1})}, \tag{5.1.2}$$

$$w_* = \frac{-Dp_*}{\rho(nm^{-1} + gK^{-1})}.$$

Consider now a horizontal layer of free fluid at rest, characterized by the dynamic viscosity μ and let us perturb this static equilibrium by a perturbation $\mathbf{v} = \mathbf{v}_0(z) \exp[i(k_x x + k_y y) + nt]$, where $\mathbf{v}_0(z) = (u, v, w)$. Then, eliminating the pressure, it is easy to deduce for w the equation [1]

$$(5.1.3) \quad D\{[\rho - \mu n^{-1}(D^2 - k^2)] Dw - n^{-1}(D\mu)(D^2 + k^2) w\} = \\ = -k^2\{gn^{-2}wD\rho + [\mu n^{-1}(D^2 - k^2) - \rho] w + 2n^{-1}(D\mu) Dw\}.$$

The condition on a horizontal impervious wall lying in the plane $z = z_i$ is

$$(5.1.4) \quad w = Dw = 0,$$

and on the frontier of separation $z = z_s$ of two fluids characterized by quantities denoted by (′) and (″), we have

$$(5.1.5) \quad w' = w'', \quad Dw' = Dw'', \quad \mu'(D^2 + k^2) w' = \mu''(D^2 + k^2) w'';$$

on the frontier $z = z_s$ we must also impose the condition of continuity for the normal tension, which, denoting $\delta(f) = f''(z) - f'(z)$, can be written as

$$(5.1.6) \quad \delta\{[\rho - \mu n^{-1}(D^2 - k^2)] Dw - n^{-1}(D\mu)(D^2 + k^2) w\} = \\ = -(kn^{-1})^2 [g\delta(\rho) - k^2 T] w' - 2k^2 n^{-1}(Dw')\, \delta(\mu),$$

where T is the surface tension on the plane $z = z_s$.

The expression of the pressure with respect to w is

$$(5.1.7) \quad k^2 p = [\mu(D^2 - k^2) - \rho n] Dw + (D\mu)(D^2 + k^2) w.$$

Consider now, in a horizontal layer the motion of an inviscid free fluid having the velocity $(U(z), 0, 0)$. Perturbing this flow with perturbations of the form $\mathbf{v} = \mathbf{v}_0 \exp i(k_x x + k_y y + nt)$ keeping unchanged the rest of the above notations, we deduce from the equation of motion the relation

$$(5.1.8) \quad (n + k_x U)(D^2 - k^2) w - k_x(D^2 U) w - \\ - \frac{gk^2 w D\rho}{\rho(n + k_x U)} + [(n + k_x U) Dw - k_x(DU) w] \frac{D\rho}{\rho} = 0.$$

The conditions

$$(5.1.9) \quad w'(n + k_x U'') = w''(n + k_x U')$$

$$(5.1.10) \quad \delta\{\rho(n + k_x U) Dw - \rho k_x(DU) w\} = \\ = gk^2 \left[\delta(\rho) - \frac{k^2 T}{g}\right] \left(\frac{w'}{n + k_x U'}\right)$$

are fulfilled on the interface $z = z_i$.

We mention that in this case the pressure has the expression

$$-k^2 p = \mathrm{i}\rho(n + k_x U)\,\mathrm{D}w - \mathrm{i}\rho k_x(\mathrm{D}U)\,w. \tag{5.1.11}$$

Let the plane $z = z_p$ separate a free fluid layer from a porous layer; the boundary conditions at $z = z_p$ are then

$$p = p_*, \quad w = w_*, \tag{5.1.12}$$

where w_* has the expression (5.1.2) and p—the expression (5.1.7) or (5.1.11).

If $z = z_*$ is the plane of separation of two porous layers, then for $z = z_*$ we impose the conditions

$$p_{1*} = p_{2*}, \quad w_{1*} = w_{2*}. \tag{5.1.12$'$}$$

In this way, the mathematical problem of the stability of fluid motions in the presence of porous media is an eigenvalue problem which, in the case of the Reyleigh-Taylor instability, consists of Equations (5.1.1), (5.1.3) and the boundary conditions corresponding to the frontiers of the given configuration and in the case of the Kelvin-Helmholtz instability of Equations (5.1.1), (5.1.8) and the corresponding boundary conditions. The eigenfunction is (w, p_*) and n is the eigenvalue. In practice, from Equations (5.1.1), (5.1.3) and (5.1.8) two, four and, respectively, two linearly-independent solutions (w, p_*) are deduced for each layer; then the general form of p_* and w is written as a linear combination with unknown coefficients and to this general solution the corresponding boundary conditions are imposed. Thus, an algebraic system of equations is obtained for coefficients; in order to have nontrivial solutions, a determinant must vanish, which yields the *characteristic equation* having n as an unknown. The sign of the real part of n (for the Rayleigh-Taylor case and of the imaginary part in the other case) indicates stability or instability. Since the conditions on the frontier between a porous layer and a free fluid as well as those between two porous media introduce an n in the characteristic equation while the conditions on an interface an n^2, one can foresee for a concrete flow, the degree of this equation.

§ 5.2. RAYLEIGH-TAYLOR INSTABILITY

Consider a horizontal layer $(L < z < \infty)$ of free fluid having constant density ρ_2, another layer of free fluid $(0 < z < L)$ having the constant density ρ_1 and a porous medium $(-\infty < z < 0)$; assuming that the density in the two adjacent layers of free fluid and respectively porous medium is the same, the fluid which saturates the porous medium must also have the density ρ_1. If the free fluid is an ideal one then from (5.1.3) we get

$$(D^2 - k^2)\, w = 0,$$

which has the general solutions $w_1 = Ae^{kz} + Be^{-kz}(0 < z < L)$ and $w_2 = Ce^{-kz}(z > L)$; the fluid being ideal, w must satisfy only the first condition (5.1.5). It follows

$$Ae^{kL} + Be^{-kL} = Ce^{-kL}. \tag{5.2.1}$$

Taking into account this relation, Condition (5.1.6) becomes

$$\rho_2 + \rho_1(Ae^{kL} - Be^{-kL})\,(Ae^{kL} + Be^{-kL}) = \\ = kn^{-2}\,[g(\rho^2 - \rho_1) - k^2T]. \tag{5.2.2}$$

Since, in the case considered, from Equation (5.1.1) we obtain $p_* = Me^{kz}$ (the other solution is physically unacceptable since it becomes unbounded as $z \to -\infty$) conditions (5.1.12) can be written in the form

$$A + B = -\rho_1^{-1} Mk(nm^{-1} + gK^{-1})^{-1}, \\ A - B = -(\rho_1 n)^{-1}\, kM. \tag{5.2.3}$$

The system (5.2.2), (5.2.3) in A, B and M has the characteristic equation

$$\rho_2 + \rho_1\, \frac{nK(1 + m \operatorname{th} kL) + mg}{nK(\operatorname{th} kL + m) + mg \operatorname{th} kL} = \frac{k}{n^2}\,[(\rho_2 - \rho_1)\, g - k^2T],$$

or, equivalently,

$$n^3 + mg((\rho_2 \operatorname{th} kL + \rho_1)\, Qn^2 - K(m + \operatorname{th} kL)\, Q\theta n - \\ - mgQ\theta \operatorname{th} kL = 0, \tag{5.2.4}$$

where $\theta = k\,[(\rho_2 - \rho_1)\, g - k^2T]$, $Q^{-1} = K[(m + \operatorname{th} kL)\, \rho_2 + (1 + m \operatorname{th} kL)\, \rho_1]$. As $kL \to \infty$, we obtain from (5.2.4) the characteristic equation which governs the Rayleigh-Taylor instability in the absence of the porous medium

$$\rho_2 + \rho_1 = kn^{-2}\,[g(\rho_2 - \rho_1) - k^2T]. \tag{5.2.4$'$}$$

The characterization (given in § 5.1) of this instability follows immediately.

When $K = 0$, i.e. when instead of the porous medium we have an impervious one, we get from (5.2.4)

$$\rho_2 + \rho_1 \operatorname{cth} kL = kn^{-2}\,[(\rho_2 - \rho_1)\, g - k^2T]. \tag{5.2.4$''$}$$

It follows that the corresponding stratification is stable if $\rho_2 < \rho_1$ and unstable if $\rho_2 > \rho_1$. In this last case the perturbations with the wave number $k \in (0, k_c)$ where, $k_c = [(\rho_2 - \rho_1)\, gT^{-1}]^{\frac{1}{2}}$, are amplified as $t \to \infty$.

When $K \to \infty$, i.e. when instead of a porous medium we have a free fluid, we also have $m \to 1$ [3], and we still obtain Equation (5.2.4)′.

In the neutral case $k = k_c$, if $\rho_2 > \rho_1$, Equation (5.2.4) has two roots $n = 0$, the third one being

$$n = \frac{-mg(\rho_2 \operatorname{th} k_cL + \rho_1)}{K[(m + \operatorname{th} kL)\, \rho_2 + (1 + m \operatorname{th} k_cL)\, \rho_1]},$$

which corresponds to the presence of the porous medium. It is easy to see that in the case considered the porous medium does not induce instability, a fact that could be physically anticipated owing to the dissipative effect of this medium. Nevertheless in § 5.3 it is shown that sometimes the presence of the porous medium can induce instability in the same way; for instance, the viscosity, which in general dissipates energy, nevertheless induces instability in plane parallel flows.

§ 5.3. THE KELVIN-HELMHOLTZ INSTABILITY

We assume now that the free fluid lying between two horizontal planes has the velocity $(U, 0, 0)$, where $U = \alpha + \beta z$, α and β are real numbers. We note that the case $\alpha = \beta = 0$ corresponds to the fluid at rest, and $\alpha \neq 0$, $\beta = 0$ — to the uniform flow whose Kelvin-Helmholtz instability is analised in [7].

For this velocity Equation (5.1.8) becomes

$$(\mathrm{D}^2 - k^2)\, w = 0, \tag{5.3.1}$$

whose solutions are of the form $w = A\mathrm{e}^{kz} + B\mathrm{e}^{-kz}$. From this equation we deduce also $p_* = M\mathrm{e}^{kz} + N\mathrm{e}^{-kz}$.

The case of two free fluids between two porous media. Assume that the planes $z = L$, $z = 0$ and $z = -H$ bound four fluid layers as follows: in the domain $L < z < \infty$ there is a free fluid having the velocity $V_2 = \alpha_2 + \beta_2 z$, density ρ_2 and the perturbation velocity $w_2 = C\mathrm{e}^{-kz}$. In the zone $0 < z < L$ there lies a free fluid characterized by $V_1 = \alpha_1 + \beta_1 z$, $w_1 = A\mathrm{e}^{kz} + B\mathrm{e}^{-kz}$; for $-H < z < 0$ we have a porous medium with $p_{1*} = M\mathrm{e}^{kz} + N\mathrm{e}^{-kz}$, and for $z < -H$ — another porous medium characterized by $p_{2*} = P\mathrm{e}^{kz}$.

Imposing Conditions (5.1.9) and (5.1.10) at $z = L$, Conditions (5.1.12) at $z = 0$ and Conditions (5.1.12)' at $z = -H$ we obtain the system

$$
(5.3.2)\quad \begin{cases}
a_1 C e^{-kL} = a_2(A e^{kL} + B e^{-kL}) - \rho_2 a_2 C k e^{-kL} - \\
- \rho_2 k_x \beta_2 C e^{-kL} - \rho_1 a_1 k A e^{kL} + \rho_1 a_1 k B e^{-kL} + \\
+ \rho_1 k_x \beta_1 A e^{kL} + \rho_1 k_x \beta_1 B e^{-kL} = -a_5 k^2 \dfrac{A e^{kL} + B e^{-kL}}{a_1}, \\
M + N = \rho_1 i k^{-2}[k_x \beta_1 A + k_x \beta_1 B - (a_1 - \\
\qquad - k_x L \beta_1)\, k(A - B)], \\
A + B = \rho_1^{-1}\, i a_3^{-1} k(M - N), \\
M e^{-kH} + N e^{kH} = P e^{-kH}, \\
a_4 k(M e^{-kH} - N e^{kH}) = a_3 k P e^{-kH},
\end{cases}
$$

where

$$a_1 = n + k_x V_1(L) = n + k_x(\alpha_1 + \beta_1 L),$$

$$a_2 = n + k_x V_2(L) = n + k_x(\alpha_2 + \beta_2 L),$$

$$a_1 - k_x L \beta_1 = n + k_x \alpha_1 = n + k_x V_1(0),$$

$$a_3 = n m_1^{-1} - i g K_1^{-1}, \quad a_4 = n m_2^{-1} - i g K_2^{-1},$$

$$a_5 = \rho(\rho_1 - \rho_2 + k^2 T g^{-1}).$$

The characteristic equation corresponding to (5.3.2) is

$$
(5.3.3)\quad \rho_2 a_2^2 + \rho_1 a_1^2 \frac{c \operatorname{th} kL + d}{c + d \operatorname{th} kL} = kg(\rho_1 - \rho_2 + k^2 T g^{-1}) + \frac{\rho_1 k_x \beta_1 a_1 - \rho_2 k_x \beta_2 a_2}{k},
$$

where $c = (a_1 - k_x L \beta_1)(a_3 + a_4 \operatorname{th} kH)$, and $d = a_3(a_3 \operatorname{th} kH + a_4) + k_x \beta k^{-1}(a_3 + a_4 \operatorname{th} kH)$. As would only be expected, for $\beta_1 = \beta_2 = 0$ we regain the characteristic equation corresponding to uniform streams [7].

Let us now suppose that $\beta_1^2 + \beta_2^2 \neq 0$. For $K_1 = 0$, (5.3.3) becomes

$$
(5.3.4)\quad \rho_2 a_2^2 + \rho_1 a_1^2 c \operatorname{th} kL = kg(\rho_1 - \rho_2 + k^2 T g^{-1}) + \frac{\rho_1 k_x \beta_1 a_1 - \rho_2 a_2 k_x \beta_2}{k},
$$

from which it can be seen that if the first porous layer is impervious ($K_1 = 0$) n no longer depends on m_2 and on K_2 as was expected. If $m_1 = m_2 = m$ and $K_1 = K_2 = K$, from (5.3.3) we get

$$\rho_2 a_2^2 + \rho_1 a_1^2 \frac{(a_1 - k_x L \beta_1) \operatorname{th} kL + a_3 + k_x \beta_1 k^{-1}}{a_1 - k_x L \beta_1 + (a_3 + k_x \beta_1 k^{-1}) \operatorname{th} kL} = \tag{5.3.5}$$

$$= kg(\rho_1 - \rho_2 + k^2 T g^{-1}) \frac{\rho_1 k_x \beta_1 a_1 - \rho_2 k_x \beta_2 a_2}{k},$$

and as $K \to 0$, we obtain (5.3.4). Suppose $K_1 \to \infty$, $K_2 \to \infty$ ($m_1 = 1$, $m_2 = 1$) and $V_1 = 0$. In this case, (5.3.3) leads to

$$\rho_2 a_2^2 + \rho_1 n^2 = kg(\rho_1 - \rho_2 + k^2 T g^{-1}) - \rho_1 a_2 k_x k^{-1} \beta_2.$$

If, in addition, $\beta_2 = 0$, $\alpha_2 = 0$, we obtain formula (24) from [7]. This last one can be also obtained if $K_1 = K_2 = V_1 = V_2 = 0$.

From (5.3.3) we can deduce the characteristic equation for the case of a free fluid in the half plane $z > 0$ and a porous medium lying in $z < 0$, taking $\alpha_1 = \alpha_2$, $\beta_1 = \beta_2 = 0$, $\rho_1 = \rho_2$, $T = 0$, $m_1 = m_2 = m$, $K_1 = K_2 = K$. Thus we have $a_3(a_1 + a_3) = 0$ whence $n_1 = [\mathrm{i}gK^{-1} - k_x\alpha_1][1 + m^{-1}]^{-1}$ and $n_2 = \mathrm{i}gmK^{-1}$.

Finally $m_1 = m_2 = m$, $K_1 = K_2 = K$ and $\beta_1 = \beta_2 = 0$ correspond to the case of a porous medium in the half plane $z < 0$ and two free fluids in the half plane $z > 0$, and the characteristic equation is

$$\rho_2(n + k_x\alpha_2)^2 + \rho_1(n + k_x\alpha_1)^2 \frac{(n + k_x\alpha_1) \operatorname{th} kL + nm^{-1} - \mathrm{i}gK^{-1}}{m + k_x\alpha_1 + (nm^{-1} - \mathrm{i}gK^{-1}) \operatorname{th} kL} =$$

$$= k[(\rho_2 - \rho_1) g - k^2 T],$$

i.e. just relation (39) from [4].

The free fluid between two porous media. We consider the plane $x0z$ divided into three strips. In the middle, from $z = L$ to $z = 0$, there is the free fluid whose basic flow is $V = \alpha + \beta z$ and the z-component of the perturbation velocity is $w = Be^{kz} + Ce^{-kz}$. In the first layer, the second porous medium having $p_{2*} = De^{-kz}$ is located between $z = \infty$ and $z = L$ and in the first strip, there is another medium between $z = 0, z = -\infty$, having $p_{1*} = Ae^{kz}$. Writing Conditions (5.1.12) for $z = L$ and $z = 0$, we get system (5.3.6).

$$\begin{cases} De^{-kL} = \rho \mathrm{i} k^{-2} [k_x\beta(Be^{kL} + Ce^{-kL}) - ak(Be^{kL} - Ce^{-kL})], \\ Be^{kL} + Ce^{-kL} = -\rho^{-1}\mathrm{i}b_2^{-1}kDe^{-kL}, \\ A = \rho \mathrm{i} k^{-2}[k_x\beta(B + C) - a_0 k(B - C)], \\ B + C = \rho^{-1}\mathrm{i}b_1^{-1}kA, \end{cases} \tag{5.3.6}$$

where $a = n + k_x(\alpha + \beta L)$, $a_0 = n + k_x\alpha$ and $b_j = nm_j^{-1} - \mathrm{i}gK_j^{-1} (j = 1, 2.)$. The characteristic equation corresponding to (5.3.6) is

$[k^2b_1b_2 + k^2aa_0 - k_x\beta k(b_1 - b_2) - k_x^2\beta^2]$ th $kL + k^2(ab_1 + b_2a_0) + k_x^2\beta^2 kL = 0$, or, equivalently,

$$(5.3.7)\qquad n^2[(m_1^{-1}m_2^{-1} + 1)\,\mathrm{th}\,kL + m_1^{-1} + m_2^{-1}] + n\{m_1^{-1}(k_x\alpha + k_x\beta L) + m_2^{-1}k_x\alpha + (2\alpha k_x + \beta L k_x)\,\mathrm{th}\,kL - \mathrm{i}g[K_1^{-1}K_2^{-1} + (K_1^{-1}m_2^{-1} + K_2^{-1}m_1^{-1})\,\mathrm{th}\,kL] + k_x k^{-1}\beta(m_2^{-1} - m_1^{-1})\,\mathrm{th}\,kL\} + k_x^2(\alpha^2 + \alpha\beta L)\,\mathrm{th}\,kL - k_x^2\beta^2k^{-2}\,\mathrm{th}\,kL + k_x^2Lk^{-1}\beta^2 - \mathrm{i}gk_x[K_2^{-1}\alpha + K_1^{-1})\,\alpha + \beta L) + \beta k^{-1}(K_2^{-1} - K_1^{-1})\,\mathrm{th}\,kL] - g^2K_1^{-1}K_2^{-1}\,\mathrm{th}\,kL = 0.$$

For $\beta = 0$ or for $k_x = 0$, we obtain analogous cases, studied in [7]. For $m_1 = m_2 = m$ and $K_1 = K_2 = K$, from (5.3.7) we get

$$(5.3.8)\qquad n^2[(1 + m^{-2})\,\mathrm{th}\,kL + 2m^{-1}] + n\{(-2\mathrm{i}gK^{-1}m^{-1} + 2k_x\alpha + k_x\beta L)\,\mathrm{th}\,kL + 2k_x\alpha m^{-1} - 2\mathrm{i}gK^{-1} + k_x\beta L m^{-1}\} + [(-g^2K^{-2} + k_x^2\alpha^2 + k_x^2\alpha\beta L - k_x^2\beta^2k^{-2})\,\mathrm{th}\,kL - \mathrm{i}gK^{-1}k_x(2\alpha + \beta L) + k_x^2\beta^2Lk^{-1}] = 0.$$

Equation (5.3.8) is of second degree, and has the form

$$(5.3.8)'\qquad an^2 + (b + \mathrm{i}c)\,n + d + \mathrm{i}f = 0,$$

where a, b and d are positive numbers and c and f are negative. Setting $n_j = n_{rj} + \mathrm{i}n_{ij}$, $j = 1, 2$, for the roots of (5.3.8)′, we have

$$D = \frac{af^2 - bfc + dc^2}{a^3} = -n_{1i}n_{2i}\,[(n_{1r} - n_{2r})^2 + (n_{1i} + n_{2i})^2].$$

Hence, when $Da^3 > 0$, motion is unstable, for $Da^3 = 0$ we obtain the boundary between stability and instability, and for $Da^3 < 0$ the motion is stable. Denoting $\delta = \dfrac{\beta L}{\alpha}$, $kL = \eta$ (δ and η being dimensionless positive numbers), we get

$$(5.3.9)\qquad Da^3 = g^2K^{-2}k_x^2\alpha^2(2 + \delta)^2\,[2m^{-1} + (1 + m^{-2})\,\mathrm{th}\,kL] - 2g^2K^{-2}k_x^2\alpha^2(2 + \delta)^2\,(m^{-1} + \mathrm{th}\,kL) - (1 + m^{-1}\,\mathrm{th}\,kL) + 4g^2K^{-2}k_x\alpha^2(1 + m^{-1}\,\mathrm{th}\,kL)^2\left[\left(1 + \delta - \frac{\delta^2}{\eta^2}\right)\mathrm{th}\,kL + \frac{\delta^2}{\eta}\right] - 4g^4K^{-4}\,\mathrm{th}\,kL\,[1 + \mathrm{th}\,kL]^2.$$

When Da^3 vanishes, by means of the dimensionless Gheorghitza's number

$$G_3^2 = g^{-2}k_x^2\alpha K^2,$$

we obtain in the plane (G_3^2, η) (for m and δ taking assigned values) the following stability curve

$$G_3^2 = \left\{\delta^2\left[\frac{1}{\eta \operatorname{th} \eta} - \frac{1}{\eta^2} - \frac{1}{4} - \frac{1 - \operatorname{th}^2 \eta}{4(m + \operatorname{th} \eta)^2}\right] - \right. \tag{5.3.10}$$

$$\left. - \delta \frac{1 - \operatorname{th}^2 \eta}{(m + \operatorname{th} \eta)^2} - \frac{1 - \operatorname{th}^2 \eta}{(m + \operatorname{th} \eta)^2}\right\}^{-1}.$$

If the expression between brackets is negative, then the motion is stable, since the right-hand side of (5.3.9) is negative. Thus, we can restrict the intervals for δ, η and m; denote

$$g(\eta) = \frac{1}{\eta \operatorname{th} \eta} - \frac{1}{\eta^2} - \frac{1}{4}, \qquad f(\eta, m) = \frac{1 - \operatorname{th}^2 \eta}{4(m + \operatorname{th} \eta)^2}$$

and consider the equation

$$\delta^2(g - f) - 4f\delta - 4f = 0, \tag{5.3.11}$$

whose roots are

$$\delta_1 = \frac{2\sqrt{f}}{\sqrt{g} - \sqrt{f}}, \qquad \delta_2 = -\frac{2\sqrt{f}}{\sqrt{g} - \sqrt{f}}. \tag{5.3.12}$$

It may be shown that f and g are positive decreasing functions. If $g = f$, then the right-hand side of (5.3.11) is negative, hence the motion is stable. To have instability, we must impose the condition $g > f$, namely

$$\frac{1}{\eta \operatorname{th} \eta} - \frac{1}{\eta^2} - \frac{1}{4} - \frac{1 - \operatorname{th}^2 \eta}{4(m + \operatorname{th} \eta)^2} > 0. \tag{5.3.13}$$

In this way, for a given m, we find the bounds between which η may vary in order to have instability. For $m = 1$, this interval is $0.6 \leqslant \eta \leqslant 2.34$, for $m = 0.5$ we have $1.03 \leqslant \eta \leqslant 2.26$ and for $m = 0.25$ it corresponds $1.33 \leqslant \eta \leqslant 2.234$. More generally, it may be shown that if m decreases, the interval for which inequality (5.3.13) holds decreases too. For $m = 0$ we have only stability and we could expect that there exists an interval of variation of m, inducing stability for every $\eta \in [0.6; 2.34]$. Elementary calculations indicate the interval $[0; 0.009]$.

So far, we have shown that the instability may take place only if $\eta \in [0.6; 2.34]$ and $m > 0.009$. Taking into account (5.3.13), it may be proved that the extra conditions $\delta_1 > 0$ and $\delta_2 < 0$ must hold. Hence the trinomial $\delta^2(g-f) - 4f\delta - 4f$ shall be positive iff $\delta > 2\sqrt{f}(\sqrt{g} - \sqrt{f})^{-1}$. For every m and η we shall find a bound for δ, and, as was to be expected, for small δ the motion is always stable. The lower bound of δ is increased when m decreases. Thus, the greatest δ for which we still have stability is $\delta = 1.9145$ for $\eta = 1.9$. Some bounds of δ for various m and η are given in Tables 1, 2 and 3.

Table 1. *Instability lower bounds for* δ *when* $m = 1$.

η	0.6	0.7	0.8	1	1.5	2	2.1	2.3	2.34
δ	500	23	11.25	5.485	2.381	1.965	2.158	2.216	21.82

Table 2. *Instability lower bounds for* δ *when* $m = 0.5$.

η	1.04	1.3	1.35	1.5	1.8	2.0	2.2
δ	231	9.453	8.011	5.611	3.936	3.96	6.757

Table 3. *Instability lower bounds for* δ *when* $m = 0.25$.

η	1.35	1.5	1.8	2.0	2.2
δ	89.966	17.38	8.081	8.123	26.16

Let us come back to the neutral curve (5.3.10). From the right-hand side, it is seen that: a) the stability domain decreases when δ increases and m is kept fixed; b) the stability domain increases when m decreases and δ is kept fixed. Denoting by $G^2_{3cr}(m, \delta, \eta_{cr})$ the smallest value of Gheorghitza's number corresponding to the dimensionless wave number η_{cr}, we have further: c) for m fixed and δ decreasing, η_{cr} increases, i.e. $\delta_2 < \delta_1$, $\eta_{cr}(\delta_2) > \eta_{cr}(\delta_1)$; d) for δ fixed and m decreasing, η_{cr} increasing, i.e. $m_1 < m_2$, $\eta_{cr}(m_1) > > \eta_{cr}(m_2)$.

The properties a) and b) show respectively that the term βz from velocity has a stabilizing effect and that porosity has the same effect. For each m and δ assigned we obtain a neutral curve, below which there is the stability domain and above which lies the instability domain. In figs 5.3.1 and 5.3.2 we have represented the curves of neutral stability for $m = 0.5$ and $m = 0.25$ when $\delta = 240$, 300 and

500. The above properties can also be seen on these figures. Thus, the case of the flow governed by Equation (5.3.8)′ is completely studied.

If now $k \to \infty$ in (5.3.8), we get

$$n_{1,2} = [2igK^{-1} - 2k_x\alpha - k_x\beta L) + k_x\beta L] (1 + m^{-1})^{-1},$$

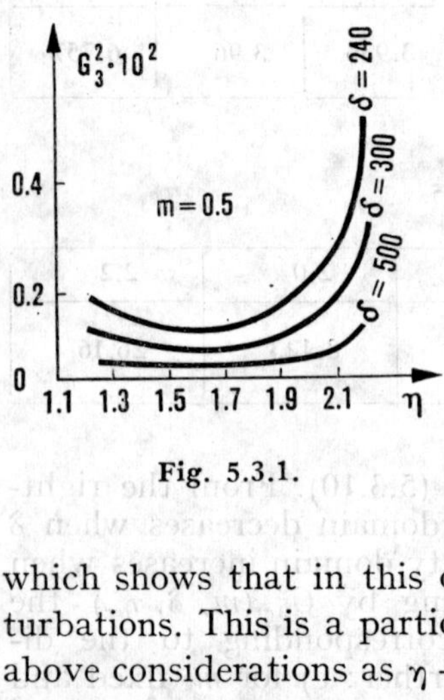

Fig. 5.3.1.

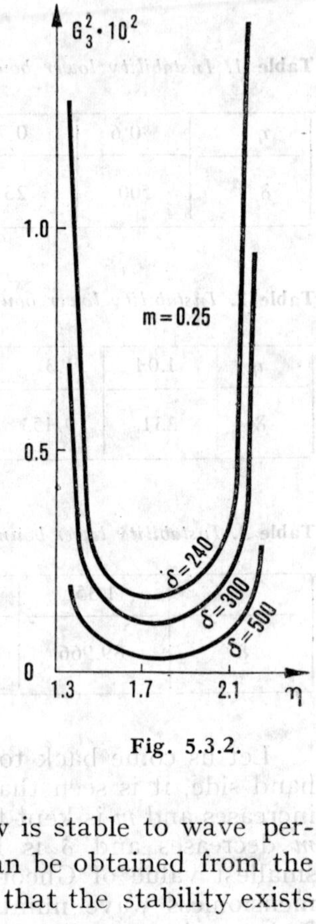

Fig. 5.3.2.

which shows that in this case the linear flow is stable to wave perturbations. This is a particular case which can be obtained from the above considerations as $\eta \to \infty$. We mention that the stability exists for every $\eta > 2.34$.

Porous layer in fluid. First let a free fluid lie between $z = -\infty$ and $z = 0$, where $w_1 = Be^{-kz}$. The first porous medium having $p_{1*} = M_1 e^{kz} + N_1 e^{-kz}$ is situated between the planes $z = 0$ and $z = -H$. The planes $z = -H$ and $z = -E$ ($E > 0$, $H > 0$, $E > H$) separate the second porous medium, characterized by $p_{2*} = M_2 e^{kz} + N_2 e^{-kz}$, from the first porous medium and from the second free fluid located between $z = -E$ and $z = -\infty$, having $w_2 = Ae^{kz}$.

Writing now the conditions (5.1.12) for $z = 0$ and $z = -E$ and (5.1.12)′ for $z = -H$, we get

$$\begin{cases} M_1 + N_1 = \rho i k^{-2}(k_x \beta_1 + a_1 k)\, B, \\ B = ik\rho^{-1} a_3^{-1}(M_1 - N_1), \\ M_1\, e^{-kH} + N_1\, e^{kH} = M_2\, e^{-kH} + N_2\, e^{kH}, \\ a_4(M_1\, e^{-kH} - N_1\, e^{kH}) = a_3(M_2\, e^{-kH} - N_2\, e^{kH}), \\ M_2\, e^{-kH} + N_2\, e^{kH} = \rho i k^{-2}(k_x\beta_2 - ka_2 A e^{-kH}), \\ A\, e^{-kH} = ik\rho a_4^{-1}(M\, e^{-kH} - N_2\, e^{kH})\,; \end{cases}$$

the characteristic equation corresponding to this system is

$$a_3(k_x\beta_1 k^{-1} + a_1 + a_3 \operatorname{th} kH)\,[(a_2 - k_x\beta_2 k^{-1}) \operatorname{th} k(E - H) + a_4] +$$
$$+ a_4[(k_z\beta_1 k^{-1} + a_1) \operatorname{th} kH + a_3][a_2 - k_x\beta_2 k^{-1} +$$
$$+ a_4 \operatorname{th} k(E - H)] = 0,$$

which, after introducing $ka_1' = ka_1 + k_x\beta_1$, $ka_2' = ka_2 - k_x\beta_2$ becomes the characteristic equation

$$a_3(a_1 + a_3 \operatorname{th} kH)\,[a_3 \operatorname{th} k(E - H) + a_4] +$$
$$+ a_4(a_1 \operatorname{th} kH + a_3)\,[a_2 + a_4 \operatorname{th} k(E - H)] = 0$$

corresponding to the case of the uniform currents, studied in [7].

The above considerations led St. I. Gheorghitza [4] to put into evidence an erroneous boundary condition, used sometimes in studies concerning the flow in the presence of porous media [14], [12], [13].

§ 5.4. THE CASE OF A VERTICAL CYLINDER

Consider a vertical circular cylinder with the radius a, having the axis in the z-direction, in the interior of which there are two horizontal layers of ideal free fluids located as follows: one in the halfplane $z > 0$, having density ρ', and the other between the planes $z = 0$ and $z = -H$, having density ρ; at $z < -H$ we have a porous medium with the filtration coefficient K, porosity m and density ρ. Suppose that under the influence of a perturbation, an irrotational flow appears, characterized in the first layer of free fluid by the potential φ, and in the second layer by the potential

φ'. If v_r, v_θ, v_z are the components of the perturbation velocity in cylindrical coordinates r, θ and z, we have $v_r = \frac{\partial \varphi}{\partial r}$, $v_\theta = \frac{1}{r}\frac{\partial \varphi}{\partial \theta}$, $v_z = \frac{\partial \varphi}{\partial z}$, where

(5.4.1) $$\Delta\varphi = \frac{\partial \varphi}{\partial r} + r\frac{\partial^2 \varphi}{\partial r^2} + \frac{1}{r}\frac{\partial^2 \varphi}{\partial \theta^2} + r\frac{\partial^2 \varphi}{\partial z^2} = 0.$$

The boundary conditions are

(5.4.2) $$\frac{\partial \varphi}{\partial r} = 0, \quad \frac{\partial \varphi'}{\partial r} = 0, \quad \text{for } r = a;$$

in addition, $\varphi' \to 0$ as $z \to +\infty$. The general solution of this equation, taking into account conditions (5.4.2), is

(5.4.3) $$\begin{cases} \varphi = (C_1 \mathrm{e}^{kz} + C_2 \mathrm{e}^{-kz})\, \mathrm{e}^{\sigma t} \cos n\theta j_n(kr), \\ \varphi' = C' \exp(\sigma t - kz) \cos n\theta j_n(kr) \end{cases}$$

where n is an integer, j_n is the n-th order Bessel function and k are the roots of the equation

(5.4.4) $$\frac{\mathrm{d}j_n(kr)}{\mathrm{d}r} = 0, \quad \text{for} \quad r = a.$$

The body forces being $(0, 0, -\rho g z)$, the perturbed pressures are

(5.4.5) $$p = -\rho\frac{\partial \varphi}{\partial t} - \rho g z, \quad p' = -\rho'\frac{\partial \varphi'}{\partial t} - \rho' g z,$$

and represent the linearized form of Bernoulli's equations. Denoting by ζ the displacement of the perturbed interface, we have

(5.4.6) $$\frac{\partial \varphi}{\partial z} = \frac{\partial \varphi'}{\partial z} = \frac{\partial \zeta}{\partial t}, \quad \text{for} \quad z = \zeta;$$

and taking into account (5.4.3), it follows that ζ has the form

(5.4.7) $$\zeta = a\, \mathrm{e}^{\sigma t} \cos n\theta j_n(kr).$$

In the linear approximation, taking into account the capilarity effects, the dynamic condition on the interface is

(5.4.8) $$p - p' = -T\left(\frac{\partial^2 \zeta}{\partial r^2} + \frac{1}{r}\frac{\partial \zeta}{\partial r} + \frac{1}{r^2}\frac{\partial^2 \zeta}{\partial \theta^2}\right),$$

where T is the surface tension. If the unperturbed interface corresponds to $z = 0$, then

$$(5.4.9) \qquad (C_1 - C_2)\, k = -C'k = \sigma a;$$

and from relations (5.4.3), (5.4.7), (5.4.5) and (5.4.8) ,we deduce

$$(5.4.10) \qquad -\sigma\rho(C_1 + C_2) + \sigma\rho' C' - a\rho g + a\rho' g = Tk^2 a.$$

Consequently, requiring the perturbation in the free fluids to satisfy the boundary condition for $z = +\infty$, and on the wall of the cylinder $z = 0$, as well as the kinematic and dynamic conditions on the interface $z = 0$, we have obtained three relations (5.4.9) and (5.4.10) for each C_1, C_2, C' and σ. We join now the perturbed motion from the free fluid with that from the porous medium where the perturbation satisfies the equation

$$\frac{1}{m}\frac{\partial \mathbf{v}_*}{\partial t} = -\frac{1}{\rho}\,\mathrm{grad}\,(p_* + \rho g z) - g\frac{\mathbf{v}_*}{K},$$

whence

$$(5.4.11) \qquad \Delta(p_* + \rho g z) = 0,$$

Δ being the Laplacian written in cylindrical coordinates. Since v_{*z} must vanish for $z \to \infty$, we have

$$(5.4.12) \qquad p_* = \mathrm{D} \exp(\sigma t + kz) \cos n\theta\; j_n(kr) - \rho g z,$$

hence

$$(5.4.13) \qquad v_{*z} = -k\mathrm{D}\rho^{-1}(\sigma m^{-1} + gK^{-1}) \exp(\sigma t + kz) \cos n\theta\; j_n(kr).$$

From the continuity condition of the normal velocity and the pressure at $z = -H$, we get

$$(5.4.14) \qquad C_1 \mathrm{e}^{-kH} - C_2 \mathrm{e}^{kH} = -\mathrm{D}\rho^{-1} \mathrm{e}^{kH} (\sigma m^{-1} + gK^{-1})^{-1}$$

and

$$(5.4.15) \qquad -\rho\sigma(C_1 \mathrm{e}^{-kH} + C_2 \mathrm{e}^{kH}) = \mathrm{D}\mathrm{e}^{kH}.$$

Eliminating C_1, C_2, C' and D in (5.4.9), (5.4.10), (5.4.14), (5.4.15) we obtain the equation in σ

$$(5.4.16) \qquad \sigma^2\rho' + \sigma^2\rho\,\frac{\widetilde{a}\,\mathrm{th}\,kH + 1}{\widetilde{a} + \mathrm{th}\,kH} + \Theta = 0,$$

where $\tilde{a} = \sigma(gK^{-1} + \sigma m^{-1})^{-1}$ and $\Theta = kg(\rho - \rho' + Tk^2g^{-1})$. The relation (5.4.16) may also be put in the form

$$(5.4.16)' \quad \sigma^2\rho' + \sigma^2\rho\frac{\rho \operatorname{th} kH + gK^{-1} + \sigma m^{-1}}{\sigma + (gK^{-1} + \sigma m^{-1}) \operatorname{th} kH} = -kg(\rho - \rho' + Tk^2g^{-1}),$$

hence for $K \to \infty (m = 1)$ this equation is that from the classical case in the absence of the porous medium.

Comparing Equation (5.4.16) with that deduced in [7] for the stability of the equilibrium of two layers of free fluids in the presence of a porous medium, and introducing in [7] $n = -i\sigma$, $V_1 = V_2 = 0$, $L = 0$, $m_1 = 1$, $K_1 = \infty$, $m_2 = m$, $K_2 = K$, $\rho_2 = \rho'$, $\rho_1 = \rho$, we get that the characteristic equation corresponding to this flow is identical to (5.4.16). Hence, the condition for the stable equilibrium of two free fluid layers and of a porous medium layer bounded by a vertical cylinder has the same form as in the case when the layers are unbounded. Nevertheless we note that in the case of unbounded layers k is arbitrary, while in the case of the cylinder k satisfies (5.4.4).

As in the case of the unbounded flows, the porosity still has a stabilizing effect for the case of the cylinder, the gravitation being the single cause of instability.

REFERENCES

[1] Gheorghitza, St. I., *On the neutral stability in homogeneous porous media,* Com. Acad. R.P.R., **10**, 12, 1069–1073 (1960). (In Romanian).

[2] Gheorghitza, Șt. I., *The marginal stability in porous inhomogeneous media,* "Proc. Cambr. Phil. Soc.", **57**, 4, 1961, 871–877.

[3] Gheorghitza, St. I., *Mathematical methods in underground hydrogazodynamics,* Ed. Acad. Bucharest, 1966. (In Romanian).

[4] Gheorghitza, St. I., *A generalization of some results concerning the stability of superposed fluids,* "Bull. Math. de la Soc. Sci. Math. de la R. S. de Roumanie" **11** (59), 4, 1967, 25–35.

[5] Gheorghitza, St. I., *On the stability of superposed fluids in the presence of inhomogenous porous media,* "J. Math. Phys. Sci.", **3**, 3, 1969, 269–279.

[6] Gheorghitza, St. I., *Sur la stabilité de l'équilibre des fluides en présence des milieux poreux non homogènes,* "Bull. Math. de la Soc. Sci. Math. de la R. S. de Roumanie", **13**, (61), 2, 1969, 157–165.

[7] Gheorghitza, St. I., *The generalization of some results concerning the Kelvin-Helmholtz instability,* "J. Sci. Eng. Res.", **13**, 1, 1969, 39–47.

[8] Gheorghitza, St. I., *On the marginal stability in porous media,* "J. Math. Phys. Sci.", **3**, 4, 1969, 386–392.

[9] Georgescu, A., St. I. Gheorghitza, *On the Kelvin-Helmholtz instability in presence of porous media*, "Rev. Roum. Math. Pures et Appl.", **14**, 1, 1971, 27–39.

[10] Georgescu, A., St. I. Gheorghitza, *Instability of two superposed liquids in a circular tube in the presence of a porous medium*, "Rev. Roum. Math. Pures et Appl.", **14**, 5, 1971, 677–680.

[11] Lapwood, E. R., *Convection of a fluid in a porous medium*, "Proc. Cambridge Phil. Soc.", **44**, 1948, 508–521.

[12] Slezkin, N. A., *On the motion of viscous flnid with free surface over a porous bottom*, "Vest. Mosk. Univ. Mat. Meh. Astr.-Fiz. Chem.", **3**, 1957 (In Russian).

[13] Tao, L. N., D. D. Joseph, "J. Appl., Mech.", 1962, 429.

[14] Taylor, G. I., J. C. P. Miller, *Fluid flow between porous rollers*, "Quart. J. Mech. Appl. Math.", **9**, 1956, 129.

APPENDICES

1. OPERATORS IN HILBERT SPACES

A real (respectively complex) Banach space H is called a *Hilbert space* if the norm $\|\cdot\|_H$ is induced by a scalar product, hence by a symmetrical (respectively hermitic) positively definite bilinear form $(\cdot\,,\cdot)$ such that

$$\|u\|_H = \sqrt{(u, u)}.$$

Then for every $u, v \in H$, we have the *Schwarz inequality*

$$|(u, v)| < \|u\| \cdot \|v\|.$$

Two vectors u and v are called *orthogonal* if $(u, v) = 0$; given a linear subspace H_1 of H, the subspace

$$H_1^{\perp} = \{\mathbf{v} \in H \mid (u, v) = 0, \ \ (u \in H_1)\}$$

is called the *orthogonal complement* of H_1 in H. From the definition, it follows that $H_1 \cap H_1^{\perp} = \varnothing$, (hence the sum $H_1 \oplus H_1^{\perp}$ is a direct sum of linear subspaces) and $H_1^{\perp} = (\overline{H}_1)^{\perp}$.

The orthogonal decomposition theorem. *If the subspace* H_1 *is closed in* H, *then* $\mathrm{H} = \mathrm{H}_1 \oplus \mathrm{H}_1^{\perp}$, *hence every element* $\mathrm{u} \in \mathrm{H}$ *can be written in an unique way as a sum of the elements* $\mathrm{u}_1 \in \mathrm{H}_1$, *and* $\mathrm{u}_2 \in \mathrm{H}_2$.

This theorem implies the existence of a linear map, called a *map of orthogonal projection* on H_1, $P: H \to H_1$, defined by the equality

$$P(u) = u_1, \text{ if } u - u_1 \in H_1^{\perp},$$

which for every $u \in H$ and $v \in H_1$, satisfies the relation

$$(u, v) = (Pu, v).$$

Since $\|Pu\| \leqslant \|u\|$, this map is bounded, its norm being 0 if $H_1 = \{0\}$ and 1 in the other cases.

Representation theorem (Riesz-Fischer). *Every bounded linear functional on a Hilbert space* H, *is of the form*

$$l(u) = (u, v_l),$$

for a certain $\mathrm{v}_l \in \mathrm{H}$.

In particular, it follows that every Hilbert space is reflexive, and the weak convergence of a sequence $\{u_n\}$ of elements from H is equivalent to the convergence of the numerical sequences $\{(u_n, v)\}$ for every $v \in H$.

A generalization of the Riesz-Fischer theorem is given by the following.

Lemma (*Lax-Millgram*). *Let* F *be a bilinear form defined on the Hilbert space* H, *which satisfies the conditions*

$$1^\circ \quad |F(u, v)| \leqslant c_1\|u\| \cdot \|v\|$$

$$2^\circ \quad c_2\|u\|^2 \leqslant F(u, u)$$

for certain strictly positive constants c_1 *and* c_2.

Then, every linear bounded functional on H *is of the form*

$$l(u) = F(u, v_l)$$

for a certain element $v_l \in H$.

A set of elements of the Hilbert space H is called *total* in H, if zero is the only element orthogonal to the set E, or, equivalently, if E is dense in H. The set $E = \{x_i\}_{i \in I}$ is called *orthonormal* if $\|x_i\| = \sqrt{(x_i, x_i)} = 1$, and $(x_i, x_j) = 0$ for $i \neq j$. For every $x \in H$, the scalars

$$c_i = (x, x_i), \quad i \in I$$

are called the *Fourier coefficients* of x with respect to E. These coefficients are all vanishing save for a countable number of them, and the series $\sum_{i \in I} c_i x_i$ converges to an element $x' \in H$ such that $x - x' \perp E$. From the above considerations, it follows that if E is a total orthonormal system of vectors in H, every element $x \in H$ may be uniquely written in the form

$$x = \sum_{i \in I} c_i x_i,$$

where c_i are the Fourier coefficients of x with respect to E. In order that the topology of the Hilbert space H be separable (i.e. that there exists a countable set dense in H), it is necessary and sufficient that there exist a finite or at most countable system, total in H. The space $L^2(\Omega)$ is separable.

A linear map A defined on a linear subspace of H denoted by $\mathfrak{D}(A)$ and taking values in H is called a *linear operator in the Hilbert space* H; the image of $\mathfrak{D}(A)$ through A is a linear subspace of H denoted by $\mathfrak{R}(A)$. If the operator $A : \mathfrak{D}(A) \to \mathfrak{R}(A)$ is injective, then it is called *invertible* since there exists the map $A^{-1}: \mathfrak{R}(A) \to \mathfrak{D}(A)$. A is invertible iff the equation $Au = 0$ has the only solution $u = 0$. B is an *extension* of A if $\mathfrak{D}(B) \supset \mathfrak{D}(A)$; in this case we write $B \supset A$.

Suppose that $\mathfrak{D}(A)$ is dense in H and consider the set

$$\mathfrak{D}^*(A) = \{v \in H \mid \exists v^* \in H, \ (Au, v) = (u, v^*), \ (u \in \mathfrak{D}(A))\}.$$

$\mathfrak{D}^*(A)$ is a linear subspace of H, and for $v \in \mathfrak{D}^*(A)$, the element v^* is uniquely determined by the above property; hence we may define a map A^* called the adjoint map of A, having $\mathfrak{D}(A^*) \overset{def}{=} \mathfrak{D}^*(A)$, such that

$$(Au, v) = (u, A^*v), (u \in \mathfrak{D}(A), v \in \mathfrak{D}(A^*)).$$

A linear operator A is called *symmetric* if for every $u, v \in \mathfrak{D}(A)$, we have $(Au, v) = (u, Av)$ or, equivalently, if $A \subset A^*$. A symmetric map is called *self-adjoint* if $A = A^*$.

Some connections between symmetric and self-adjoint operators are given by the following *theorems* due to *von Neumann*: 1°. A symmetric operator, defined on the whole Hilbert space H is self-adjoint. 2°. A symmetric operator having $\mathfrak{R}(A) = H$ is self-adjoint. 3°. A symmetric operator with $\overline{\mathfrak{R}(A)} = H$ is invertible and A^{-1} is also symmetric. 4°. A is self-adjoint iff A^{-1} is self-adjoint.

A symmetric operator A is called *positive definite* if there exists a number $m > 0$ such that

$$(Au, u) \geqslant m(u, u), \text{ for every } u \in \mathfrak{D}(A).$$

A symmetric operator A is called *positive* if $(Au, u) \geqslant 0$ for every $u \in \mathfrak{D}(A)$, and $(Au, u) = 0$ iff $u = 0$.

The class of orthogonal projections coincides with the class of self-adjoint, idemponents $(P^2 = P)$ defined everywhere.

If A is a positive definite operator, then

$$[u, v]_A \overset{def}{=} (Au, v)$$

is a new scalar product on $\mathfrak{D}(A)$, and the norm $\|\cdot\|_A$ associated with this product is stronger than the norm introduced by the scalar product of H. Then, completing the normed space $\mathfrak{D}(A)$, with respect to the norm $\|\cdot\|_A$, we obtain the Hilbert space H_A, called the *energetic space* of the map A, which, owing to the relation between $\|\cdot\|_A$ and $\|\cdot\|_H$, can be identified with the subspace of H, consisting of elements which are limits in H of sequences from $\mathfrak{D}(A)$ which are Cauchy in $\|\cdot\|_A$. H_A is separable iff H is separable. It is easy to see that the restriction $\tilde{A}$ of the operator A^* to the subspace $H_1 = \mathfrak{D}(A^*) \cap H_A$ is a positive definite, self-adjoint operator, defined in the Hilbert space H. Since $H_1 \supset \mathfrak{D}(A)$, and A^* is an extension of A, then $\tilde{A}$ is a self-adjoint operator called the *Friedrichs extension* of A. Moreover $H_A = \mathfrak{D}(\tilde{A}^{\frac{1}{2}})$, where $\tilde{A}^{\frac{1}{2}}$ is a self-adjoint positive definite operator called the square root of $\tilde{A}$.

Let Ω be a domain of $\mathbb{R}^n$; a linear 1-st order differential operator (defined in $C_0(\Omega)$) is a sum of the form

$$A = \sum_{0 \leqslant |\alpha| \leqslant l} a_\alpha \mathrm{D}^\alpha,$$

where $\mathrm{D}^\alpha = \dfrac{\partial^{|\alpha|}}{\partial x_1^{\alpha_1} \ldots \partial x_n^{\alpha_n}}$, and $a_\alpha \in C^{|\alpha|}(\Omega)$ are the coefficients of the operator A. Since $\mathfrak{D}(A) = C_0^l(\Omega)$ is a dense subspace of $H =$

$= L^2(\Omega)$ it follows that A may be regarded at as an operator in $L^2(\Omega)$ (of course the convention at the beginning of Ch. 2 still holds) namely $\mathfrak{D}(A) = \{u \in C_0^l(\Omega) \mid u \in L^2(\Omega), Au \in L^2(\Omega)\}$. Define the linear operator $A^+ : C_0^l(\Omega) \to L^2(\Omega)$ such that for every $u, v \in C_0^l(\Omega)$ the following equality holds

$$(1.1) \qquad (Au, v) = (u, A^+v)$$

where

$$A^+v = \sum_{0 \leqslant |\alpha| \leqslant l} (-1)^{|\alpha|} D^\alpha(\bar{a}_\alpha v).$$

The bar indicates the complex conjugate. If H is a real space then instead of $\bar{a}_\alpha$ we shall have a_α. The operator A^+ is called the *formal adjoint of A*, or, the *adjoint of* A *in Lagrange's sense.* If $A(\cdot) = A^+(\cdot)$ then A is called formally selfadjoint. Analogous definitions can be given for differential operators defined in $C_0^l(\Omega, \mathbb{R}^m)$.

Let A be a linear l-th order differential operator defined on $\mathfrak{D}(A) = \{u \in C^l(\Omega) \cap C^{l-1}(\bar{\Omega}) \mid Au(x) = \sum_{0 \leqslant |\beta| \leqslant l-1} b_\beta D^\beta u(x) = 0$ for $x \in \partial\Omega$ where $b_\beta \in \mathbb{R}\}$. Then $\mathfrak{D}(A)$ is dense in $L^2(\Omega)$. If the elements of $\mathfrak{D}(A)$ have some symmetry properties then $\overline{\mathfrak{D}(A)}^{\|\,\|}$ is a Hilbert subspace of $L^2(\Omega)$ where those properties hold.

Let $A : \mathfrak{D}(A) \subset H \to H$ be a linear operator. We say that A is *continuous* if $u_n, u \in \mathfrak{D}(A)$, $\lim_{n\to\infty} u_n = u$ implies $\lim_{n\to\infty} Au_n = Au$. A is called *bounded* if there exists a constant $C \geqslant 0$ such that $\|Au\| \leqslant C\|u\|$ for every $u \in \mathfrak{D}(A)$.

In applications a symmetric operator is sometimes called self-adjoint. In reality these two concepts are identical only if the operator is bounded. An operator which is not bounded is called *unbounded*. Since H is a Hilbert space the boundedness and continuity of a linear operator are equivalent (§ 2.1). By definition the set $G = \{(u, Au) \in X \times X \mid u \in \mathfrak{D}(A)\}$ is called the graph of A. We say that A is *closed* if from $u_n \in \mathfrak{D}(A)$, $\lim_{n\to\infty} u_n = u$, $\lim_{n\to\infty} Au_n = v$ it follows $u \in \mathfrak{D}(A)$ and $v = Au$. The closed graph theorem in Hilbert spaces states: a linear operator is closed iff its graph is closed. As every linear continuous (and, hence, bounded) operator is closed it follows that the notion of closedness generalizes the boundedness. A closed linear operator is in general not bounded, but if $\mathfrak{D}(A) = H$ then the closed operator is bounded. The adjoint A^* is a closed operator, consequently every self-adjoint operator is closed. Generally the self-adjoint operators are not bounded. If $\mathfrak{D}(A) = H$ the self-adjoint linear operator being closed is also bounded (Hellinger-Toeplitz theorem). The linear differential operators defined on strict subsets of $L^2(\Omega)$ are unbounded. If A is a closed operator and B is a bounded operator than $A + B$ is closed.

Consider the equation $Au = f, f \in L^2(\Omega)$ where A is the above defined linear differential operator. It may have no solution in $\mathfrak{D}(A)$ but it may have in a larger class. But for this larger class the operator is no longer defined, consequently, it has to be extended. We obtain an extension by closing the operator A. We say that the linear operator $A: \mathfrak{D}(A) \subset H \to H$ has a closed extension $\bar{A}$ if the graph of $\bar{A}: \mathfrak{D}(\bar{A}) \subset H \to H$ is the closure of the graph of A. A has a closed extension iff from $\{u_n\} \subset \mathfrak{D}(A)$, $\lim_{n\to\infty} u_n = 0$ and $\lim_{n\to\infty} Au_n = v$ it follows that $v = 0$. Let us show that the linear differential operator A of order l introduced above has a closed extension. In fact, take a sequence $u_n \to 0$, $u_n \in \mathfrak{D}(A)$; for every $\varphi \in C_0^l(\Omega)$ equality (1.1) gives $(A_u, \varphi) = (u_n, A^+\varphi)$. By the continuity of the scalar product in the limit we have as $n \to \infty$ $(v, \varphi) = (0, A^+\varphi) = 0$ for every $\varphi \in C_0^l(\Omega)$. Due to the density of $C_0^l(\Omega)$ in H it follows that $v = 0$, where $v = \lim_{n\to\infty} Au_n$. By the above mentioned theorem the existence of the closed extension $\bar{A}$ of A follows. In particular the closed extention $\bar{A}$ of the operator D^α from $\mathfrak{D}(D^\alpha)$ to $\mathfrak{D}(\bar{A}) \subset L^2$ is called the *generalized derivative* and satisfies

$$(1.2) \qquad \int_\Omega (\bar{D}^\alpha u)(x)\, \varphi(x)\, dx = \int_\Omega u(x)\, ((D^\alpha)^+ \varphi)(x)\, dx,$$

for every $\varphi \in C_0^\infty(\Omega)$.

Relation (1.2) shows that the generalized derivative is the adjoint of the operator formally adjoint to the operator D^α of usual differentiation.

Proposition 2.1.1 expresses the fact that the generalized derivative in $L^p(\Omega)$ is a closed operator.

For linear elliptic operators $A: \mathfrak{D}(A) \to L^2(\Omega)$ we say that $u \in \mathfrak{D}(\bar{A})$ is a *generalized solution* of $Au = f$, $f \in L^2(\Omega)$ if u satisfies the equality $(f, \varphi) = (u, A^+\varphi)$ for every $\varphi \in C_0^\infty(\Omega)$. Other types of generalized solutions were introduced in Ch. 2 for the Navier-Stokes equations.

It can be proved that $\bar{A}$ is the minimal closed extension of A i.e. every closed extension B of A is an extension of $\bar{A}$ (hence $B \supset \bar{A}$). A^* is the maximal symmetric extension of A. When $\bar{A}$ exists it follows that $(\bar{A})^* = A^*$. If A^{**} exists then $A \subset A^{**}$ whence $\bar{A}$ exists. If $\mathfrak{D}(A) = H$ and $\bar{A}$ exists then A^{**} exists too and $\bar{A} = A^{**}$. If $A = A^{**}$ it follows that A is closed ($A = \bar{A}$).

An operator which is not linear is called a nonlinear mapping. A *nonlinear mapping* $T: E \to F$ where E and F are two Banach spaces is *bounded* if it transforms every bounded set of E in a bounded set of F. T is said continuous if $u_n \to u$ strongly in E implies $Tu_n \to Tu$ strongly in F. Let E and F be two reflexive Banach spaces and $T: E \to F$ a nonlinear mapping. We say that T is *completely contin-*

uous if for every $\{u_n\}_{n\in\mathbb{N}}, u_n \in E, u_n \to u \in E$ it follows that $\{T(u_n)\}_{n\in\mathbb{N}}$ is convergent (in norm) towards $T(u) \in F$. A nonlinear mapping $T: E \to F$ is *compact* if 1° it is continuous; 2° for every bounded $M \subset E$ then $T(M) \subset F$ is relatively compact (i.e. $\overline{T(M)}$ is compact). If T is linear it is called compact if 2° holds (because 1° will follow from 2°). Due to the reflexivity of E every completely continuous mapping is compact. This property fails if E is not reflexive, since then the unit ball of E is no longer weakly compact. Generally a compact mapping is not completely continuous, but the two definitions are equivalent for linear mappings. This allowed us to use the concept of completely continuity for linear operators which satisfy property 2° (section 2.1.6). Since every linear compact operator is completely continuous it transforms every weakly convergent sequence into a strongly convergent sequence. This is why the existence theorem for solutions of the Navier-Stokes equations requires a suitable use of inequalities between various norms. Indeed, in Ch. 2 the stationary Navier-Stokes equations were put in the form $Tu = f$ where T is a compact mapping defined on some subspace of the Hilbert (hence reflexive) space $L^2(\Omega)$. The reflexive Banach spaces $L^p(\Omega)$ $(1 < p < \infty)$ were used too. The Navier-Stokes equations linearized about the null solutions are called the Stokes equations and can be written as $F_0'u = f$ where F_0' is the Fréchet derivative of F at the point 0. In branching theory linearizations about nontrivial basic solutions u_0 of the Navier-Stokes equations are made and consequently F_{u_0}' must be studied. F_0' and F_{u_0}' are completely continuous linear operators.

If A is a positive operator then the only solution of the equation $Au = 0$ is $u = 0$ since otherwise $u_0 \neq 0$, $Au_0 = 0$ would imply $(Au_0, u_0) = 0$ but $(Au_0, u_0) = 0$ implies $u_0 = 0$, a contradiction. Since every positive definite operator is a positive operator too it follows that the only solution only of $Au = 0$ where A is positive definite is the null solution.

Consider the equation

$$Au = f, \; f \in H, \tag{1.3}$$

$u \in \mathfrak{D}(A)$ where H is a Hilbert space and $A: \mathfrak{D}(A) \subset H \to H$ is a positive operator. From the uniqueness of the null solution for $Au = 0$ it follows that (1.3) has at most one solution. Generally (1.3) may have no solution. This is due to the fact that A is defined on a narrow class and therefore A is not surjective on H. Let us extend A up to a new surjective operator $\tilde{A}$ on a Hilbert space ($\mathfrak{R}(\tilde{A}) = H_1$). This extension can always be done if A is positive definite. Indeed $\tilde{A}$ will be the Friedrichs extension. Since $\tilde{A}$ is linear self-adjoint positive definite operator then $\tilde{A}^{-1}$ exists and is a bounded linear operator $\tilde{A}^{-1}: H_1 \to H_1$ i.e. $\mathfrak{D}(\tilde{A}^{-1}) = \mathfrak{R}(\tilde{A}^{-1}) = H_1$. It follows that the equation

$$\tilde{A}u = f \tag{1.3$'$}$$

has a solution $u_0 = \tilde{A}^{-1} f$ and this solution is unique. Since generally the solution u_0 does not satisfy (1.3) it is called a generalized solution of (1.3). As the operator $\tilde{A}$ is closed it follows that this second extension of the meaning of the solution was obtained by means of a closed extention of A. The first extension was obtained considering the functional $(u, A^+\varphi)$. In the second extension the energy functional $F: \mathfrak{D}(A) \to \mathbb{R}$, $Fu = (Au, u) - (u, f) - (f, u)$ was used . A theorem can be proved which asserts that every solution of (1.2) is a minimum point for F and conversely. If (1.3) has no solution it follows that F does not attain its minimum on $\mathfrak{D}(A)$. It means that the space in which the minimum of F is sought is too narrow and it must be enlarged. The possibility of extending F is immediate since, although F was defined initially on $\mathfrak{D}(A)$ it has a meaning also on H_A. In H_A every term of F is a linear bounded functional of u; therefore, by Riesz-Fischer theorem, there exists a unique element $u_0 \in H_A$ such that $F(u) = \|u - u_0\|^2_{H_A} - \|u_0\|^2_{H_A}$. This formula shows that u_0 is a minimum point of F. The corresponding extension $\tilde{A}$ of A will enjoy the property that $\tilde{A}u_0 = f$ (i.e. u_0 is a solution of (1.3)').

For operators A which are only positive H_A may contain elements which are not from H such that the above arguments do not hold.

If A is a positive definite operator then the imbedding of H_A in H is completely continuous since $\mathfrak{D}(\tilde{A}^{\frac{1}{2}}) = H_A$ and $\mathfrak{R}(\tilde{A}^{\frac{1}{2}}) = H$ and $\tilde{A}^{\frac{1}{2}}$ is compact. For instance if $H = N(\Omega)$, Ω is a bounded set of $\mathbb{R}^2$ or $\mathbb{R}^3$, $A = -P\Delta: N^2 \to N$ (Section 2.2.2) is positive definite. Then $H_A = N^1$ and the imbedding of N^1 into N is completely continuous (see also Corollary of Theorem 2.1.4).

If A is a compact operator and B is a bounded operator then AB and BA are compact. A finite sum of compact linear operators is is compact. Every nonlinear compact mapping is bounded. Therefore in Banach spaces the compactness ensures the boundedness of the nonlinear mappings while for linear operators the only continuity is sufficient to induce their boundedness.

REFERENCES

[1] Cristescu, R., *Functional analysis*, Ed. Didactică și Pedagogică, Bucharest, 1979. (In Romanian).

[2] Maurin, K., *Methods of Hilbert spaces*, Izd. Mir., Moscow, 1965. (In Russian).

[3] Mihlin, S. G., *Variational methods in the mathematical physics*, Nauka, Moskow, 1970. (In Russian).

[4] Nicolescu, M., *Real functions and elements of topology*, Ed. Didactică și Pedagogică, Bucharest, 1968. (In Romanian).

2. SEMIGROUPS OF OPERATORS IN BANACH SPACES

A family $\{B_t\}_{t \in [0, \infty)}$ of bounded linear operators in a Banach space X is called a *strongly continuous semigroup of operators* if the following relations hold

1) $B_s B_t = B_{s+t}, \ s, t \geqslant 0,$

2) $B_0 = I,$

3) For every $x \in X$, $t \to B_t(x)$ defines a continuous map from $[0, \infty)$ to X.

The semigroup is called *continuous in the uniform topology* if instead of 3) it satisfies the stronger condition

3') The map $t \to B_t$ is a continuous map from $[0, \infty)$ to $(X \to X)$ endowed with the topology given by the norm of the operators.

For instance, if B is a bounded linear operator, for every given t we may consider the operator

$$e^{tB} \overset{def}{=} \sum_{n=0}^{\infty} \frac{(tB)^n}{n!},$$

where the series of the right-hand side is absolutely convergent in the Banach algebra $(X \to X)$ of the bounded linear operators from X to X, and its limit satisfies the relation

$$\| e^{tB} \| \leqslant \sum_{n=0}^{\infty} \frac{|t|^n \| B \|^n}{n!} = e^{|t| \, \|B\|}.$$

For $s, t \in [0, \infty)$ it is easily seen that $\{e^{tB}\}$ defines a semigroup of operators continuous in the uniform topology and that B is the *infinitesimal generator* of this semigroup, i.e.

$$B = \lim_{h \to 0} \frac{e^{tB} - I}{h} \cdot$$

Conversely, it can be shown that if $\{B_t\}$ is a semigroup continuous, in the uniform topology, then there exists a bounded linear operator B, defined by the equality

$$B = \lim_{t \to 0} \frac{B_t - I}{t},$$

such that $B_t = e^{tB}$ for $t \geqslant 0$.

A (not necessarily bounded) linear operator B defined on a dense subspace $\mathfrak{D}(B) \subset X$ is called the infinitesimal generator of the strongly continuous semigroup of operators $\{B_t\}$ if, for every $x \in \mathfrak{D}(B)$, it satisfies the relation

$$Bx = \lim_{t \to 0} \frac{B_t x - x}{t} \cdot$$

The following theorem gives a criterion for the operator B to generate a strongly continuous semigroup of operators.

Theorem (*Hille-Yosida*) [1]. *The necessary and sufficient condition in order that a closed linear operator, defined on a dense subspace of* X, *be the infinitesimal generator of a strongly continuous semigroup of operators* B_t *with the property* $\|\mathrm{B}_t\| \leqslant e^{\omega t}$ *for a certain* $\omega \in \mathbb{R}$, *is that* $(\mathrm{B} + \lambda \mathrm{I})^{-1}$ *exist for* $\lambda > \omega$ *and satisfies the inequality*

$$\| B - \lambda I \| \leqslant (\lambda - \omega)^{-1}, \; \lambda > \omega.$$

3. SPECTRAL THEORY OF LINEAR OPERATORS

Let $A : \mathfrak{D}(A) \subset B \to B$ be a linear continuous operator defined on a complex Banach space B, λ a complex number, and I the identity operator. The linear operator $A_\lambda = (A - \lambda I)^{-1}$ is called the *resolvent* of A. We say that λ is a *regular point* of A if A_λ exists and is densely defined and continuous on $\mathfrak{D}(A_\lambda) \subset B$. The set $\rho(A)$ of all regular points of A is called the *resolvent set* of A. The resolvent set of a closed operator is an open set. The resolvent set of a continuous operator is not empty. The complementary set $\sigma(A)$ of $\rho(A)$ is called the *spectrum* of A. The spectrum consists of three sets: the *point spectrum* $\sigma_p(A)$, the *continuous spectrum* $\sigma_c(A)$ and the *residual spectrum* $\sigma_r(A)$. A point of $\sigma_p(A)$ is called an *eigenvalue* of A and has the property that A_λ does not exist. $\sigma_c(A)$ consists of points for which A_λ exists, is densely defined on B but it is not continuous. $\sigma_r(A)$ is formed by points λ for which A_λ exists, is continuous but it is not densely defined. If A is defined on the whole space B and it is bounded or it is closed then $A_\lambda(A)$ is defined on the whole space X and it is bounded. Generally the point spectrum is not discrete. If λ is an eigenvalue of A the equation $Au = \lambda u$ has nontrivial solutions called *eigenfunctions* or *eigenvectors*. The dimension of the space spanned by all the eigenfunctions corresponding to a given eigenvalue λ is called the *rank* of λ. The nontrivial solutions of $(A - \lambda I)^n$ $n \in \mathbb{N}$ corresponding to a given eigenvalue λ of A are called *generalized eigenfunctions* of A corresponding to λ. The dimension of the space spanned by all the eigenfunctions and generalized eigenfunctions of A corresponding to λ is called the *multiplicity* of λ. If an eigenvalue has the multiplicity 1 it is called *simple eigenvalue.*

Most of the flows investigated in Ch. 3 have the property that the least eigenvalue of the Stokes operator F_0' is simple. Then, by Leray-Schauder degree theory (or, equivalently by Krasnosel'skii theorem) it follows that λ_0 is a branch point. If λ_0 is not simple

[1] N. Dunford, J. T., Schwartz, *Linear operators*. Interscience Publishers, New York, London, 1968.

we may consider the problem in a suitable subspace where λ_0 is simple. This is the case with some convection problems (Section 3.4.1). Generally the linearized operations $F'_{\tilde{u}}$ from hydrodynamic stability theory are not symmetric. Nevertheless, for some motions posessing symmetry of rotation $F'_{\tilde{u}}$ may be symmetrized by means of a bounded invertible operator T [18] (from Ch. 3) (i.e. $T^{-1} F'_{\tilde{u}} T$ is symmetric and it can be made self-adjoint with respect to an appropriate scalar product of $W^{l,2}$). When F'_0 is not symmetrizable then in order to deduce the multiplicities of the eigenvalues, the eigenvalue problem for F'_0 is reduced to an integral equation with oscillatory kernel in Krein's sense.

All the eigenvalues λ of self-adjoint operators have rank 1. So, to prove that λ is simple we must prove that to λ corresponds a single eigenfunction. By Leray-Schauder on Krasnosel'skii theorem a sufficient condition for λ to be a branch point for the solutions of $F(u, \lambda) = 0$ is that λ 1° is a *real* (since it is a physical parameter) eigenvalue of F'_0 and 2° have odd multiplicity. If the spectrum of F'_0 is not real then branching does not occur as for instance in the case of the rigid rotation of fluids treated by Yudovich. In Section 3.4.3 we gave the Odqvist sufficient condition for nonbranching.

The reciprocal λ^{-1} of a nonvanishing eigenvalue λ of A is called a *characteristic value* of A. The spectrum of a continuous operator (which is not the null operator) is not empty. By definition the *spectral radius* is $r_\sigma(A) = \sup\limits_{\lambda \in \sigma(A)} |\lambda|$. It is used in Hopf bifurcation. If A is a compact linear operator then its spectrum is compact and $r_\sigma(A) \leqslant \leqslant \|A\|$. This property holds even for bounded linear operators. If A is normal (i.e. $AA^* = A^*A$) then $r_\sigma(A) = \|A\|$. If $\lambda \in \sigma_r(A)$ then $\lambda \in \sigma_p(A^*)$. If A is symmetric (hence $A \subset A^*$) then $\sigma_p(A) \subset \sigma_p(A^*)$. The eigenvalues of a symmetric operator are real. The eigenfunctions of a symmetric operator corresponding to distinct eigenvalues are orthogonal. They can be made orthogonal by the Gram-Schmidt orthonormalization procedure. If A is self-adjoint then $\sigma_r(A) = \emptyset$. The system of eigenfunctions of a positive operator is orthogonal with respect to the energy norm. We say that the symmetric operator A is bounded from below if there exists a real number k such that $(Au, u) \geqslant k\|u\|^2$ for every $u \in \mathfrak{D}(A)$.

Let A *be bounded from below operator and denote by* d *the greater lower bound of the value of the functional (called the Rayleigh quotient)* $\dfrac{(Au, u)}{(u, u)}$ *where* $(\cdot,\cdot)$ *is the scalar product in the Hilbert space* H *and* $\mathrm{A}\colon \mathfrak{D}(\mathrm{A}) \subset \mathrm{H} \to \mathrm{H}$. *If there exist* $\mathrm{u}_0 \in \mathfrak{D}(\mathrm{A})$ *such that* $\dfrac{(\mathrm{A}u_0, u_0)}{(u_0, u_0)} = \mathrm{d}$ *then* d *is the smallest eigenvalue of* A *and* u_0 *is the corresponding eigenfunction.*

Let $\lambda_1 \leqslant \lambda_2 \leqslant \ldots \leqslant \lambda_n$ *be the first* n *eigenvalues of the operator* A *bounded from below and denote by* $u_1, u_2, \ldots, u_n$ *the corresponding orthogonal eigenfunctions. Suppose that there exists* $u = u_{n+1} \neq 0$ *such that* $\frac{(Au, u)}{(u, u)}$ *takes its minimum value at* u_{n+1} *in the class of functions which satisfy the following constraints* $(u, u_1) = (u, u_2) = \ldots = (u, u_n) = 0$. *Then* u_{n+1} *is an eigenvector of* A *and corresponds to the eigenvalue* $\lambda_{n+1} = (Au_{n+1}, u_{n+1})/(u_{n+1}, u_{n+1})$. In Appendix 1 the solution of a boundary value problem was reduced to finding the minimum of the energy functional. The last two theorems reduce the solution of an eigenvalue problem to finding the minimum of the functional $(Au, u)/(u, u)$. In Appendix 7 we deal with operators to which these theorems do not apply; and their eigenvectors are stationary (not necessary minimum) points of a functional obtained from (Au, u) by suitable by parts integrations.

If A is a self-adjoint positive definite operator, $(Au, u) \geqslant k(u, u)$, $k \geqslant 0$, then the interval $(-\infty, k)$ of the real axis belongs to $\rho(A)$ and therefore if $\lambda \in \sigma_p(A)$ then $\lambda \geqslant k$. If A is a closed operator then $\sigma(A) = \sigma_p(A)$. If A is a compact operator then its spectrum is discrete and $\sigma_p(A)$ is a countable set. Therefore the spectrum of a compact(linear operator) consists of at most a countable set of discrete eigenvalues. From von Neuman's theorems it follows that if A is symmetric and $\mathfrak{D}(A) = H$ then its boundedness is equivalent with its self-adjointness. So, we have

Theorem 3.1. *If* $A : H \to H$ *is compact and positive (therefore it is self-adjoint) then there exists an infinite orthogonal set of eigenfunctions* u_i $(i = 1, 2, \ldots)$ *total in* H *and the corresponding eigenvalues form a monotone decreasing sequence*

$$\lambda_1 \geqslant \lambda_2 \geqslant \lambda_3 \geqslant \ldots \to 0 \tag{3.1}$$

where $\lambda = 0$ *is the only possible point of accumulation.*

From this theorem it follows that every $u \in H$ may be written in the form of a Fourier series $u = \sum_{i=1}^{\infty} a_i u_i$ where $a_i = (u, u_i)$, $a_i \in \mathbb{R}$. This result will be applied in Appendix 7 to derive some total sets in $L^2(-0.5; 0.5)$.

A bounded self-adjoint operator $A : H \to H$ on a Hilbert space $H \neq \{0\}$ is said to have a *pure point spectrum* or *purely discrete spectrum,* if A has an orthonormal set of eigenvectors which is total in H. If A is not compact it may happen that $\sigma_c(A) \neq \varnothing$.

For other types of operators σ_p is not discrete or it may happen that $\sigma(A) = \sigma_c(A)$. Then every function of B may be represented by integrals involving solutions of $Au = \lambda u$ where $\lambda(x) \in \sigma_c(A)$. The above-mentioned representantion for the solutions of $Au = f$ in terms of the corresponding solutions of $Au = \lambda u$ are basic in §§ 2.3, 2.4. The § 2.3 is in fact the spectral analysis of the Fréchet derivative $F'_{\bar{u}}(u, \lambda)$ of the Navier-Stokes equations $F(u, \lambda) = 0$ corresponding to the linearization about the basic flow $\bar{u}$. In linear hydrodynamic stability theory the dependence on t (time) is taken into acocunt through the normal modes $e^{\sigma t}$ where σ belongs to $\sigma(F'_{\bar{u}})$. This choice holds only when $\sigma(F'_{\bar{u}}) = \sigma_p(F'_{\bar{u}})$ is discrete and the set of corresponding eigenvectors and generalized eigenvectors is total in the Hilbert space H of the problem. The theorems of § 2.3 establish the completeness of the normal modes in H. As a consequence the solution of the (nonstationary) Navier-Stokes equations will be written in the form of a Fourier series upon this total set. For some fluid flows (as for instance plane Couette motion) $F'_{\bar{u}}$ may not have a discrete point spectrum or it may happen that $\sigma(F'_{\bar{u}}) = \sigma_c(F'_{\bar{u}})$ as in the case of the Rayleigh equation (1.1.48). In this case the solution of the Navier-Stokes equations will no longer be a series but an integral and hence the Laplace transform with respect to t must be taken. The theorems on completeness of the normal modes are based on some hypotheses: so, for a concrete motion we have to prove that these hypotheses hold.

According to Mikhlin [3] (from Appendix 1) an operator $A : \mathfrak{D}(A) \subset H \to H$ is said to have a *discrete spectrum* if it has an infinite set of eigenvalues

$$0 < \lambda_1 \leqslant \lambda_2 \leqslant \dots \leqslant \lambda_n \leqslant \dots \to \infty \tag{3.2}$$

and the corresponding eigenvectors form a system total in H as well as in H_A.

Theorem 3.2. *If* $\mathrm{A} : \mathfrak{D}(\mathrm{A}) \subset \mathrm{H} \to \mathrm{H}$ *is a positive definite operator such that the imbedding of* H_A *into* H *is compact then it has a discrete spectrum.*

The operator $-P\Delta$ defined in Section 2.2.2 satisfies the condition of Theorem 3.2. The projections of the space N (introduced in Section 2.1.8) on the finite dimensional spaces spanned by eigenvectors of $\tilde{\Delta}$ (i.e. the Fredholm self-adjoint extension of $-P\Delta$) were of fundamental importance in proving the finite genericity of the set of solutions of Navier-Stokes equations (§ 4.4).

The converse to theorem 3.2 holds too.

Theorem 3.2′. *If* $\mathrm{A} : \mathfrak{D}(\mathrm{A}) \subset \mathrm{H} \to \mathrm{H}$ *is a positive definite operator which has a discrete spectrum then the imbedding of* H_A *into* H *is compact.*

We also have

Theorem 3.3. *If* A *is a positive definite operator which satisfies the hypotheses of Theorem 3.2′ then* A^{-1} *is compact.*

But if A is positive then A^{-1} is positive and from Theorem 3.3 it follows that A^{-1} is positive. By Theorem 3.1 we have that if A is a positive definite operator and A^{-1} is compact then it has a purely discrete spectrum and $\lambda_1 \geqslant \lambda_2 \geqslant \dots \to 0$ as $n \to \infty$ where λ_i are the eigenvalues of A^{-1}. Since $\mu_i = \lambda_i^{-1}$ where μ_i are the eigenvalues of A it follows (3.2).

Let us now mention some spectral properties of compact operators useful in alternative problems.

The above theorems make evident the strong relationship between the compactness of a linear operator and the existence of a point spectrum for it. This can be immediately seen from another equivalent definition for linear compact operators. We say that a linear compact operator $A_1 : \mathfrak{D}(A_1) \subset H \to H$ is degenerate if it can be represented in the form $A_1 u = \sum_{k=1}^{n} (u, \psi_k)\, \varphi_k$ where $\psi_k, \varphi_k \in H$. A is called compact if for every $\varepsilon > 0$, Au can be written as $Au = A_1 u + A_2 u$ where A_1 is a degenerate operator and $\|A_2\| < \varepsilon$. In other words A compact is a sum of an operator defined on a finite dimensional space H_1 and an invertible operator on $H \setminus H_1$. Taking into account that the product of a compact operator by a bounded operator is compact it follows that a compact operator may be written as a sum of a closed operator and the identity operator. Let H be a Hilbert space and $C : H \to H$ be a closed linear operator. Then $\mathcal{R}(C) \perp \perp \mathcal{N}(C^*)$ and $H = \mathcal{R}(C) \oplus \mathcal{N}(C^*)$. So, if A is a compact operator and λ_0 is a nonvanishing eigenvalue of A then $A - \lambda I$ is closed; and then for every $u \in H$ we have $(A - \lambda_0 I)\, u \in \mathcal{R}(A - \lambda_0 I) = (\mathcal{N}(A^* - \lambda_0 I^*))^{\perp}$. It follows that a necessary condition for the equation $(A - \lambda I) u = f$, $f \in H$, A compact, to have a solution is that $f \in (\mathcal{N}(A^* - \lambda_0 I^*))^{\perp}$ i.e. f must be orthogonal to all the eigenvectors of A^*. It can be proved that this condition is also sufficient [104] (from Ch. 2). If $\lambda_0 \notin \sigma_p(A)$, as A is compact, it follows that $\lambda_0 \in \rho(A)$; hence the resolvent exists, is bounded and defined on all of H. Then the equation $(A - \lambda_0 I)\, u = f$, $f \in H$ has the unique solution $u = \mathrm{A}_{\lambda_0} f$. Summarizing we have that if A is a compact operator then the equation $(A - \lambda_0 I)\, u = f$, $f \in H$ has a unique solution if the corresponding homogeneous equation $(A - \lambda_0 I)\, u = 0$ has the only null solution (i.e., λ_0 is not an eigenvalue of A); if $(A - \lambda_0 I)\, u = 0$ has nontrivial solutions $(A - \lambda_0 I)\, u = f$ has some solutions (i.e. λ_0 is an eigenvalue of A) iff f is orthogonal on the right-hand side of the adjoint equation $(A^* - \lambda_0 I^*)\, u^* = f^*$ (i.e. $f \in (\mathcal{N}(A^* - \lambda_0 I^*))^{\perp}$). Briefly, if A is compact then $A - \lambda I$ satisfies the Fredholm alternative. A linear operator which satisfies the Fredholm alternative is called a *Fredholm operator* (Appendix 5). So, if A is compact then $A - \lambda I$ is Fredholm. The essential properties of a compact operator A which imply this assertion are: 1) $\lambda \neq 0$, $\lambda \in \sigma_p(A)$ then $\dim \mathcal{N}(A - \lambda I) < \infty$; 2) $\mathcal{R}(A - \lambda I)$ is closed. This property allows to obtain $H = \mathcal{R}(A - \lambda I) \oplus \mathcal{N}(A^* - \lambda I^*)$ where $\mathcal{N}(A^* - \lambda I^*) = (\mathcal{R}(A - \lambda I))^{\perp}$ is closed; 3) $\dim \mathcal{N}(A - \lambda I) = \dim \mathcal{N}(A^* - \lambda I^*) < \infty$ (i.e. A and A^* have the same number of linearly independent eigenvectors. If $A : H \to H$ is only linear then the eigenvectors corresponding to different eigenvalues are linearly independent. This property will be used in Appendix 7). We saw that every eigenvalue of a compact operator has finite multiplicity. It follows that there exists $r > 0$ such that for every $n > r$ we have $\mathcal{N}(A - \lambda I)^n \subset \mathcal{N}(A - \lambda I)^r$. It can be proved that (*) $H = \mathcal{N}(A - \lambda I)^r \oplus \mathcal{R}(A - \lambda I)^r$ [1]. In order to solve some nonlinear problems $Fu = f, F : H_1 \to H_2$ the so-called *alternative methods* are used. These methods extend the Fredholm alternative and the Liapunov-Schmidt Lemma to more general operators F and spaces H_1, H_2. The basic idea of these approaches is to reduce the study of solutions of $Fu = f$ to a lower dimensional problem. Namely perform the splittings $H_1 = H_{11} \oplus H_{12}$, $H_2 = H_{21} \oplus H_{22}$ where $\dim H_{11} = n < \infty$, $\dim H_{12} < \infty$ and put $u = u_1 + u_2$ where $u_1 \in H_{11}$ is a linear combination (with the undetermined coefficients ξ_i) of n known functions of H_{11}. Projecting $Fu = f$ on H_{22} an equation is obtained from which $u_2 = u_2(\xi_i)$ may be uniquely determined. Projecting $Fu = f$ on H_{21} we obtain a finite system of algebraic equations for the determination of ξ_i called the *bifurcation equation* or *determining equation* [2].

This Appendix mainly concerns compact operators. A spectral analysis of a large class of concrete problems of practical interest is badly needed in stability and bifurcation of fluid flows.

REFERENCES

[1] Kreyszig, E., *Iutroductory functional analysis with applications*, John Wiley & Sons, New York, 1978.

[2] Hale, J. K., *Generic bifurcation with applications*, Research Notes in Mathematics **17**, Pitman, London, 1977, 59–157.

4. CALCULUS OF VARIATIONS

To solve a direct variational problem means: given a space S of functions and a functional $j : S \to R$ find the function $u \in S$ which makes j stationary i.e. $\delta j(u) = 0$ where δj is the first Fréchet variation of j. If $\delta j(u_0) = 0$ then u_0 is called a stationary or critical point of j. The point u for which j takes a minimum (maximum) value in a neighbourhood $U \ni u$ is called a minimum (maximum) point of j and it is a stationary point. Usually in a variational problem minimum points are looked for. If u is a stationary point of j then u is a solution of an operator equation $Tu = 0$, where $T : S \to S$ is deduced from $\delta j(u) = 0$. To solve the inverse problem of the calculus of variations means: given the space S and the equation $Tu = 0$ where $T : S \to S$ is a (generally nonlinear) map find a functional $j : S \to R$ whose stationary points are solutions of $Tu = 0$.

Classical setting [1]. S is a set of continuous functions characterized by: smoothness conditions, boundary conditions and other restrictions. In the simplest case S is a subset of $C^2[a, b]$ consisting of functions $u : [a, b] \to \mathbb{R}$ which satisfy the boundary conditions $u(a) = u_1$, $u(b) = u_2$. j is the functional $j = \int_a^b F(x, u, u')\, dx$, ($u' = du/dx$), where F is a continuous function in all the three variables x, u, u'. A direct problem is to find the function which is two times continuously differentiable on $[a, b]$ and which passes through the points (a, u_1), (b, u_2) and for which j takes a minimum value. We say that u_0 is a relative minimum of j if $j(u) \geqslant j(u_0)$ for every $u \in U_1 = \{u \in S \mid |u - u_0| < \varepsilon,\ \varepsilon > 0$ is a small number$\}$. A finer definition corresponds to a k-th order neighbourhood $U_k = \{u \in S \cap C^k(a, b) \mid |u^{(m)}(x) - u_0^{(m)}(x)| < \varepsilon, m = 0, 1, \ldots, k\}$, where $u^{(m)} = \dfrac{d^m u}{dx^m}$.

If $S_1 \subset S$ then $\min\limits_{u \in S_1} j \geqslant \min\limits_{u \in S} j$. If functions u are extended to $[a_1, b_1] \supset [a, b]$ and their corresponding space is S_1 then $S_1 \supset S$ and therefore $\min\limits_{u \in S_1} j \leqslant \min\limits_{u \in S} j$. Similarly if $[a, b]$ is situated between two intervals $[a_2, b_2] \subset [a, b] \subset [a_1, b_1]$ then

$$\min_{u \in S_2} j \geqslant \min_{u \in S} j \geqslant \min_{u \in S_1} j. \tag{4.1}$$

This inequality holds for domains $\Omega \subset \mathbb{R}^n$ of definition of u. Put $\Omega_2 \subset \Omega \subset \Omega_1$ and define three variational problems corresponding to Ω_2, Ω, Ω_1. Then the variational problem corresponding to Ω is intermediate between those corresponding to Ω_2 and Ω_1. The inequalities (4.1) yield bounds for the minimum of the given functional. The procedure of obtaining these bounds is called the *Weinstein method of intermediate problems* and it is extremely useful, especially when the boundaries of Ω are complicated. The Weinstein method was used in the derivation of Serrin's criteria for arbitrary domains (Section 1.3.1) by means of variational problems where Ω is a sphere or a cube.

Let B_1 and B_2 be two Banach spaces and $U \subset B_1$ an open set; a (generally nonlinear) map $T : B_1 \to B_2$ is called Fréchet differentiable at $\tilde{u} \in U$ if there exists a linear continuous operator $T'_{\tilde{u}} : B_1 \to B_2$ such that for every $u \in B_1$ and $\tilde{u} + u \in B_1$ we have

$$\lim_{\|u\|_{B1} \to 0} \frac{\| T(u + \tilde{u}) - T(u) - T'_{\tilde{u}}(u) \|_{B2}}{\| u \|_{B_1}} = 0.$$

The operator $T'_{\tilde{u}}$ is called the Fréchet derivative of T at the point $\tilde{u}$; so it depends on $\tilde{u}$ and acts on u. The Fréchet derivative of a nonlinear map T at $\tilde{u}$ is the linearization of T around $\tilde{u}$. In fluid dynamics the Stokes operator is the linearization of the Navier-Stokes equations about origin hence it is T'_0; the Oseen operator is $T_{U\infty} \cdot T'_{\tilde{u}}$ depends on $\tilde{u}$ and therefore its spectrum will depend on $\tilde{u}$. The points of primary bifurcation belong to $\sigma_p(T'_{\tilde{u}})$ where $\tilde{u}$ is the basic flow. For the motions where it is possible to calculate the spectrum of $T'_{\tilde{u}}$ a bifurcation study is successful. For example in Bénard steady convection we have $\tilde{u} = 0$. For nonstationary convection which is a *secondary branching* (i.e. the branching from a bifurcating solution) $\tilde{u}$ will be the velocity in the Bénard cells. Since it is difficult to calculate the spectrum of a linear differential operator with variable coefficients which depend on a function $\tilde{u}$ which is generally known only numerically, the required calculation for the second bifurcation may seldom be carried out.

The Fréchet derivative may be defined in the context of manifolds of maps (Appendix 5). If $B_1 = B_2 = \mathbb{R}$ then T is a function of a real variable and the concept of Frechet derivative coincides with that of usual derivative. If $B_1 = \mathbb{R}^n$ and $B_2 = \mathbb{R}^m$, $T : \mathbb{R}^n \to \to \mathbb{R}^m$, the Fréchet derivative is the Jacobian of T.

Let $u \in S$; the *variation* δu is by definition a linear operator $\delta u : S \to S$, $\delta u(\tilde{u}) = \tilde{u} - u$. Usually by variation δu of u we mean the value of δu at $\tilde{u}$. The variation $\delta j(u)$ of j is by definition the Fréchet derivative of j at the point u (i.e. j'_u). In the simplest case we hawe $\delta j_u(\tilde{u}) = \int_a^b \left(\frac{\partial F}{\partial u} - \frac{\mathrm{d}}{\mathrm{d}x}\frac{\partial F}{\partial u'}\right) \delta u(\tilde{u})\, \mathrm{d}x$. Since $\delta j(u)$ must vanish for every $\tilde{u} \in S$, by Lagrange's Lemma, we have

$$\frac{\partial F}{\partial u} - \frac{\mathrm{d}}{\mathrm{d}x}\frac{\partial F}{\partial u'} = 0, \tag{4.2}$$

which is called the *Euler equation*. The solutions $u \in S$ of (4.2) are stationary points of j and conversely. Some of these stationary points may be minimum points. A sufficient condition for $u \in S$ be a minimum point is that $\delta^2 j_u(\tilde{u}) > 0$ for every $\tilde{u} \in S$, where $\delta^2 j_u$ is the second Fréchet variation (derivative) of j at u.

If F is a quadratic function of u, u' then (4.2) is a linear second order ordinary differential equation. If F depends on higher order derivatives up to $u^{(k)}$ then (4.2) will have order $2k$. If u is a vector then instead of (4.2) we have a system of Euler equations for each component of u. If x is a vector then the corresponding Euler equation will be a partial differential equation. In all these cases similar definitions for $S, F, j, \delta j_u, \delta u$ etc. can be introduced. We say that a *variational problem* is *isoperimetric* if the function of S must satisfy some constraint $\Gamma(u) = 0$. Then the stationary points of j in S are the stationary points of $j - \lambda\Gamma$ in S' where S' is the space of the functions of S which do not satisfy $\Gamma(u) = 0$. λ is an unknown number if Γ is a constraint of integral type and λ is a function if the constraint is differential. λ is called the *Lagrange multiplier*. Eigenvalue probblems are isoperimetric problems in which the eigenvalue λ is a Lagrange multiplier. In hydrodynamic stability theory functionals which are the quotient of two integrals occurs frequently. For example $j(u) = \int_a^b F_1(x, u, u', u'')\,dx \Big/ \int_a^b F_2(x, u, u')\,dx$ is such that $\sqrt{j_2}$ (where $j_2 \equiv \int_a^b F_2(x, u, u')\,dx$) has the properties of a norm and $j_1(u) \equiv \equiv \int_a^b F_1(x, u, u', u'')\,dx$ is homogeneous of order two in u and u'. In this case put $y = u\,(j_2(u))^{-2}$; it follows that $j(u) = j_1(y)$ and $j_2(y) = 1$. So, instead of calculating the minimum of j in the class S we have to calculate min $j_1(y)$ for $y \in S$ with the constraint $j_2(y) - 1 \equiv 0$. Thus we obtain an isoperimetric problem which is easier to calculate than the given variational problem. This scheme is basic in obtaining isoperimetric inequalities (as for instance (1.3.28)) which play an essential role in global stability. They were used extensively by D.D. Joseph and his school. The application of energy methods in linear hydrodynamic stability theory leads to variational problems of all the above-mentioned types as it can be seen in Ch. I. Isoperimetric problems will be treated in Appendix 7 too. In order to shorten the calculations in Appendix 7 the boundary value problems or eigenvalue problems were written in the variational form. The reduction of a variational problem to a differential one is convenient when the solution of the Euler equation can be explicitly written as is the case with the isoperimetric inequalities.

Generalized setting. Consider the equation

$$Au = f, \quad f \in H, \tag{4.3}$$

where $A : \mathfrak{D}(A) \subset H \to H$ is a map and H is a Hilbert space. It if of interest to find a variational formulation for (4.3) i.e. to find a fsnctional whose minimum is attained at points u which are solutions ou (4.3). This is the inverse problem similar to that in the classical case. In Appendix 1 we indicated two ways of introducing generalized solutions of (4.3) by extending the operator A via extending the functional j. The operator and the associated variational formulations were equivalent. Weak formulations of operator equations whose generalized solutions are minimum points for some functionals are called *variational problems*. Thus the problems satisfied by the generalized solutions of Ch. 2 are of variational form [2].

Geometrical setting. Global (variational) analysis is the geometrical (topological) extension of the calculus of variations from functionals on linear spaces of functions to functionals defined on manifolds of maps. It originates in Morse theory of the calculus of variations on Riemannian manifolds of curves joining two points; the stationary points are the geodesics [4], [5]. For manifolds M_1, M_2

of maps $T: M_1 \to M_2$ the bifurcation points are the critical points of the Fréchet derivative defined on tangent spaces of M_1 and M_2. The catastrophe theory of R. Thom dealt with bifurcations of gradient dynamical systems. In Appendix 7 we show that the catastrophe curve for the characteristic equation of the eigenvalue problem governing the linear stability of a fluid flow bounds the neutral hypersurfaces. The concepts and results of global analysis extend the corresponding ones from the generalized setting. Some of them are presented in Appendix 5. Many recent studies on global bifurcation, Hopf bifurcation etc. are written in the geometric language; we mention some of them in Ch. 4.

REFERENCES

[1] Lavrentiev, M. A., L. A. Liusternik, *Lectures Notes on calculus of variations*, Ed Tehnică, Bucharest, 1955 (Romanian).

[2] Temam, R., *Navier-Stokes equations, theory and numerical analysis*, North-Holland, Amsterdam, 1977.

[3] Guirault, V., P. A. Raviart, *Finite element approximation of the Navier-Stokes equations*, Springer Lecture Notes **749**, Berlin, 1979.

[4] Morse, M., *Global variational analysis; Weierstrass integrals on a Riemannian manifold*, Princeton University Press, 1976.

[5] Smale, S., *The mathematics of time*, Springer-Verlag, New York, 1980.

5. GEOMETRIC METHODS IN BRANCHING THEORY

The definition of a branch (or bifurcation) point given in Section 2.5.3 may be also stated as follows.

Definition 5.1. Let X and Y be two Banach spaces, $F: X \times \mathbb{R} \to Y$ a C^∞ nonlinear mapping with $F(0, \lambda) = 0$ for every $\lambda \in \mathbb{R}$, $S = \{(u, \lambda) \in X \times \mathbb{R} \mid F(u, \lambda) = 0, u \neq 0\}$. $\lambda_0 \in \mathbb{R}$ is called a branch point for the problem

$$F(u, \lambda) = 0, \ (u, \lambda) \in X \times \mathbb{R}$$

if $(0, \lambda_0)$ belongs to the closure of S in $X \times \mathbb{R}$.

This definition is of local type and it refers to the behaviour of subsets of S around a given point $(0, \lambda_0)$. We present now some recent results on *global branching* theory which study the global structure of the branches: the boundedness of the components of the set of solutions i.e. maximal connected subsets. This theory originates in two independent papers of Rabinowitz and Turner from 1971 [4], where it was shown that if λ_0 is a characteristic value of odd multiplicity of the Fréchet derivative F_0' of the mapping F then there exists a component of the nontrivial solutions of this equation whose closure contains the point $(0, \lambda_0)$. If this component is a bounded subset of $X \times \mathbb{R}$ then its closure contains an element $(0, \lambda) \in X \times \mathbb{R}$ where λ is a characteristic value of F_0' and $\lambda \neq \lambda_0$. In other words, a branch which appears at $(0, \lambda_0)$ passes through another bifurcation point $(0, \lambda)$, all its points on the response diagram being at finite distance.

In the global theory important results have been obtained by C. A. Stuart. Here we mention one of them [5]. Let X and Y be two real Banach spaces,

$$F(u, \lambda) = 0, \text{ for } (u, \lambda) \in X \times \mathbb{R},$$

$F : X \times \mathbb{R} \to Y$ is a continuous mapping such that $F(0, 0) = 0$. The component of S which contains the point $(0, 0)$ is called a principal component of S and is denoted by $\mathcal{C}$. If $u = 0$ is the only solution of $F(u, 0) = 0$, $\mathcal{C}^+ = \{(u, \lambda) \in \mathcal{C} \mid \lambda \geqslant 0\}$ and $\mathcal{C}^- = \{(u, \lambda) \in \mathcal{C} \mid \lambda \geqslant 0\}$ are both connected and, if the usual compactness assumptions are made, then these sets are unbounded. However, if the compactness assumptions are relaxed then $\mathcal{C}^+$ and $\mathcal{C}^-$ are bounded. Their global behaviour is studied by means of topological degree and ordered Hilbert space theory.

More general definition of the branch point are those which do not take into account the known branch; in the following we give three of them. The methods used to study the corresponding branchings are based on theories in differential geometry (e.g. singularity of mappings, transversality, catastrophes, structural stability).

Definition 5.2[3]. Let X and Y be two Banach spaces, τ a family of mappings from X to Y and suppose there is a norm on the members of τ. Let $T \in \tau$ and assume there exists $u_0 \in X$ such that $Tu_0 = 0$. The operator T is said to be a branch point for τ at u_0 if for every neighbourhood U of T and V of u_0 there is an $S \in U$ and $u_1, u_2 \in V$ $(u_1 \neq u_2)$ such that $Su_1 = Su_2 = 0$.

Some connection can be established between the definition 5.1 and 5.2. if we recall that the mappings from a neighbourhood of T are equal to T up to higher-order terms. Therefore in every neighbourhood of the known solution u_0 there are two approximate solutions u_1 and u_2 of the equation $Tu = 0$.

Definition 5.3 (Smale). If $h: X \to Y$ is a continuous map between topological spaces then $y_0 \in Y$ is a global branch point of h if, for every neighbourhood U of y_0, not all the sets $h^{-1}(y)$, $y \in U$ are homeomorphic (i.e. $h^{-1}(y)$ changes topological type at y_0). The point $u_0 \in X$ is a local bifurcation point for h if for every neighbourhood U of $v_0 = h(u_0)$ and V of u_0 the sets $h^{-1}(v) \cap V$, $v \in U$, are not all homeomorphic (i.e. $h^{-1}(v)$ locally changes topological type at (u_0, v_0)).

Topological methods are based on last characterization of branch points (u_0, v_0) namely the change of the topological type at (u_0, v_0).

Let X, Y be two smooth manifolds and $h: X \to Y$ a differentiable map (i.e. it has Fréchet derivatives of every order). We say $u_0 \in X$ is a *regular point* of h if the linear map (called the Fréchet differential of h at u_0) $h'_{u_0}: T_{u_0}X \to T_{h(u_0)}Y$ has the maximum rank. By definition, the rank of h at u_0 is the dimension of the image through h'_{u_0} of the space $T_{u_0}X$. Obviously the maximum rank of h at u_0 can be min (dim $T_{u_0}X$, dim $T_{h(u_0)}Y$) which is equal to min (dim X, dim Y).

It can be proved that if u_0 is a regular point of h then there exists a neighbourhood U of $v_0 = h(u_0)$ such that all the sets $h^{-1}(v)$, $v \in U$ are just singletons i.e. they are homeomorphic. Finally it can be shown that $u_0 \in X$ is a *regular point* of h if h'_{u_0} is injective. If a point is not regular it is called a *singular point*. Therefore from definition 5.3 it follows that bifurcation points are among the singularities of mappings. On the other side u_0 is regular iff h is *transversal* on $\{u_0\}$ ($h \pitchfork \{u_0\}$) [6]; these facts show the connection between branching theory on one hand and singularity theory and transversality on the another one. The connection between the definitions 5.1 and 5.3 can be obtained by letting $h: S \to \Lambda$, $(x, \lambda) \in \Lambda$ where S is the set of all the solutions of $F(u, \lambda) = 0$.

Finally we give another definition of branch point which together with the definitions 5.2, 5.3 may be applied to problems with no parameters.

Definition 5.4 (Thom). A vector field is a bifurcation point (of a family) if it is in the complement of structurally stable vectors fields (relative to the given family).

If now in definition 5.1 we let λ belong to $\mathbb{R}^n$ then the set of all bifurcation points will be an $n - 1$ manifold called the *bifurcation surface* or *the catastrophe surface*. Recently the theory of catastrophes was applied to equations governing the motion of some fluids [34].

A basic theorem in nonlinear functional analysis (and consequently in branching theory) is the implicit function theorem which gives, among others, the existence of solutions, the points where an equation has a unique solution branch. Prior to stating it for Banach spaces X, Y and for smooth mappings $F: X \times \mathbb{R} \to Y$ we shall present an intuitive sketch of it in the case $X = Y = \mathbb{R}$, $F(x, y) = 0$. In this case we can determine $y = y(x)$ in a neighbourhood of a point (x_0, y_0) whenever the tangent to the curve $F(x, y) = 0$ at (x_0, y_0) is not vertical or, since

$$y' = -\frac{\partial F}{\partial x}\Big/\frac{\partial F}{\partial y}, \quad \frac{\partial F}{\partial y}\Big|_{(x_0,\, y_0)} = 0.$$

Let X and Y be two Banach spaces (in general of infinite dimension) and $F: X \times \mathbb{R} \to Y$ a smooth mapping such that $F(u_0, \lambda_0) = 0$. The implicit function theorem states that if the Fréchet derivative $F'_u(u_0, \lambda_0) \cdot$ is invertible, then there exists a neighbourhood U of λ_0 and a continuous mapping $\lambda \to u(\lambda)$ such that $F(u(\lambda), \lambda) = 0$

for every $\lambda \in U$, $u(\lambda_0) = u_0$ and the mapping u is differentiable in U and $u_\lambda = -F'_u \circ F_\lambda$. Taking into account the above characterization of the singular and regular points, with the aid of the Fréchet derivative and by implicit function theorem it follows: a necessary condition for a point (u_0, λ_0) to be a branch point for $F(u, \lambda) = 0$ is that (u_0, λ_0) be a singular point of F, or, equivalently, λ_0 is the eigenvalue of $F'_u(u_0, \lambda_0) - \lambda I$ where $F(u_0, \lambda) = 0 \ \forall \ \lambda > 0$.

A sufficient condition that (u_0, λ_0) bc a branch point for $F(u, \lambda) = 0$ is that the multiplicity of λ_0 be odd. This may be proved by topological degree methods. The above necessary condition and sufficient condition hold for compact F. For more general noncompact mappings we quote [1]. We mention that in recent years the bifurcation from eigenvalues of $F'_u(u_0, \lambda_0) - \lambda I$ of even multiplicity has been developed.

In short, the Liapunov-Schmidt procedure analised in § 3.3 consists of the following. Let X and Y be two Banach spaces and $F: X \times \mathbb{R} \to Y$ a C^k map, $k \geqslant 1$. The Fréchet derivative $F'_x(x, \lambda)$: $: X \to Y$ of F with respect to x at the point (x, λ) is a continuous linear map. This derivative represents the linearization of F around the point (x, λ). Let (x_0, λ_0) be a solution of the nonlinear equation $F(x, \lambda) = 0$ and assume that $F'_x(x_0, \lambda_0)$ is a Fredholm operator. Then, taking into account the eigenspace of $F'_x(x_0, \lambda_0) + \lambda_0 I$ (i.e. the null space of $F'_x(x_0, \lambda_0)$), the spaces H and Y are split into two parts $X = X_1 \oplus X_2$ where $X_1 = \mathfrak{N}(F'_x(x_0, \lambda_0))$, $\dim X_1 < \infty$ and $Y = Y_1 \oplus Y_2$ where $Y_1 = \mathfrak{R}(F'_x(x_0, \lambda_0))$ $\dim Y_2 < \infty$. Correspondingly, by means of the projection $P: Y \to Y_1$, the equation $F(x, \lambda) = 0$ splits into two equations

$$(5.1) \qquad PF(x_1 + x_2, \lambda) = 0, \ x = x_1 + x_2, \ x \in X, \ x \in X_1, \ x_2 \in X_2$$

$$(5.2) \qquad (I - P) F(x_1 + u(x_1, \lambda), \lambda) = 0.$$

By the implicit function theorem, $x_2 = u(x_1, \lambda)$ is the unique solution near (x_0, λ_0) of (5.1). The equation (5.2) is the bifurcation equation of $\dim Y_2$ equations in $\dim X_1$ unknowns. In this way, the determination of the multiple solutions x of $F(x, \lambda) = 0$ has been reduced to the solution of (5.2).

Recently [2] this procedure has been expreseed in terms of differential topology and some extensions have been obtained in this framework. We present here only the simplest case when X and Z are Banach manifolds and Y is a Banach space. Assume that $F: X \times Z \to Y$ is a C^1 map. Define the Fréchet derivative $F'_x(x_0, \lambda_0)$: $T_x X \times \mathbb{R} \to T_{F(x_0, \lambda_0)} Y$. Again we split $T_{x_0} X = X_1 \oplus X_2$ and $T_{F(x_0, \lambda_0)} Y = Y_1 \oplus Y_2$. Since for fixed λ_0 the map F is transversal to the submanifold Y_2 at (x_0, λ_0) then, by transversality in a neighbourhood of (x_0, λ_0) the space $S_P = \{(x, \lambda) \mid PF(x, \lambda) = 0\}$ (i.e. the space of the solutions of (5.1)) is a smooh manifold of $X \times Z$ tangent to $X_1 \times T\lambda_0 Z$ at (x_0, λ_0). If $F_P = F|_{S_P}$ then, obviously, $F(x, \lambda) = 0$ iff

$(I - P) F_P(x, \lambda) = 0$ iff $F_P(x, \lambda) = 0$. This last equation represents the geometric version of the bifurcation equation (5.2). In the above $T_{x_0}X$ stands for the tangent space of X at x_0 and we recall that the mapping $g : X \to Y$, X, Y smooth manifolds, is *transversal* to a submanifold $W \subset Y$, at $x \in X$ if: a) either $g(x) \notin W$ or 2) $g(x) \in W$ and it satisfies the equality $g'_x(T_xX) + T_{g(x)}W = T_{g(x)}Y$ where $g'_x : T_xX \to \to T_{g(x)}Y$ [6].

Let us now give some notions used in § 4.4. Let B_1 and B_2 be two Banach spaces and $A : B_1 \to B_2$ a continuous linear operator. A is called a *Fredholm operator* if: 1) $\dim \mathcal{N}(A) < \infty$; 2) the range of A is closed; 3) coker $A = B_2/$ range A has finite dimension. Let M_1 and M_2 be two differentiable manifolds (locally like Banach spaces). A nonlinear mapping $F : M_1 \to M_2$ of class C^1 is called a *Fredholm mapping* if for every $u \in M_1$ the Fréchet derivative $F'_u : T_uM_1 \to \to T_{F(u)}M_2$ is a Fredholm operator. Its *index* is, by definition, the index of F'_u (which does not depend on $u \in M_1$), index $F'_u = = \dim \mathcal{N}(F'_u) - \dim \operatorname{coker} (F'_u)$.

Sard theorem. *Let* U *be an open set of* $\mathbb{R}^p$ *and* $F : U \to \mathbb{R}^q$ *be a* C^s *mapping where* $s > \max(p - q, 0)$. *Then the set of singular values of* F *in* $\mathbb{R}^q$ *has measure zero (i.e. the regular values of* F *are almost all the points of* $\mathbb{R}^q$).

Sard-Smale theorem. *Let* M_1 *and* M_2 *be two differentiable manifolds and* $F : M_1 \to M_2$ *a* C^q *mapping with* $q > \max$ *(index* F, 0*). Then the regular values of* F *are almost all the points of* M_2 *(i.e. are every point of* M_2 *with the exception of a rare closed subset of* M_2).

A *regular value* of F is its value $F(u)$ where u is a regular point of F. $F(u)$ is called a *singular value* of F if u is a singular point of F.

REFERENCES

[1] Zeidler, E., *Some recent results in bifurcation theory*, Mathematical Physics and Physical Mathematics, K. Maurin, R. Raczka (eds.), PWN-Polish Scientific Publishers, Warszawa, 1976, 185–202.

[2] Marsden J., *On the geometry of the Liapunov-Schmidt procedure*, LNM **755**, Springer, Berlin, 1979, 77–82.

[3] Marsden, J., *Qualitative methods in bifurcation theory*, Bull. Amer. Math Soc., **84**, 6 (1978) 1125–1148.

[4] Rabinowitz, P. H., *Some global results for nonlinear eigenvalue problems*, J. Funct. Anal., **7**, (1971) 487–513.

[5] Stuart, C. A., *Three fundamental theorems in bifurcation*, Seminario di Analisi Funzionale ed applicazione, Università degli Studi della Calabria, Sept. 1977.

[6] Choudary, A. D. R., *Teorema de transversalitate și aplicațiile sale*, Doctoral Dissertation, Bucharest, 1980.

6. NEW METHODS FOR SOLVING THE ORR-SOMMERFELD EQUATION

The Orr-Sommerfeld equation (O–S eq.) is the most famous in hydrodynamic stability theory (h.s.th.) since it governs the linear stability of stationary plane-parallel flows of the viscous fluids against two-dimensional perturbations. So, it corresponds to the simplest widely physically acceptable model. The O-S eq. is obtained as a first step in every investigation which takes into account extra effects (three-dimensionality, the curvature of the walls, the nonlinearity, the nonstationarity, the temperature, the magnetic and electrical fields, the compressibility, the nonlinearity of constitutive equations etc.). This explains its intensive and sometimes dramatic investigation from its very derivation. But due to its many unpleasant properties, the study of this equation is a very difficult task. The O-S equation is a nonselfadjoint fourth order ordinary differential equation with variable (complex-valued) coefficients which depend on the velocity profile of the basic flow. Its spectral analysis is not yet complete even for simple velocity profiles. The O-S eq. is couched in singular asymptotic perturbation theory due to the fact that the parameter $(\alpha \text{ Re})^{-1}$ in it is small and the first approximation of the regular O-S eq. is the singular Rayleigh equation. As a result, to overvome these difficulties a large number of methods have been developed [1]. Among them the most suitable are the *asymptotic methods* (mainly developed by Heisenberg, Tollmien, Lin, Reid) whose principles were sketched in Ch. 1. By these methods the solutions of the Orr-Sommerfeld equation which are continuous functions are approximated by means of (generally divergent) asymptotic series in the powers of the small parameter $\varepsilon = (\alpha \text{ Re})^{-1}$. These series are continuous for $\varepsilon > 0$ and discontinuous for $\varepsilon = 0$. This is why two asymptotic phenomena arise: the (mathematical) boundary layer and the Stokes phenomenon. The Stokes lines emerge from the turning point y_c (i.e. $U(y_c) = c_r$ (Section 1.2.1)). The boundary layer phenomenon takes place near $y = 0$ where, due to the fact that in passing from the O-S eq. to the Rayleigh equation two degress are lost, two boundary conditions are also lost. The rigorous analysis of the *turning point* was initiated by W. Wasow. We mention the basic contribution of W. H. Reid in studying (for the O-S eq.) the case with more than one turning point. In order to exploit the standard asymptotic theory for the second order differential equations he used Airy functions insead of traditional Hankel functions.

Although, at the time being, there are a lot of new asymptotic methods used to study the O-S eq. which improve in many respects the early heuristic approaches by Heisenberg and Tollmien, nevertheless the main difficulties persist due to the two above-mentioned asymptotic phenomena. A very good discussion on old and new asymptotic methods from hydrodynamic stability theory may be found in [1]. Here we want to mention the *multiple scale method* (or homogenization) which in about 1965 began to be used in deriving the macroscopic model for compounds media with periodic microstructure [2], [3]. By this method it was shown [4] that for almost parallel basic flows the perturbations are, in the first approximation, the very Tollmien-Schlichting waves whose amplitude and phase have a slow evolution governed by some equations whose left-hand side is just the Orr-Somerfeld equation. The homogenization was applied to turbulence by L. Tartar [5].

The use of Fourier series methods upon orthogonal sets of functions are customarily applied to solve equations with constant coefficients. In the next Appendix we present them in detail.

When applied to equations with variable coefficients (like O-S eq.) they lead to cumbersome calculations. Nevertheless even for this type of equations substantial simplifications can be obtained if the expansion functions do not satisfy all the boundary conditions of the problem (e.g.$\varphi = D\varphi = 0$ at $y = 0,1$ for the eigenvalue problem for the O-S eq. (1.1.25), (1.1.26)). This method, called by us the *Budianski-DiPrima method*, in spite of its elegance and efficiency is less known. Instead, methods of Chandrasekhar-Galerkin type where the expansion functions satisfy all the boundary conditions of the problem are widely used. It is precisely to emphasize the advantages of the Budianski-DiPrima method that we gave such a large emphasis to Appendix 7.

Since expansion functions φ_n which satisfy an equation very close to the O-S eq. will be close to the eigenfunctions of the O-S eq. φ_n are chosen for instance to be the eigenfunctions of the problem

$$\begin{cases}(D^2 - \alpha^2)^2 \varphi_n + \lambda_n (D^2 - \alpha^2) \varphi_n = 0, \\ \varphi_n = D\varphi_n = 0 \text{ at } y = 0,1\end{cases}$$

or

$$\begin{cases}(D^2 - \alpha^2)^2 \varphi_n - \lambda_n \varphi_n = 0 \\ \varphi_n = D\varphi_n = 0 \text{ at } y = 0,1,\end{cases}$$

or Chebyshev polynomials $T_n(z) = \cos (n \cos^{-1} z)$ $(n = 0, 1, 2, \ldots)$ [6]. It remains a task for future work to decide if the Budianski-DiPrima method applied to Orr-Sommerfeld equation leads to less calculations.

As will be shown in Appendix 7 the expansion methods lead to exactly the same secular equation as in the *variational methods* if the same expansion functions are used. A variational method based on O-S eq. as well as its adjoint (see Part *C* of our Appendix 7) was applied by L. H. Lee and W. C. Reynolds [7], namely if $L\varphi = 0$ stands for the problem (1.1.25), (1.1.26) and $L^*\varphi^* = 0$ is its adjoint then their variational equation is $\delta \int_0^1 \varphi^* L\varphi \, dy = 0$. In [8] by means of the Lebon-Lambermont variational criterion for hydrodynamics another variational form of the O-S eq. is deduced. For other variational methods see [1] and [8].

The *finite-diference method* was used by L. H. Thomas to solve the Orr-Sommerfeld equation for the case of the plane Poiseuille flow as early as 1953 [78] (from Ch. 1). Subsequently, due to the facilities offered by high speed computers many numerical methods were applied to O-S eq. directly or to various equivalent forms. Thus, in the *initial-value methods (shooting)* one writes the O-S eq. in the form of a first order system

$$\boldsymbol{\psi}' = \mathbf{N}\boldsymbol{\psi} \tag{6.1}$$

where, for instance, $\boldsymbol{\psi} = [\varphi, \varphi', \varphi'' - \alpha^2\varphi, \varphi''' - \alpha^2\varphi']^T$,

$$\mathbf{N} = \begin{pmatrix} 0 & 1 & 0 & 0 \\ \alpha^2 & 0 & 1 & 0 \\ 0 & 0 & 0 & 1 \\ -i\alpha\,\mathrm{R}U'' & 0 & \alpha^2 + i\alpha\,\mathrm{R}(U - c) & 0 \end{pmatrix}.$$

If $\boldsymbol{\psi}_1$ and $\boldsymbol{\psi}_2$ are two linearly independent solutions which satisfy the initial conditions $\boldsymbol{\psi}_1(0) = [1, 0, 0, 0]^T$, $\boldsymbol{\psi}_2(0) = [0, 0, 1, 0]^T$, then the general solution of (1.1.25) will be $\boldsymbol{\varphi} = A_1\boldsymbol{\varphi}_1 + A_2\boldsymbol{\varphi}_2$, which when introduced in the boundary conditions at $y = 1$, yield the secular equation. The numerical solution of this equation will give the neutral curve. There are many other choices for $\boldsymbol{\psi}$ and, correspondingly, many initial-value methods adapted to solve the O-S eq. We present the so-called *compound matrices method* which gives better estimates [9], [10]. It uses $\boldsymbol{\varphi} = [\varphi, \varphi', \varphi'', \varphi''']^T$ such that (1.1.25) becomes

$$\boldsymbol{\varphi}' = A(y)\,\boldsymbol{\varphi} \tag{6.2}$$

where $\mathbf{A}(y) =$

$$= \begin{pmatrix} 0 & 1 & 0 & 0 \\ 0 & 0 & 1 & 0 \\ 0 & 0 & 0 & 1 \\ -\{\alpha^4 + i\alpha\,\mathrm{Re}[\alpha^2(U - c) + U'']\} & 0 & 2\alpha^2 + i\alpha\,\mathrm{Re}(U - c) & 0 \end{pmatrix}.$$

Let $\boldsymbol{\varphi}_1$ and $\boldsymbol{\varphi}_2$ be two solutions of (6.2) which satisfy the initial conditions $\boldsymbol{\varphi}_1(0) = [0, 0, 1, 0]^T$, $\boldsymbol{\varphi}_2(0) = [0, 0, 0, 1]^T$ and put

$$\boldsymbol{\Phi}(y) = \begin{pmatrix} \varphi_1 & \varphi_2 \\ \varphi_1' & \varphi_2' \\ \varphi_1'' & \varphi_2'' \\ \varphi_1''' & \varphi_2''' \end{pmatrix}, \tag{6.3}$$

$$z_1 = \varphi_1\varphi_2' - \varphi_1'\varphi_2, \qquad z_4 = \varphi_1'\varphi_2'' - \varphi_1''\varphi_2',$$

$$z_2 = \varphi_1\varphi_2'' - \varphi_1''\varphi_2, \qquad z_5 = \varphi_1'\varphi_2''' - \varphi_1'''\varphi_2',$$

$$z_3 = \varphi_1\varphi_2''' - \varphi_1'''\varphi_2, \qquad z_6 = \varphi_1''\varphi_2''' - \varphi_1'''\varphi_2''$$

so that that the identity $z_1 \cdot z_6 - z_2 \cdot z_5 + z_3 \cdot z_4 = 0$ takes place. Differentiating (6.3), taking into account (6.2), and putting $\mathbf{Z} = [z_1, \ldots, z_6]^T$ we obtain

$$\mathbf{Z}' = \mathbf{B}(y)\,\mathbf{Z}, \tag{6.4}$$

where

$$\mathbf{B}(y) = \begin{Vmatrix} 0 & 1 & 0 & 0 & 0 & 0 \\ 0 & 0 & 1 & 1 & 0 & 0 \\ 0 & a_2 & 0 & 0 & 1 & 0 \\ 0 & 0 & 0 & 0 & 1 & 0 \\ -a_4 & 0 & 0 & a_2 & 0 & 1 \\ 0 & -a_4 & 0 & 0 & 0 & 0 \end{Vmatrix},$$

$a_2 = 2\alpha^2 + \mathrm{i}\alpha\,\mathrm{Re}(U - c)$, $a_4 = \alpha^4 - \mathrm{i}\alpha\,\mathrm{Re}[\alpha^2(U - c) + U'']$. Thus $\mathbf{Z}$ is the second component of $\mathbf{\Phi}$ and $\mathbf{Z}(0) = [0, 0, 0, 0, 0, 1]^T$. Imposing the boundary conditions at $y = 1$ we obtain the secular equation [1].

In these initial-value methods the boundary condition at $y = 0$ were used to write two linearly independent solutions and then, after some algebra the other boundary conditions at $y = 1$ were imposed to derive the equation of the neutral curve. The numerical applications concerned frequently the plane Poiseuille flow and boundary layer type flows. The same remarks are valid for the following *method which uses a second type Volterra equation* [11]. (In fact, all the initial value methods are related to some integral equations). Putting $f = \varphi'' - \alpha^2\varphi$ such that since $\varphi(0) = \varphi'(0) =$ we have

$$\varphi = \frac{1}{\alpha}\int_0^y f(p)\sinh\alpha(y - p)\,\mathrm{d}p. \tag{6.5}$$

Denoting $\Phi = f'' - \alpha^2 f$ we obtain

$$f = \frac{1}{\alpha}\int_0^y \Phi(p)\sinh\alpha\,(y - p)\,\mathrm{d}p + A\cosh\alpha y + B\sinh\alpha y \tag{6.6}$$

where Φ is the solution of the Volterra integral equation

$$\Phi(y) + \mathrm{i}\alpha\,\mathrm{Re}\int_0^y K(y, p)\,\Phi(p)\,\mathrm{d}p = -\mathrm{i}\,F(y), \tag{6.7}$$

where $K(y, p) = [c - U - U''(2\alpha^2)^{-1}]\sinh\alpha\,(y - p) + U''(2\alpha)^{-1} \times (y - p)\cosh\alpha(y - p)$, $F(y) = \alpha A\,\mathrm{Re}[(c - U)\cosh\alpha\,y + U''(2\alpha)^{-1} \times y\sinh\alpha y] + \alpha B\,\mathrm{Re}\{[c - U - U''(2\alpha^2)^{-1}]\sinh\alpha y + U''(2\alpha)^{-1} y\cosh\alpha y$. For a given velocity profile $U(y)$ then equation (6.7) has a unique solution which depends on the undetermined constants A and B.

From (6.5) and (6.6) we have

$$\varphi = \frac{1}{2\alpha^2}\int_0^y [(y-p)\cosh\alpha(y-p) - \frac{1}{\alpha}\sinh\alpha(y-p)] \times \tag{6.8}$$

$$\times \Phi(p)\,\mathrm{d}p + \frac{A}{2\alpha}\,y\,\sinh\alpha y + \frac{B}{2\alpha}\left(y\cosh\alpha y - \frac{1}{\alpha}\sinh\alpha y\right).$$

Now requiring φ from (6.8) to satisfy $\varphi(1) = \varphi'(1) = 0$ we obtain the neutral curve.

We mention the existence of a large number of other numerical methods to study the O-S eq. Many of these methods are of finite difference or finite element type.

REFERENCES

[1] Drazin, P., W. H. Reid., *Hydrodynamic stability*, Cambridge University Press, 1981.

[2] Bensoussan, A., J.—L. Lions, G. Papanicolau, *Asymptotic analysis for periodic structure*, Amsterdam, North-Holland, 1978.

[3] Sanchez-Palencia, E., *Comportements local et macroscopique d'un type de milieux physique hétérogènes*, Int. J. Eng. Sci., 12 (1974) 331—351.

[4] Bouthier, M., *Stabilité lineaire des écoulements presque parallèles par la méthode des échelles multiples*, C. R. Acad. Sci. Paris, **273 A** (1971) 1101—1104.

[5] Tartar, L., *Homogenization of turbulence*, Singular perturbations and boundary layer theory, Lecture Notes in Mathematics **594**, Springer-Verlag, Berlin, 1977.

[6] Orszag, S. A., *Accurate solution of the Orr-Sommerfeld stability equation*, J. Fluid Mech., **50**, (1971) 689—703.

[7] Lee, L. H., W. C. Reynolds, *On the approximate and numerical solution of Orr-Sommerfeld problems* Q. J. Mech. Appl. Math., **20**, 1—22 (1967).

[8] Lebon, G., H. Nguyen, *Hydrodynamic stability by variational methods*, Inst. J. Heat Mass Transfer, **17**, (1974) 655—667.

[9] Lakin, W. D., B. S. Ng, W. H. Reid, *Approximations to the eigenvalue relation for the Orr-Sommerfeld problem*, Phil. Trans. of the Royal Soc. London, **289**, nr. 1360 (1978) 347—371.

[10] Ng, B.S., W. H. Reid, *An initial value method for eigenvalue problems using compound matrices*, J. Computational Physics, **30**, 1 (1979) 125—136.

[11] Panaitescu, V., *Asupra stabilității lineare a unei curgeri laminare plane*, St. Cerc. Mec. Apl., **30**, 3 (1971) 549—558.

7. ANALYTICAL METHODS TO SOLVE SOME EIGENVALUE PROBLEMS IN HYDRODYNAMIC AND HYDROMAGNETIC STABILITY THEORY

Most of the eigenvalue problems occurring in linear hydrodynamic and hydromagnetic stability theory consist of an ordinary differential (usually nonsymmetric) equation (or system) of order $n \geqslant 6$ with constant coefficients and of n homogeneous boundary conditions.

The coefficients depend on one or more real positive parameters (as for instance the Reynolds number, the Rayleigh number, the Prandtl number, wave numbers etc.). The unknown functions represent perturbations of the velocity, temperature, magnetic field etc. are the components of the eigensolution $\mathbf{v} = (v_1, ..., v_k)$. The parameters are the components of the eigenvalue $\mathbf{R} = (R_1, ..., R_m)$. In the neutral case these eigenvalue problems are of the form

$$\mathbf{A}(\mathbf{R})\mathbf{v} = 0, \quad x \in (a, b) \tag{7.1}$$

$$\mathbf{B}(\mathbf{R})\mathbf{v} = 0, \quad x = a, b, \tag{7.2}$$

where $v_i(x) : [a, b] \to \mathbb{R}$, $i = 1, ..., k$, $\mathbf{A}$ is a $l \times k$-matrix whose entries A_{ij} are linear differential operators. The order of $A_{1j_1}, ... A_{lj_l}$ is less than or equal to n for every mutually disjoint $j_1, ..., j_l \leqslant k$. $\mathbf{B}$ are the boundary conditions which involve p-th order derivatives, $p \leqslant n - 1$. Since x is a scalar these problems correspond to plane parallel or almost parallel motions; for the three-dimensional case x must be replaced by $\mathbf{x} = (x_1, x_2)$. Similar results with those derived in this section will hold for the three-dimensional case.

The problem (7.1), (7.2) generates a linear operator $L : \mathfrak{D}(L) \subset \subset (L^2(a, b))^k \to (L^2(a, b))^k$, $\mathfrak{D}(L) = \{\mathbf{v} \in (C^\infty[a, b])^k \mid \mathbf{B}(\mathbf{R})\ \mathbf{v} = 0$ at $x = a, b\}$. We have $\mathfrak{D}(L) = (L^2(a, b)^k$ (Appendix 1). Then the problem (7.1), (7.2) may be written as

$$L\mathbf{v} = 0, \ \mathbf{v} \in \mathfrak{D}(L).$$

Generally $\mathfrak{D}(L)$ consist of functions characterized by smoothness properties; boundary conditions and, sometimes, some special requirements concerning their oddness, periodicity etc. Thus, in general $\overline{\mathfrak{D}(L)}^{\|\ \|} = H \subset (L^2(a, b))^k$.

The main problem of interest for stability theory is the determination, in the space $(R_1, ..., R_m)$, of the neutral hypersurface NS which separates the domain of linear stability from that of linear instability i.e. the derivation of the smallest eigenvalue $\mathbf{R}$ of (7.1), (7.2) (that is the smallest R_i for $R_1, ..., R_{i-1}, R_{i+1}, ..., R_m$ kept fixed). This eigenvector has the origin (0, 0, ..., 0) and its vertex lies on NS. The equation of NS, say $D_g = 0$, may be obtained either by a) direct or by b) variational methods or c) by requiring the general solution of (7.1) (deduced by means of the characteristic equation) to satisfy (7.2) or d) by other methods. In the following we shall present several analytical (i.e. not numerical) methods of types a), b) and c) to derive $D_g = 0$ for some typical eigenvalue problems. To solve $D_g = 0$ generally numerical methods must be used.

The aim of this appendix is to emphasize the advantages of the Budiansky-DiPrima type methods [8] (from Ch. 1) as compared with Chandrasekhar-Galerkin type approaches [4] and to show how the solution based on the characteristic equation leads to bounds for the neutral hypersurfaces. These bounds are the very catastrophe (bifurcation) surface CS of the characteristic equation.

This last solution is suitable for eigenvalue problems where many physical parameters appear [3]. In [5] the CS for all the typical problems in [5] (from Ch. 1) were derived.

Prior to presenting the above-mentioned methods we make some remarks basic to the sequel.

Remark 7.1. $v_i \in C^\infty[a, b]$ because every solution of (7.1), (7.2) is of the form $P(x)\cosh \lambda x + Q(x)\sinh \lambda x$ where $P(x), Q(x)$ are polynomials of a degree smaller or equal to $n - 1$ and λ are (generally) complex numbers. This remark will enable us to derive some extra boundary conditions from equation (7.1), which facilitate the derivation of the variational principles. The reduction of (7.1), (7.2) to an equivalent eigenvalue problem with a (single) scalar eigensolution also use Remark 7.1.

Remark 7.2. Suppose that (7.1) was reduced to a single even order equation and in (7.2) the even and odd derivatives occur separately. Then the new problem will have even and odd solutions ($\mathbf{v}_e$ and $\mathbf{v}_0$ respectively). In this case $D_g = D_e \cdot D_0$ where $D_e = 0$ and $D_0 = 0$ are the equations of NS corresponding to even and odd solutions respectively. Therefore instead of solving (7.1), (7.2) in the class of general functions (i.e. neither even nor odd) we solve (7.1), (7.2) in the class of even and odd functions separately. As a consequence the smallest eigenvalue $\mathbf{R}_{min}$ will be the solution of $D_e = 0$ or $D_0 = 0$ and the corresponding eigensolution of (7.1), (7.2) will be even or odd. If for some particular choice of the parameters $\mathbf{R}_{min}$ corresponds to even solutions, say, then this situation holds for any other values of those parameters. The order of the determinants D_e or D_0 is half of the order of D_g. Thus this remark reduces the calculation of a n-th order determinant to that of a $\frac{n}{2}$-th order determinant. Since in the following Fourier series will be used extensively we recall some facts from their theory.

Remark 7.3. Inappropriate choices of Fourier series may introduce extra periodicity properties on $[a, b]$. For instance, if we want to expand $\mathbf{v}_e$ into a Fourier cosine series from the set $\{\cos 2ln\, \pi x\}_{n \in \mathbb{N}}$, and if $a = -0.5$, $b = 0.5$, then for $l > 1$ $\mathbf{v}_e$ will have the extra period $\frac{1}{l}$ which is the period of all the functions of this set. $\mathbf{v}_e$ was supposed to be a scalar even function on $[-0.5; 0.5]$.

Remark 7.4. For every function of L^2 we may write many expansions in Fourier series upon total sets in L^2. In this section we shall frequently use the following orthonormal sets in $L^2(-0.5; 0.5)$: $\{E_1, E_3, E_5, ...\}$, $\{F_1, F_3, F_5, ...\}$, $\{1, E_2, E_4, E_6, ...\}$, $\{F_2, F_4, ...\}$. The orthogonal set $\{1, E_1, E_3, \ldots\}$ is total in $L^2(-1,1)$. These sets are total in the subpaces of L^2 consisting of even or odd functions. By definition $E_n = \sqrt{2} \cos n\pi x$ and $F_n = \sqrt{2} \sin n\pi x$. This notation differs slightly from that of Section 1.1.3.

Remark 7.5. Since generally termwise differentiation of a Fourier series fails, to give the Fourier series of the derivative of a function we use the so-called backward integration technique. It is an integration by parts. As a result the Fourier coefficients of the derivative are given in terms of the Fourier coefficients of the function itself. For example if $v = \sum_{n=1}^{\infty} v_{2n-1} E_{2n-1}$ and $a = -0.5$, $b = 0.5$ then we put $\frac{dv}{dx} = \sum_{n=1}^{\infty} v'_{2n-1} F_{2n-1}$. Obviously $v'_{2n-1} = \int_{-0.5}^{0.5} \frac{dv}{dx} F_{2n-1}(x)\, dx =$

$$= 2[v \sin(2n-1)\pi x]_{-0.5}^{0.5} - \sqrt{2} \int_{-0.5}^{0.5} (2n-1)\,\pi \cdot v(x) \cos(2n-1)\,\pi x\, dx =$$

$= 2(-1)^{n+1} \sqrt{2} v(0.5) - (2n-1)\pi v_{2n-1}$. Hence, $v'_{2n-1} = 2\sqrt{2}(-1)^{n+1} v(0.5) - (2n-1)\,\pi v_{2n-1}$. (We see that v is termwise differentiable only if $v(0.5) = 0$). Similarly $v^{(2k)}_{2n-1} = (2n-1)\pi v^{(2k-1)}_{2n-1}$, $v^{(2k+1)}_{2n-1} = 2\sqrt{2}(-1)^{n+1} D^{2k} v(0.5) - (2n-1)\pi v^{(2k)}_{2n-1}$ where $v^{(m)}_{2n-1}$ stands for the Fourier coefficients of $D^m v$. Thus by applying step by step the backward integration technique we finally obtain the Fourier coefficients of every derivative of v in terms of the Fourier coefficients of v.

A. Direct Fourier series technique; the expansion functions satisfy all the boundary conditions of the problem. This method, method A, consists in using backward integration and the boundary conditions (7.1) so expansions in Fourier series are written for the unknown function **v** as well as for their derivatives up to the n-th order. Introducing these expansions into (7.1) we obtain an infinite system of linear equations having as unknowns the (constant) coefficients of the above mentioned expansions. The unknown function is non-vanishing iff not all these coefficients vanish i.e. iff the corresponding Cramer determinant vanishes. This will be precisely the equation of the neutral curve of (7.1), (7.2). As illustration let us consider the eigenvalue problem

$$\left.\begin{aligned} &(7.3)\quad (D^2 - k^2)\theta + v = 0, \\ &(7.4)\quad (D^2 - k^2)^2 v - k^2 R(1 + El)\,\theta - k^2 R\, El D\varphi = 0, \end{aligned}\right\} x \in (-0.5; 0.5)$$

$$(D^2 - k^2)\,\varphi + D\theta = 0, \tag{7.5}$$

$$v = D^2 v = \theta = \varphi = 0 \text{ at } x = \pm 0.5, \tag{7.6}$$

which governs the neutral linear stability of a viscous fluid layer heated from below and subjected to a dielectrophoretic force. R is the Rayleigh number, El is a dimensionless number proportional to the squared electric field, $D = \frac{d}{dx}$, $v(x)$, $\theta(x)$, $\varphi(x) \in C^\infty[-0.5; 0.5]$ are the perturbations of the velocity, temperature and electrical potential respectively [1]. If El $= 0$, equation (7.5) ceases to have physical meaning and thus problem (7.1)—(7.6) reduces to (7.3), (7.4) (7.6). In (7.6) we no longer consider the functions φ. Apply now the method A to the problem obtained

$$\left.\begin{array}{ll}(7.3) & (D^2 - k^2)\theta + v = 0 \\ (7.4)' & (D^2 - k^2)^2 v - k^2 R\, \theta = 0\end{array}\right\} x \in (-0.5;\ 0.5)$$

$$(7.6)' \qquad v = D^2 v = \theta = 0 \text{ at } x = \pm 0.5.$$

Taking into account (7.6)′ and Remark 7.4 and using backward integration for even solutions (v, θ) of (7.3), (7.4)′, (7.6)′ we write

$$(7.7) \quad v^{(1)}_{2n-1} = -(2n-1)\pi v_{2n-1},\ v^{(2)}_{2n-1} = -(2n-1)^2\pi^2 v_{2n-1},\ v^{(4)}_{2n-1} = (2n-1)^4\pi^4 v_{2n-1},\ \theta^{(2)}_{2n-1} = -(2n-1)^2\pi^2\theta_{2n-1}.$$

Introducing the corresponding expansions in $\{E_1, E_3, ...\}$ into (7.3), (7.4)′ and denoting $A_n = (2n-1)^2\pi^2 + k^2$ we obtain

$$\begin{cases} -A_n\,\theta_{2n-1} + v_{2n-1} = 0, \\ A_n^2 v_{2n-1} - k^2 R\,\theta_{2n-1} = 0, \end{cases}$$

which leads to the following equation for the neutral curve:

$$\begin{vmatrix} -A_1 & 1 & 0 & 0 & 0\ldots \\ -k^2R & A_1^2 & 0 & 0 & 0\ldots \\ 0 & 0 & -A_2 & 1 & 0\ldots \\ 0 & 0 & -k^2R & A_2^2 & 0\ldots \end{vmatrix} = 0.$$

This equation has an infinity of solutions $R_n = k^{-2}A_n^3$. The smallest solution is $R_1 = k^{-2}A_1^3$ and, thus, the equation of the neutral surface is $R = k^{-2}(4\pi^2 + k^2)$.

Let us now treat the case El $\neq 0$. For this purpose the even functions v and θ will be expanded in Fourier series in $\{E_1, E_3, ...\}$ and the odd functions φ in $\{F_2, F_4, ...\}$. We have $D^{2k}v = \sum_{n=1}^{\infty} v^{(2k)}_{2n-1} E_{2n-1}(x)$, $(k = 0, 1, 2)$, $D^{2k}\theta = \sum_{n=1}^{\infty} \theta^{(2k)}_{2n-1} E_{2n-1}(x)$, $(k = 0, 1)$,

$D\theta = -\sum_{n=1}^{\infty} (2n-1)\pi\theta_{2n-1}F_{2n-1}$, $D^{2k}\varphi = \sum_{n=1}^{\infty} \varphi^{(2k)}_{2n} F_{2n}(x)$, $(k = 0, 1)$,

$D\varphi = \varphi_0' + \sum_{n=1}^{\infty} \varphi^{(1)}_{2n} E_{2n}(x)$, where $v^{(2k)}_{2n-1}$ and $\theta^{(2k)}_{2n-1}$ are given by (7.7)

and $\varphi_{2n}^{(1)} = 2n\pi\varphi_{2n}$, $\varphi_{2n}^{(2)} = -(2n\pi)^2 \varphi_{2n}$, $\varphi_0' = \int_{-0.5}^{0.5} \mathrm{D}\varphi(x)\,\mathrm{d}x = \varphi(0.5) - \varphi(-0.5) = 0$. Introducing these series into (7.3)–(7.5) we obtain

$$\sum_{n=1}^{\infty} (A_n \theta_{2n-1} + v_{2n-1}) E_{2n-1} = 0, \tag{7.8}$$

$$\sum_{n=1}^{\infty} \{A^2 v_{2n-1} - k^2 \mathrm{R}(1 + \mathrm{El})\theta_{2n-1}\} E_{2n-1} - k^2\mathrm{R}\,\mathrm{El}\sum_{n=1}^{\infty} (2n\pi)\,\varphi_{2n} E_{2n} = 0, \tag{7.9}$$

$$\sum_{n=1}^{\infty} (-B_n)\,\varphi_{2n} F_{2n} - \sum_{n=1}^{\infty} (2n-1)\,\pi\theta_{2n-1} F_{2n-1} = 0. \tag{7.10}$$

Multiplying a function by E_{2n-1} and integrating the result we obtain its Fourier coefficient of E_{2n-1}. If this function is identically zero then all its Fourier coefficients with respect to $\{E_{2n-1}\}_{n\in\mathbb{N}}$ will vanish. As $\{E_{2n-1}\}_{n\in\mathbb{N}}$ is a total set it follows that the converse is also true. Therefore since the function $(\mathrm{D}^2 - k^2)\theta + v$ vanishes (by equation (7.3)) it follows that its Fourier coefficients $A_n\theta_{2n-1} + v_{2n-1}$ upon $\{E_{2n-1}\}_{n\in\mathbb{N}}$ vanish. Hence,

$$-A_n\theta_{2n-1} + v_{2n-1} = 0, \; n = 1, 2, \ldots \tag{7.8$'$}$$

Equation (7.8) may be obtained by imposing to (7.8) to be orthogonal on E_{2n-1}. Impose similarly to (7.4) and (7.5) to be orthogonal on E_{2m-1} and E_{2m}. We find

$$A^2 v_{2m-1} - k^2\mathrm{R}(1 + \mathrm{El})\,\theta_{2m-1} - k^2\mathrm{R}\,\mathrm{El}\sum_{n=1}^{\infty} (2n\pi)\,\varphi_{2n}(E_{2n}, E_{2m-1}) = 0, \; m = 1, 2, \ldots . \tag{7.9$'$}$$

$$B_m\varphi_{2m} + \sum_{n=1}^{\infty} (2n-1)\,\pi\theta_{2n-1}(F_{2n-1}, F_{2m}) = 0, \; m = 1, 2, \ldots . \tag{7.10$'$}$$

By the Galerkin procedure we mean: expansion of the solution u of an equation $f(u) = 0$ as well as its derivatives in an orthogonal set of functions $\{l_n\}_{n\in\mathbb{N}}$, introduction of these series into the equation $f(u) = 0$ and then requiring the equation obtained to be orthogonal to $\{l_n\}_{n\in\mathbb{N}}$. This requirement provide an infinite system of algebraic equations for the determination of the Fourier coefficients of u. Generally Galerkin's approach is not exact and the expansion of u upon $\{l_n\}$ does not converge. Nevertheless when the set $\{l_n\}$ is total the method is exact and reduces to requiring the Fourier coefficients of the left-hand side of $f(u) = 0$ to vanish.

So, by Galerkin's method we obtain the system (7.8)′—(7.10)′. From it, taking into account that $(E_{2n}, E_{2m-1}) = 4(-1)^{n+m}(2m-1)[(2n)^2-(2m-1)^2]$, $(F_{2n}, F_{2k-1}) = 8n(-1)^{n-k+1}[(2n)^2-(2k-1)^2]$, we have, for $m = 1, 2, \ldots$,

$$[A_m^3 - k^2\,\mathrm{R}(1+\mathrm{El})]\theta_{2m-1} + \\ +k^2\,\mathrm{R}\,\mathrm{El}\sum_{n=1}^{\infty}\sum_{k=1}^{\infty}\frac{(-1)^{m-k+1}64n^2\pi^2(2m-1)(2k-1)}{[(2n\pi)^2+k^2][(2n)^2-(2m-1)^2]-[(2n)^2-(2k-1)^2]}\,\theta_{2k-1}=0. \tag{7.11}$$

The condition that the system (7.11) has a nonvanishing solution is that the infinite determinant D of the coefficients of θ_{2m-1} vanish. Therefore $D = 0$ is equivalent to the NS equations.

The method of expanding the unknown function f and its derivatives in Fourier series which satisfy termwise the boundary conditions of f will be called the Chandrasekhar-Galerkin method. Method A is a direct method of Chandrasekhar-Galerkin type. It was applied extensively in hydromagnetic stability theory by Chandrasekhar [5] (from Ch. 1).

The method A has two disadvantages: 1) the equation of NS derived by this method has the form of a determinant of infinite order which gives difficulties in its approximate solution. 2) the examples treated by us are very simple. More complicated cases require the construction of total sets of orthonormal functions which satisfy a large number of boundary conditions. Since the eigensolutions of self-adjoint operators densely defined in a Hilbert space H are total sets of H we have the possibility of constructing expansion functions with the desired boundary conditions. Nevertheless their form is complicated and the calculations based on them are cumbersome. For instance, a set of even orthonormal functions which together with their first derivatives vanish at $x = \pm 0.5$ are $C_n(x) = \cosh\lambda_n x\left(\cosh\frac{\lambda_n}{2}\right)^{-1} - \cos\lambda_n x\left(\cos\frac{\lambda_n}{2}\right)^{-1}$ where λ_n are the positive roots of $\tanh\frac{\lambda}{2} + \tan\frac{\lambda}{2} = 0$. The functions C_n are the eigenvectors and λ_n^4 are eigenvalues of the real self-adjoint operator $L = \mathrm{D}^4$, $L;\ \mathfrak{D}(L) \subset L^2(-0.5;\ 0.5) \to L^2(-0.5;\ 0.5)$, $\mathfrak{D}(L) = \{u \in C^\infty[0.5;\ 0.5] \mid u(\pm 0.5) = \mathrm{D}u(\pm 0.5) = 0\}$ and hence they form an orthonormal basis for $\overline{\mathfrak{D}(L)}^{\|\ \|_{L^2}} = H_e \subset L^2(-0.5;\ 0.5)$ where H_e consists of all the even functions of $L^2(0.5;\ 0.5)$. For a more detailed analysis of C_n and of some other expansion functions used in hydrodynamic stability theory the reader is sent to [2], [4], [5] (from Ch. 1).

B. Direct Fourier series technique; the expansion functions do not satisfy all the boundary conditions of the problem. The only difference between methods A and B consists in the fact that in method B only part of the boundary conditions (7.2) are satisfied by the expansion functions and hence only part of the involved series vanish termwise at $x = \pm 0.5$. The remaining boundary conditions (7.2) introduce some constraints on the coefficients of the series; i.e. although the expansion functions do not satisfy the corresponding boundary conditions nevertheless the series as a whole satisfies them.

Thus consider the problem (7.3)—(7.6) where v and θ are given by (7.7) and the odd function φ and its derivatives have the form

$$D^{2k}\varphi = \sum_{n=1}^{\infty} \varphi_{2n-1} F_{2n-1}(x), \quad (k = 0, 1), \ D\varphi =$$

$$= \sum_{n=1}^{\infty} (2n-1)\, \pi\varphi_{2n-1} E_{2n-1}(x), \text{ where } \varphi^{(2)}_{2n-1} = 2\sqrt{2}(-1)^{n-1}\alpha -$$

$$- (2n-1)^2 \pi^2 \varphi_{2n-1}, \ \alpha = D\varphi(\pm 0.5). \tag{7.12}$$

Introducing (7.7) and (7.12) into (7.3)—(7.5) we obtain

$$\begin{cases} -A_n\theta_{2n-1} + v_{2n-1} = 0, \\ A_n^2 v_{2n-1} - k^2 R(1+El)\theta_{2n-1} - k^2 R\, El\,(2n-1)\,\pi\varphi_{2n-1} = 0, \\ 2\sqrt{2}(-1)^{n+1}\alpha - A_n\varphi_{2n-1} - (2n-1)\,\pi\theta_{2n-1} = 0. \end{cases} \tag{7.13}$$

The first three boundary conditions (7.6) are termwise satisfied. The last condition (7.6) is not satisfied term by term because the F_{2n-1} do not vanish at $x = \pm 0.5$. This condition introduces the constraint $\Gamma = \sum_{n=1}^{\infty} \sqrt{2}(-1)^{n+1}\varphi_{2n-1} = 0$.

Solving (7.13) and putting $N_n = A_n^4 - k^2 R A_n - k^4 R\, El$, we have $\varphi_{2n-1} = 2\sqrt{2}(-1)^{n+1}[A_n^3 - k^2 R(1 + El)]\alpha N_n^{-1}$. Introducing this expression in the constraint $\Gamma = 0$ and taking into account that α is arbitrary we obtain the following equations for the NS

$$\sum_{n=1}^{\infty} [A_n^3 - k^2 R(1+El)]\, N_n^{-1} = 0. \tag{7.14}$$

Let us now choose other sets of expansion functions total in $L^2(-0.5; 0.5)$ namely $\{1, E_2, \dots\}$, $\{F_2, F_4, \dots\}$. We now show that the calculations corresponding to these sets are more complicated due to the fact that in this case three constraints are introduced. For the solution of (7.3)—(7.5) we have

$$\begin{cases} D^{2k}v = v_0^{(2k)} + \sum_{n=1}^{\infty} v_{2n}^{(2k)} E_{2n}(x), \ (k=0, 1, 2), \ D^{2k}\theta = \theta_0^{(2k)} + \\ + \sum_{n=1}^{\infty} \theta_{2n}^{(2k)} E_{2n}(x), \ (k=0, 1), \ D\theta = \sum_{n=1}^{\infty} (-2n\pi)\, \theta_{2n} F_{2n}(x), \\ D^{2k}\varphi = \sum_{n=1}^{\infty} \varphi_{2n}^{(2k)} F_{2n}(x), \ (k = 0,1), \ D\varphi = \varphi_0^{(1)} + \\ + \sum_{n=1}^{\infty} (2n\pi)\, \varphi_{2n} E_{2n}(x), \end{cases} \tag{7.15}$$

where $v_{2n}^{(2)} = 2\sqrt{2}(-1)^n \alpha - (2n\pi)^2 v_{2n}$, $v_0^{(2)} = 2\alpha$, $\alpha = \mathrm{D}v(0.5)$, $v_{2n}^{(4)} = 2\sqrt{2}(-1)^n \beta - (2n\pi)^2 2\sqrt{2}(-1)^n \alpha + (2n\pi)^4 v_{2n}$, $v_0^4 = 2\beta$, $\beta = \mathrm{D}^3 v(0.5)$, $\theta_{2n}^{(2)} = 2\sqrt{2}(-1)^n \gamma - (2n\pi)^2 \theta_{2n}$, $\theta_0^{(2)} = 2\gamma$, $\gamma = \mathrm{D}\theta(0.5)$, $\varphi_{2n}^{(2)} = -(2n\pi)^2 \varphi_{2n}$, $\varphi_0^{(1)} = \int_{-0.5}^{0.5} \mathrm{D}\varphi(x)\mathrm{d}x = \varphi(0.5) - \varphi(-0.5) = 0$. Since $F_{2n}(\pm 0.5) = 0$, $\varphi = 0$ is satisfied termwise. The rest of the boundary conditions (7.6) introduce the constraints

$$
(7.16)\quad \begin{cases} \Gamma_1 \equiv v_0 + \sum_{n=1}^{\infty} (-1)^n \sqrt{2} v_{2n} = 0, \\ \Gamma_2 \equiv v_0^{(2)} + \sum_{n=1}^{\infty} (-1)^n \sqrt{2}[2\sqrt{2}(-1)^n \alpha - (2n\pi)^2 v_{2n}] = 0, \\ \Gamma_3 \equiv \theta_0 + \sum_{n=1}^{\infty} (-1)^n \sqrt{2} \theta_{2n} = 0. \end{cases}
$$

Introducing the series (7.15) into (7.3)—(7.5) and denoting $B_n = (2n\pi)^2 + k^2$ we obtain

$$
\begin{aligned}
&2\gamma - k^2\theta_0 + v_0 = 0, \\
&2\beta - 4k^2\alpha + k^4 v_0 - k^2\mathrm{R}(1 + \mathrm{El})\theta_0 = 0, \\
&2\sqrt{2}(-1)^n \gamma - B_n\theta_{2n} + v_{2n} = 0, \\
&2\sqrt{2}(-1)^n \beta - (B_n + k^2)\, 2\sqrt{2}(-1)^n \alpha + B_n^2 v_{2n} - k^2\mathrm{R}(1 + \\
&\quad + \mathrm{El})\theta_{2n} - k^2\mathrm{R}\,\mathrm{El}(2n\pi)\varphi_{2n} = 0, \\
&-B_n\varphi_{2n} - 2n\pi\theta_{2n} = 0.
\end{aligned}
$$

The solution of this system is

$$
(7.17)\quad \begin{aligned}
&\theta_0 = \Delta(4k^2\alpha - 2\beta + 2k^4\gamma), \quad v_0 = \Delta[4k^4\alpha - 2k^2\beta + 2k^2\mathrm{R}(1 + \mathrm{El})\gamma, \\
&\theta_{2n} = M_n 2\sqrt{2}[B_n^3\gamma + (B_n^2 + k^2 B_n)\alpha - B_n\beta](-1)^n, \quad v_{2n} = \\
&= 2\sqrt{2}(-1)^n M_n[(B_n^3 + k^2 B_n^2)\alpha - \beta B_n^2 + \gamma(k^2\mathrm{R}\ B_n + k^4\mathrm{R}\,\mathrm{El})], \\
&\varphi_{2n} = 2\sqrt{2}(-1)^n M_n(2n\pi)[-(B_n + k^2)\alpha + \beta - B_n^2\gamma],
\end{aligned}
$$

where $\Delta = [k^6 - k^2\mathrm{R}(1 + \mathrm{El})]^{-1}$, $M_n^{-1} = B_n^4 - k^2\mathrm{R}\,B_n - k^4\mathrm{R}\,\mathrm{El}$.

Introducing (7.17) in (7.16) and denoting $\Delta_k = \sum_{n=1}^{\infty} M_n B_n^k$ we find the following system having α, β and γ as unknowns

$$\alpha(2k^4\Delta + 2\Delta_3 + 2k^2\Delta_2) + \beta(-k^2\Delta - 2\Delta_2) + $$
$$+ \gamma[k^2\mathrm{R}(1 + \mathrm{El})\,\Delta + 2k^2\mathrm{R}\,\Delta_1 + 2k^4\mathrm{R}\,\mathrm{El}\Delta_0] = 0,$$

$$\alpha[1 + 2(k^4\Delta_2 - k^2\mathrm{R}\,\Delta_1 - k^4\mathrm{R}\,\mathrm{El}\Delta_0) +$$
$$+ \beta(2\Delta_3 - 2k^2\Delta_2) + \gamma(-2\,k^2\mathrm{R}\,\Delta_2 - 2\,k^4\mathrm{R}\,\mathrm{El}\,\Delta_1 +$$
$$+ 2k^4\mathrm{R}\,\Delta_1 + 2k^6\mathrm{R}\,\mathrm{El}\Delta_0) = 0,$$

$$\alpha(2k^2\Delta + 2\Delta_2 + 2k^2\Delta_1) + \beta(-\Delta - 2\Delta_1) + \gamma(k^4\Delta + 2\Delta_3) = 0.$$

It follows that the equation of NS writes

$$\begin{vmatrix} 2k^4\Delta+2\Delta_3+2k^2\Delta_2 & -k^2\Delta-2\Delta_2 & \begin{matrix} k^2\mathrm{R}(1+\mathrm{El})\,\Delta+2k^2\mathrm{R}\Delta_1+ \\ +2k^4\mathrm{R}\,\mathrm{El}\Delta_0 \end{matrix} \\ 1+2k^4\Delta_2-2k^2\mathrm{R}\Delta_1-2k^4\mathrm{R}\,\mathrm{El}\,\Delta_0 & 2\Delta_3-2k^2\Delta_2 & \begin{matrix} -2k^2\mathrm{R}\Delta_2-2k^4\mathrm{R}\,\mathrm{El}\Delta_1+ \\ +2k^4\mathrm{R}\,\Delta_1+2k^6\mathrm{R}\,\mathrm{El}\,\Delta_0 \end{matrix} \\ 2k^2\Delta+2\Delta_2+2k^2\,\Delta_1 & -\Delta-2\Delta_1 & k^4\Delta+2\Delta_3, \end{vmatrix} = \tag{7.18}$$
$$= 0.$$

Equality (7.18) is another form of the equation of the neutral surface for the eigenvalue problem (7.3)–(7.6).

The method of expanding the unknown function f in Fourier series which do not termwise satisfy the boundary conditions of f will be called the Budiansky-DiPrima method. Hence method B is a direct method fo Budiansky-DiPrima type.

In applying method B we obtain the equation for NS in the form of an infinite series equal to zero.

The terms of this series contain the eigenvalue. In passing from the n-th approximation of this eigenvalue to its $n + 1$-st approximation then instead of n terms we must consider $n + 1$ terms. Recall that in order to obtain the same effect when applying the method A we have to consider an $n + 1$-st order determinant instead of a n-th order one. This shows the first advantage of method B.

We have also to note that, unlike in method A, in method B the expansion functions are very simple. Nevertheless to avoid complicated calculations these functions must be chosen so as to satisfy as many boundary conditions as possible, or, in other words, to introduce fewest possible constraints $\Gamma_1 = 0, \ldots, \Gamma_k = 0,\ k > 0$. This can be seen from the two treatments of the problem (7.3)–(7.6).

C. *Variational principles for nonsymmetric operators.* By a variational principle we mean a theorem which establishes the equatity between the set of solutions of a boundary value problem and the set of stationary points of a corresponding functional. In this section we present such principles stated in terms of the given problem and its adjoint.

Let a and b be two real numbers and $L: \mathfrak{D}(L) \subset C^\infty[a, b] \to C^\infty[a, b]$ a linear n-th order differential operator with constant coefficients. Put $\overline{\mathfrak{D}(L)}^{\|\ \|_{L^2}} = H$ (i.e. H is the Hilbert space obtained by completing $\mathfrak{D}(L)$ in $L^2(a, b)$ norm), and denote by L^* the adjoint of L; $L^*: \mathfrak{D}(L^*) \subset \subset H \to H$. For every $v \in \mathfrak{D}(L)$ and $u \in \mathfrak{D}(L^*)$ we have $(Lv, u) = (v, L^*u)$, where $(\cdot, \cdot)$ is the scalar product in $L^2(a, b)$ $\left(\text{i.e. if } u, v, \in L^2(a, b) \text{ then } (u, v) = \int_a^b u\, v\, \mathrm{d}x\right)\cdot$ Therefore for $v, u: [a, b] \to \mathbb{R}$, it follows $(Lv, u) = \int_a^b Lv(x)u(x)\, \mathrm{d}x$. Integrate by parts and suppose that the boundary conditions (which enter the definition of $\mathfrak{D}(L)$ and $\mathfrak{D}(L^*)$) are such that we can put (Lv, u) in the form of a new functional

$$(Lv, u) = \int_a^b L_1u(x)\, L_2u(x)\, \mathrm{d}x \text{ where } L_1: \mathfrak{D}(L) \to H,\ L_2: \mathfrak{D}(L^*) \to H.$$

Further integrations lead to $\int_a^b L_1v(x)\, L_2u(x)\, \mathrm{d}x = \int_a^b v(x)L^*u(x)\, \mathrm{d}x$. Conversely, from $\int_a^b L_1v(x)\, L_2u(x)\, \mathrm{d}x$ we can go back to $\int_a^b Lv(x)u(x)\, \mathrm{d}x$. On the other hand, denoting by $j(v, u)$ the functional $j: \mathfrak{D}(L) \times \times \mathfrak{D}(L^*) \to \mathbb{R}$, $j(v, u) = \int_a^b L_1v(x)\, L_2u(x)\, \mathrm{d}x$ we have its variation $\delta j = \int_a^b [\delta(L_1v)\, L_2u + L_1v\delta(L_2u)]\, \mathrm{d}x$. By exactly the same integrations as in the above form $\int_a^b L_1v(x)\delta(L_2u(x))\, \mathrm{d}x$ we can go back to $\int_a^b L_1v\delta u\, \mathrm{d}x$ while from $\int_a^b \delta(L_1\, v)L_2u\, \mathrm{d}x$ we obtain $\int_a^b (\delta v)\ (L^*u)\, \mathrm{d}x$.

Therefore

$$\delta j = \int_a^b Lv\delta u \,\mathrm{d}x + \int_v^b L^* u \delta v \,\mathrm{d}x. \tag{7.19}$$

From (7.19) we have immediatly the following variational principle which states that v, u are solutions of the direct and adjoint problem iff (v, u) is a stationary point of the functional j.

Theorem 7.1 $Lv = 0$, $L^*u = 0$, *iff* $\delta j(v, u) = 0 \quad \forall\, v \in \mathfrak{D}(L)$, $u \in \mathfrak{D}(L^*)$.

As an illustration consider the equations (7.3)—(7.5) and the boundary conditions (7.20)

$$v = \mathrm{D}v = \theta = \varphi = 0 \text{ at } x = \pm 0.5. \tag{7.20}$$

For El $= 0$, the equation (7.5) loses its physical meaning so we obtain the equation (7.3), (7.4)′ and the boundary conditions

$$v = \mathrm{D}v = \theta = 0 \text{ at } x = \pm 0.5. \tag{7.6)''$$

Eliminating v in (7.3), (7.4)′ we obtain the Couette case where v, T from (1.1.42), (1.1.43) are replaced by θ, R from (7.3), (7.4)′, (7.6)″. Hence let us consider the problem

$$(\mathrm{D}^2 - k^2)^3\, \theta = - k^2 \mathrm{R}\, \theta,\ x \in (-0.5\,; 0.5) \tag{1.1.42''}$$

$$\theta = (\mathrm{D}^2 - k^2)\, \theta = \mathrm{D}(\mathrm{D}^2 - k^2)\, \theta = 0 \text{ at } x = \pm 0.5. \tag{1.1.43''}$$

By theorem 7.1 this problem is equivalent to finding the stationary points of

$$j(\theta, \theta^*) = \int_{-0.5}^{0.5} (\theta^{\mathrm{IV}}\, \theta^{*\prime\prime} - 3k^2 \theta''\, \theta^{*\prime\prime} - 3k^4 \theta'\, \theta^{*\prime} - k^6 \theta\, \theta^* + k^2 \mathrm{R}\, \theta\, \theta^*)\, \mathrm{d}x$$

where $\theta^* \in \mathfrak{D}(L^*)$. The adjoint problem L^* is

$$(\mathrm{D}^2 - k^2)^3\, \theta^* = -k^2 \mathrm{R} \theta^*,\ x \in (-0.5\,; 0.5) \tag{1.1.42'''}$$

$$\theta^* = \mathrm{D}\theta^* = \mathrm{D}^2(\mathrm{D}^2 - 2k^2)\, \theta^* = 0 \text{ at } x = \pm 0.5. \tag{1.1.43'''}$$

To solve the variational problem for j we use either expansion series as in method A or B. Nevertheless since the expansion functions used in method A (which would satisfy the conditions (1.1.43)″ and (1.1.43)‴) are complicated and they introduce no additional idea compared with the expansion functions involved in method B, we use these last functions namely $\mathrm{D}^{2m}\theta = \sum_{k=1}^{\infty} \theta_{2k-1}^{(2m)} E_{2k-1}(x)$ $(m =$

$= 0, 1, 2)$, $D^{2m-1}\theta = \sum_{k=1}^{\infty} \theta_{2k-1}^{(2m-1)} F_{2k-1}(x)$ $(m = 1, 2)$, where $\theta_{2k-1}^{(1)} = -(2k-1)\pi\theta_{2k-1}$, $\theta_{2k-1}^{(2)} = -(2k-1)^2\pi^2\theta_{2k-1}$, $\theta_{2k-1}^{(3)} = (2k-1)^3\pi^3\theta_{2k-1}$, $\theta_{2k-1}^{(4)} = (2k-1)^4\pi^4\theta_{2k-1}$ and the same expansions for $D^m\theta^*$ where $\theta_{2k-1}^{*(1)} = -(2k-1)\pi\theta_{2k-1}$, $\theta_{2k-1}^{*(2)} = -(2k-1)^2\pi^2\theta_{2k-1}^*$. The functional j becomes a function of the numbers θ_{2k-1} and θ_{2k-1}^*

$$\begin{aligned} j(\theta_{2n-1}, \theta_{2n-1}^*) &= [-(2n-1)^6\pi^6 - 3k^2(2n-1)^4\pi^4 - \\ &- 3k^4(2n-1)^2\pi^2 - k^6 + k^2\mathrm{R}]\,\theta_{2n-1}\theta_{2k-1}^* = [-A_n^3 + \\ &+ k^2\mathrm{R}]\,\theta_{2n-1}\theta_{2n-1}^*. \end{aligned} \tag{7.21}$$

The boundary conditions $D(D^2 - k^2)\theta = 0$ and $D\theta^* = 0$ introduce the constraints $\Gamma_1 = \sum_{n=1}^{\infty} (2n-1)\pi A_n \theta_{2n-1} = 0$, $\Gamma_2 = \sum_{n=1}^{\infty} (2n-1)\pi\theta_{2n-1}^* = 0$. The stationary points of j in the constraints $\Gamma_1 = 0$, $\Gamma_2 = 0$ are, consequently, given by the system

$$\frac{\partial(j - \mu_1\Gamma_1)}{\partial\theta_{2n-1}} = \frac{\partial(j - \mu_2\Gamma_2)}{\partial\theta_{2n-1}^*} = 0, \; \Gamma_1 = 0, \; \Gamma_2 = 0.$$

These first two equations can be written as

$$(-A_n^3 + k^2\mathrm{R})\,\theta_{2n-1}^* - \mu_1 A_n(2n-1)\pi = 0, \tag{7.22}$$

$$(-A_n^3 + k^2\mathrm{R})\,\theta_{2n-1} - \mu_2(2n-1)\pi = 0. \tag{7.23}$$

From (7.22) we have $\theta_{2n-1}^* = -\dfrac{\mu_1 A_n(2n-1)\pi}{A_n^3 - k^2\mathrm{R}}$ which introduced in $\Gamma_2 = 0$ gives

$$\sum_{n=1}^{\infty} A_n(A_n - k^2)\,(A_n^3 - k^2\mathrm{R})^{-1} = 0,$$

which is precisely the neutral curve equation (1.1.47). Deriving θ_{2n-1} from (7.23) and introducing it in $\Gamma_1 = 0$ we obtain (1.1.47).

This method is used especially in applied studies when the direct and the adjoint equations are the same and, consequently, the corresponding functional j is symmetric. We mention that in these kinds of studies the symmetric operators are called selfadjoint.

Generally $(Lv, u) = [B_1(v)\, B_2(u)]_{-0.5}^{0.5} - j(v, u) = [B_1(v)\, B_2(u)]_{-0.5}^{0.5} - [B_1(u)\, B_2(v)]_{-0.5}^{0.5} + (v, L^*u)$ where $[B_1(v)\, B_2(u)]_{-0.5}^{0.5}$ stands for the boundary values obtained as a result of integration by parts. Since we supposed $L(v, u) = j(v, u)$ it means that this method is successful only if these boundary values vanish at some stage of integration. So, we can write such a variational principle whenever such a vanishing term is obtained. As a consequence, for a given problem we may find several variational principles.

The expressions for the Fourier coefficients of the derivatives of the unknown functions become more complicated as the order is increased. So, the best variational principle is that one whose corresponding functional contains the lowest order derivatives possible. Moreover, the best variational principle is one in which the functional is symmetric. By definition the functional $j(v, u)\colon \mathfrak{D}(L) \times \mathfrak{D}(L^*) \to \mathbb{R}$ is symmetric, if for every $v, u \in \mathfrak{D}(L)$ then $j(v, u) = j(u, v)$.

D. *Variational principles for symmetric operators.* Let L be the operator introduced at the beginning of this appendix and suppose that it is symmetric, therefore of order $2l$, $l \in N$. Integration by parts l times leads for every $v, u \in \mathfrak{D}(L)$, to $(Lv, u) = [B_1v\, B_2u]_a^b - j(v, u) = [B_1v\, B_2u]_a^b - [B_1u\, B_2v]_a^b + (v, Lu)$. From this formula it can be seen that if L is symmetric then the functional $j\colon \mathfrak{D}(L) \times \mathfrak{D}(L) \to \mathbb{R}$ is symmetric. Letting $v = u$ in (7.17), denoting $j(v) = j(v, v)$ and taking the variation of the functional obtained we have

$$\delta j(v) = 2(Lv, \delta v) - 2[B_1(v)\, B_2(v)]_a^b. \tag{7.24}$$

Assume that for $v \in \mathfrak{D}(L)$ we have $[B_1(v)\, B_2(v)]_a^b = 0$ where $[B_1(v)\, B_2(v)]_a^b$ is a sum of vanishing products $[B_1^k(v)\, B_2^k(v)]_a^b$ of boundary values of v and its derivatives. Then from (7.24) it follows that if $Lv = 0$, $v \in \mathfrak{D}(L)$ then $\delta j = 0$. It may happen that $[B_1(v)\,\delta B_2(v)]_a^b = 0$ for $v \in \mathfrak{D}(j) \supset \mathfrak{D}(L)$. Indeed suppose that $B_2^k(a)$ is not fixed (i.e. arbitrary). Then in order for the product $B_1^k(a)\, \delta B_2^k(a)$ to vanish for arbitrary $\delta B_2^k(a)$ it is neccessary that $B_1^k(a) = 0$. Hence $\delta j = 0$ for all $B_2^k(a)$ implies $B_1^k(a) = 0$ although the variation of j was done without any assumption on $B_1^k(a) = 0$. But $B_1^k(a) = 0$ is a boundary condition satisfied by v of $\mathfrak{D}(L)$ because $B_1^k(a)\, B_2^k(a) = 0$ (by definition) and $B_k^2(a) \neq 0$ by assumption. It follows that the variation of j was made in a larger class than $\mathfrak{D}(L)$. In this way the following variational principle holds: $Lv = 0$, $v \in \mathfrak{D}(L)$ iff $\delta j(v) = 0$, $v \in \mathfrak{D}(j)$. Note that this variational principle for L symmetric, unlike the nonsymmetric case, does not contain L^*. However the fact that δj is taken in $\mathfrak{D}(j) \supset \mathfrak{D}(L)$ is connected with the fact that $\mathfrak{D}(L^*) \supset \mathfrak{D}(L)$. (Generally there is no connection between $\mathfrak{D}(L)$ and $\mathfrak{D}(L^*)$ but when L is symmetric we have $\mathfrak{D}(L^*) \supset \mathfrak{D}(L)$ and L is the restriction of L^*).

The boundary conditions which enter the definition of $\mathfrak{D}(L)$ but which are not satisfied by the elements of $\mathfrak{D}(j)$ are called natural boundary conditions.

In applications it is more convenient to use the symmetric form of the eigenvalue problems (see for instance their spectral properties (Appendix 3) and the simple form of the variational principle just established). Given the nonsymmetric operator L we always have the possibility of defining the symmetric operator L^*L such that the equations $Lu = 0$ and $L^*Lu = 0$ are equivalent. Since this general approach leads to complicated calculations based on variational formulations we mention other simpler procedures which are strongly dependent on the boundary conditions: 1° splitting of equations by introducing new functions [8] (from Ch. 1); 2° derivation of extra boundary conditions from the very equations; 3° introduction of new functions which are linear combinations of the former functions and their derivatives; 4° multiplication of some of the equations of the system by suitable constants; 5° permutation of the indices of the components of the unknown functions. Only those new functions can be used for which boundary conditions (derived from the former ones or by 2°) may be assigned. These procedures are efficient when the system of equations (7.1) contains only even order derivatives. However, if (7.1) contains odd order derivatives then the main problem in obtaining an equivalent symmetric problem is that the terms which contain even derivatives occur symmetrically in the matrics A. This is why we adapted the procedures 1°—5° to the case when odd derivatives are present [1]. In [1] we applied these methods to the system (7.3)—(7.5) subjected to four sets of boundary conditions. Although the equations were the same in all four eigenvalue problems quite different symmetric eigenvalue problems and corresponding variational principles were nevertheless found. These differences were due to the differences in the boundary conditions.

The problem (7.3)—(7.6) is not symmetric but it becomes symmetric if the new function $u = -(\mathrm{D}^2 - k^2)v$ is introduced and, correspondingly, the equation (7.4) is split into two parts (1°) and the equation (7.5) is multiplied (4°) by $-k^2$ R El. The boundary condition for u is deduced (2°) from the equation (7.4)''. If $f = (f_1 f_2 f_3, f_4)$ and $f_1 = \theta$, $f_2 = v$, $f_3 = u$, $f_4 = \varphi$ then the corresponding problem is still nonsymetric. It becomes symmetric if, in addition, we put $f = (u, v, \theta, \varphi)$ hence a permutation of indices is done (5°). Therefore the symmetric form of (7.3)—(7.6) may be written

(7.3) $$(\mathrm{D}^2 - k^2)\,\theta + v = 0,$$

(7.4)'' $$(\mathrm{D}^2 - k^2)\,v + u = 0,$$

(7.4)''' $$(\mathrm{D}^2 - k^2)\,u + k^2\mathrm{R}(1 + \mathrm{El})\,\theta + k^2\mathrm{R}\,\mathrm{El}\,\mathrm{D}\varphi = 0,$$

(7.5) $$(\mathrm{D}^2 - k^2)\,\varphi + \mathrm{D}\theta = 0,$$

(7.6)'' $$u = v = \theta = \varphi = 0, \text{ at } x = \pm 0.5.$$

The following variational principle holds: $L_1 f = 0$, $f \in \mathfrak{D}(L_1)$ iff $\delta j_1(v) = 0, v \in \mathfrak{D}(L_1)$ where $\mathfrak{D}(L_1) = \{f = (u, v, \theta, \varphi) \in (C^\infty[-0.5;0.5])^4 |$ f satisfies (7.6)''$\}$, I is the identity operator on $\mathfrak{D}(L_1)$,

$$L_1 = \begin{pmatrix} D^2 - k^2 I & I & 0 & 0 \\ 0 & D^2 - k^2 I & I & 0 \\ k^2 R(1+El)\, I & 0 & D^2 - k^2 I & k^2 R\, El\, D \\ D & 0 & 0 & D^2 - k^2 I \end{pmatrix}$$

$$j_1(v) = \int_{-0.5}^{0.5} [-2D\theta Du - 2k^2\theta u + 2uv - (Dv)^2 - k^2v^2 + k^2R(1+El)\theta^2 -$$

$$- 2k^2 R\, El\, \varphi D\theta + k^2 R\, El\, (D\varphi)^2 + k^4 R\, El \varphi^2\, dx.$$

In order to solve the variational problem $j_1 = 0$ the function f and its derivatives which enter the expresion of j_1 are expanded in Fourier series. j_1 will become a function of the unknown coefficients, say, $j(a_n)$. Then $\dfrac{\partial j(a_n)}{\partial a_n} = 0$ will yield the NS equation. If the constraint $\Gamma(a_n) = 0$ occurs then the variational problem is isoperimetric. Introducing the Lagrange multiplier μ the NS equation will be $\dfrac{\partial (j_1 - \mu\Gamma)}{\partial a_n} = 0$, $\Gamma = 0$. If the expansion functions are $\{1, E_2, E_4, ...\}$ j_1 becomes $j_1(u_0, v_0, \theta_0, u_{2n}, v_{2n}, \theta_{2n}, \varphi_{2n}) = -2u_0\theta_0 - 2(2n\pi)^2\theta_{2n}u_{2n} - 2k^2\theta_0 u_0 - 2k^2\theta_{2n}u_{2n} - (2n\pi)^2 v_{2n}^2 - k^2v_0^2 - k^2v_{2n}^2 + k^2R(1+El)\,\theta_0^2 + k^2R(1+El)\,\theta_{2n}^2 + 2k^2R\,El(2n\pi)\,\theta_{2n}\varphi_{2n} + k^2R\,El(2n\pi)^2\,\varphi_{2n}^2 + k^4R\,El\,\varphi_{2n}^2$ and the following three constraints are introduced

$$\Gamma_1 = \theta_0 + \sum_{n=1}^{\infty} (-1)^n \sqrt{2}\theta_{2n}, \quad \Gamma_2 = v_0 + \sum_{n=1}^{\infty} (-1)^n \sqrt{2} v_{2n}, \quad \Gamma_3 = u_0 + \sum_{n=1}^{\infty} (-1)^n \sqrt{2} u_{2n} = 0.$$

Denoting by $G = j_1 - \mu_1\Gamma_1 - \mu_2\Gamma_2 - \mu_3\Gamma_3$ the variational problem is written

$$\frac{\partial G}{\partial u_0} = \frac{\partial G}{\partial v_0} = \frac{\partial G}{\partial \theta_0} = \frac{\partial G}{\partial u_{2n}} = \frac{\partial G}{\partial v_{2n}} = \frac{\partial G}{\partial \theta_{2n}} = \frac{\partial G}{\partial \varphi_{2n}} = 0 \tag{7.25}$$

$$\Gamma_1 = 0, \; \Gamma_2 = 0, \; \Gamma_3 = 0. \tag{7.26}$$

From (7.25) we have

$$
(7.27)\quad \begin{cases}
-\mu_1 - 2k^2u_0 + 2k^2\mathrm{R}(1+\mathrm{El})\,\theta_0 = 0,\\
-\mu_2 - 2k^2v_0 + 2u_0 = 0,\\
-\mu_3 - 2k^2\theta_0 + 2v_0 = 0,\\
-\dfrac{\mu_1}{2}\sqrt{2}(-1)^n - (2n\pi)^2u_{2n} - k^2u_{2n} + k^2\,\mathrm{R}(1+\mathrm{El})\,\theta_{2n} +\\
+ k^2\mathrm{R}\,\mathrm{El}\,2n\pi\varphi_{2n} = 0,\\
-\dfrac{\mu_2}{2}\sqrt{2}(-1)^n - (2n\pi)^2\,v_{2n} - k^2v_{2n} + u_{2n} = 0,\\
-\dfrac{\mu_3}{2}\sqrt{2}(-1)^n - (2n\pi)^2\,\theta_{2n} + v_{2n} = 0,\\
-k^2\mathrm{R}\,\mathrm{El}[-(2n\pi)^2\varphi_{2n} - k^2\varphi_{2n} - (2n\pi)\,\theta_{2n}] = 0.
\end{cases}
$$

If one expands in the Fourier series $\{1, E_2, E_4, \ldots\}$ instead then by applying method B to the system (7.3)—(7.6) the same system (7.27) would be obtained but with $\mu_1 = -4\mathrm{D}\,u(0.5) = -4(\beta - k^2\alpha)$, $\mu_2 = -4\alpha$, $\mu_3 = -4\gamma$. Therefore the same NS will be found as when the direct method B is applied. If u_0 and u_{2n} are eliminated between the first, the second, the fourth and the fifth equation of (7.27) and the system obtained is solved we get (7.17) Finally we obtain the NS equation in the form (7.18).

In order to derive NS of a nonsymmetric problem we may use six methods based on a given set of expansion functions: methods A, B, C applied to the nonsymmetric problem and the methods A, B, D applied to its symmetric form. As a result the same form of NS is found. In [1] and [4] many examples are worked out. In [4] the case where natural boundary conditions arise is also studied.

The convergence of the series as well as the smoothness properties of the functions have been analysed in [9].

The direct and variational methods A—D based on the expansions in Fourier series were extensively applied by us in [1], [4], [6]—[8] to many problems in hydrodynamic and hydromagnetic stability theory. The main conclusion was that the direct introduction of the series in the equation leads to the same NS as when the series are used in an equivalent variational formulation.

Due to the boundary conditions the coefficients of the Fourier series for high order derivatives become more and more complicated. Since the variational formulations do not contain the derivatives of the higest orders which occur in the equations it follows that the algebra involved in using the variational principle is simpler than in using the direct methods.

The splitting procedure introduced already in [71] (from Ch. 2) was used by us to obtain a symmetric variational formulation. Nevertheless this procedure itself leads to lower order equations and consequently to fewer calculations. Hence the split equations may be treated by direct methods A and B. If the variational formulation is based on these lower order split equations an extra simplification of the calculations is obtained.

Sometimes for a given eigenvalue problem more than one variational principle may be derived. The most convenient among them is that one in which the functional is symmetric because the highest order derivatives in this functional is half the order of the given problem.

There are many possibilities of choosing expansion functions. The best expansion functions are those which lead to the fewest possible constraints $\Gamma_k = 0$, $k = 1, 2, \ldots$. The case $k = 0$ (i.e. no constraint) known as the Chandrasekhar-Galerkin typs method involves more calculations when compared with the cases $k > 0$ (Budiansky-DiPrima type methods). Indeed, when no constraint occurs this means that the expansion functions satisfy all the boundary conditions of the problem and the neutral surface is an infinite determinant. If the expansion functions satisfy only part of these conditions then the rest of boundary conditions introduce constraints which lead to the neutral surface in the form of an (infinite) series. In this last case the variational problem associated to the given eigenvalue problem is of isoperimetric type.

When Fourier series in expansion functions which do not satisfy all the boundary conditions of the problem are used the Lagrange multiplier from the variational formulation may be expressed in terms of the boundary values of the derivatives of orders higher than those which enter the functional associated with that formulation. In the case of nonsymmetric operators the Lagrange multipliers μ_k corresponding to the constraint $\Gamma_k = 0$ for the given problem are boundary values of such higher order derivatives of the functions from the adjoint problem; and the converse assertion holds as well.

The best among the methods A—D is the variational Budiansky-DiPrima method. If consists in associating with the given eigenvalue problem a variational formulation and then solving this last one by means of Fourier series upon expansion functions which 1° do not satisfy all the boundary conditions of the initial problem and 2° introduce the fewest possible constraints $\Gamma_1 = 0$, $\Gamma_2 = 0, \ldots,$ $\Gamma_k = 0$, $k > 0$.

E. Direct approach based on the characteristic equation. Suppose that the system (7.1) has been put in the form of an equation $Lw = 0$ of order n and (7.2) are written as n homogeneous boundary conditions. Since this equation has constant coefficients let us look for its solutions in the form $e^{\lambda x}$. Then λ satisfies the algebric characteristic equation.

Case A The solutions of the characteristic equation are simple (i.e. have the multiplicity equal to 1). Then the general solution of $Lw = 0$ will be a linear combination $w_g = \sum_{i=1}^{n} A_i e^{\lambda_i x}$, where λ_i are solutions of the characteristic equation $f(\lambda_i) = 0$. Introducing w_g into the boundary conditions we obtain a system of algebraic equations having A_i as unknowns. The condition that the determinant of the coefficients of A_i vanish is just the NS equation.

For example, the system (7.3)—(7.5) may be written as [7]

$$(D^2 - k^2)^4 v + k^2 R (D^2 - k^2) v - k^4 R \, El \, v = 0. \tag{7.28}$$

Its characteristic equation is

$$(\lambda^2 - k^2)^4 + k^2 R(\lambda^2 - k^2) - k^4 R \, El = 0. \tag{7.29}$$

Put $\lambda^2 - k^2 = \mu$; then (7.29) becomes

$$\mu^4 + k^2 R \mu - k^4 R \, El = 0. \tag{7.29$'$}$$

Denoting by $\mu_i (i = 1, 2, 3, 4)$ the solutions of (7.29)$'$ we have

$$\mu_1 + \mu_2 + \mu_3 + \mu_4 = 0, \tag{7.29$''_1$}$$

$$\mu_1\mu_2 + \mu_1\mu_3 + \mu_1\mu_4 + \mu_2\mu_3 + \mu_2\mu_4 + \mu_3\mu_4 = 0, \tag{7.29$''_2$}$$

$$\mu_1\mu_2\mu_3 + \mu_1\mu_2\mu_4 + \mu_1\mu_3\mu_4 + \mu_2\mu_3\mu_4 = - k^2 R, \tag{7.29$''_3$}$$

$$\mu_1\mu_2\mu_3\mu_4 = - k^4 R \, El. \tag{7.29$''_4$}$$

Let us prove that μ_i are mutually distinct. Suppose $\mu_1 = \mu_2 = \mu_3 = \mu_4 = \eta$. From (7.29)$''_1$ it follows $\eta = 0$ but then in (7.29)$'$ we must have $R = 0$. Hence this assumption does not hold. Assume now that $\mu_1 = \mu_2 = \mu_3 = \eta$, $\eta \neq \mu_4$. Then from (7.29)$''_1$ and (7.29)$''_2$ we find $3\eta + \mu_4 = 0$ and $3\eta(\eta + \mu_4) = 0$ which give $\eta = \mu_4 = 0$. Suppose then $\mu_1 = \mu_2 = \eta$, $\eta \neq \mu_3$, $\mu_3 \neq \mu_4$, $\eta \neq \mu_4$. From (7.29)$''_1$ we have $-2\eta = \mu_3 + \mu_4$ and from (7.29)$''_2$ $\mu_3\mu_4 = 3\eta^2$ such that (7.29)$''_3$ becomes $4\eta^3 = - k^2 R$ and (7.29)$''_4$ is $4\eta^4 = - k^4 R \, El$. These last two equalities give $\eta = El \, k^2 > 0$; it follows that $4\eta^3 = - k^2 R$ cannot be satisfied. As a consequence all the $\mu_i (i = 1, 2, 3, 4)$ are simple.

Define $\lambda_1 = \sqrt{\mu_1 + k^2}$, $\lambda_2 = \sqrt{\mu_2 + k^2}$, $\lambda_3 = \sqrt{\mu_3 + k^2}$, $\lambda_4 = \sqrt{\mu_4 + k^2}$, $\lambda_5 = -\lambda_1$, $\lambda_6 = -\lambda_2$, $\lambda_7 = -\lambda_3$, $\lambda_8 = -\lambda_4$. The solutions $e^{\lambda_i x}$ are linearly independent. If only even solutions are considered then

$v = \sum_{i=1}^{4} A_i \cosh y_i x$. Since θ and φ satisfy (7.27) too then similarly $\theta = \sum_{i=1}^{4} B_i \cosh \lambda_i x$, $\varphi = \sum_{i=1}^{4} C_i \cosh \lambda_i x$. From (7.3) and (7.5) we have $B_i = A_i(k^2 - \lambda_i^2)^{-1}$, $C_i = \lambda_i(k^2 - \lambda_i^2)^{-2}$. In this way if the boundary conditions (7.6)'' are considered we obtain

$$\sum_{i=1}^{4} A_i \cosh \frac{\lambda_i}{2} = 0; \sum_{i=1}^{4} B_i \cosh \frac{\lambda_i}{2} = 0; \sum_{i=1}^{4} \lambda_i A_i \sinh \frac{\lambda_i}{2} = 0;$$

$$\sum_{i=1}^{4} C_i \sinh \frac{\lambda_i}{2} = 0,$$

or, equivalently

$$\sum_{i=1}^{4} A_i \cosh \frac{\lambda_i}{2} = 0; \sum_{i=1}^{4} A_i (k^2 - \lambda_i^2)^{-1} \cosh \frac{\lambda_i}{2} = 0;$$

$$\sum_{i=1}^{4} \lambda_i A_i \sinh \frac{\lambda_i}{2} = 0; \sum_{i=1}^{4} \lambda_i A_i (k^2 - \lambda_i^2)^{-2} \sinh \frac{\lambda_i}{2} = 0;$$

it follows that the NS equation is

$$\begin{vmatrix} \cosh \frac{\lambda_1}{2} & \cosh \frac{\lambda_2}{2} & \cosh \frac{\lambda_3}{2} & \cosh \frac{\lambda_4}{2} \\ (k^2-\lambda_1^2)^{-1} \cosh \frac{\lambda_1}{2} & (k^2-\lambda_2^2)^{-1} \cosh \frac{\lambda_2}{2} & (k^2-\lambda_3^2)^{-1} \cosh \frac{\lambda_3}{2} & (k^2-\lambda_4^2)^{-1} \cosh \frac{\lambda_4}{2} \\ \lambda_1 \sinh \frac{\lambda_1}{2} & \lambda_2 \sinh \frac{\lambda_2}{2} & \lambda_3 \sinh \frac{\lambda_3}{2} & \lambda_4 \sinh \frac{\lambda_4}{2} \\ \lambda_1(k^2-\lambda_1^2)^{-2} \sinh \frac{\lambda_1}{2} & \lambda_2(k^2-\lambda_2^2)^{-2} \sinh \frac{\lambda_2}{2} & \lambda_3(k^2-\lambda_3^2)^{-2} \sinh \frac{\lambda_3}{2} & \lambda_4(k^2-\lambda_4^2)^{-2} \sinh \frac{\lambda_4}{2} \end{vmatrix} = 0 \tag{7.30}$$

Case B The solutions λ_i of the characteristic equation have multiplicities $m_i \geqslant 1$ where for al least one i we have $m_i > 1$. As an example consider the problem

$$\{(D^2 - k^2)^2 + Qk^2\}^2 v = - Tk^2(D^2 - k^2) v, \quad x \in (-0.5; 0.5) \tag{7.31}$$

$$Dv = (D^2 - k^2)v = \{(D^2 - k^2)^2 + Qk^2\} v = D\{(D^2 - k^2)^2 + Qk^2\} v = 0 \text{ at } x = \pm 0.5 \tag{7.32}$$

(where $Q > 0$ is a magnetic dimensionless number) which governs the neutral linear stability of an electrically conducting fluid subjected to the influence of an axial magnetic field [3]. The characteristic equation has the form

$$(\lambda^2 - k^2)^4 + 2Qk^2(\lambda^2 - k^2)^2 + Q^2k^4 + Tk^2(\lambda^2 - k^2) = 0, \tag{7.33}$$

or, letting $\mu = \lambda^2 - k^2$, equivalently

$$\mu^4 + 2Qk^2\mu^2 + Tk^2\mu + Q^2k^4 = 0. \tag{7.33$'$}$$

If can be proved that the solutions of (7.33)′ do not have the multiplicity 4 or 3 for any value of the parameters. But when $T = 16(3\sqrt{3})^{-1} \times \times a\; Q\sqrt{Q}$ this equation has a double solution $\mu_1 = \mu_2 = -a\sqrt{Q}(\sqrt{3})^{-1}$ and two other distinct solutions. It follows that for $T \neq 16(3\sqrt{3})^{-1} \times \times a\; Q\sqrt{Q}$ the hypersurface is onesheeted while at

$$T = 16(3\sqrt{3})^{-1}\; a\; Q\sqrt{Q} \tag{7.34}$$

two of its sheets coincide. In the parameter space (T, Q, k), (7.34) represents a curve which is therefore the catastrophe (bifurcation) curve C of the characteristic hypersurface (7.33)′.

For (T, Q, k) which does not belong to C the general even solution of (7.31) has the form

$$v = \sum_{i=1}^{4} A_i \cosh \lambda_i x, \tag{7.35}$$

where $\lambda_i = \sqrt{\mu_i + a^2}$, $i = 1, 2, 3, 4$.

Introducing (7.35) into (7.32) we obtain the following NS equation

$$D_e(T, Q, k) = \tag{7.36}$$

$$= \begin{vmatrix} \lambda_1 \sinh \dfrac{\lambda_1}{2} & \cdots \lambda_4 \sinh \dfrac{\lambda_4}{2} \\ (\lambda_1^2 - k^2) \cosh \dfrac{\lambda_1}{2} & \cdots (\lambda_4^2 - k^2) \cosh \dfrac{\lambda_4}{2} \\ [(\lambda_1^2 - k^2)^2 + Qk^2] \cosh \dfrac{\lambda_1}{2} & \cdots [(\lambda_4^2 - k^2)^2 + Qk^2] \cosh \dfrac{\lambda_4}{2} \\ \lambda_1[(\lambda_1^2 - k^2)^2 + Qk^2] \sinh \dfrac{\lambda_1}{2} & \cdots \lambda_4[(\lambda_4^2 - k^2)^2 + Qk^2] \sinh \dfrac{\lambda_4}{2} \end{vmatrix} = 0.$$

If the point (T, Q, k) belongs to C then instead of (7.35) we have

$$v = A_1' \cosh \lambda_1 x + A_2' x \sinh \lambda_1 x + A_3 \cosh \lambda_3 x + A_4 \cosh \lambda_4 x \tag{7.35$'$}$$

where $\lambda_i = \sqrt{\mu_i + a^2}$, $(i = 1, 2, 3, 4)$, $\mu_1 = \mu_2 = -(\sqrt{3})^{-1}\, a\sqrt{Q}$, $\mu_{3,\,4} = -(\sqrt{3})^{-1}\, a\sqrt{Q}\,(1 \mp 2i\sqrt{2})$.

The corresponding NS equation will be

$$(7.36)' \qquad D_{ec}(T, Q, k) =$$

$$= \begin{vmatrix} \lambda_1 \sinh \frac{\lambda_1}{2} & \sinh \frac{\lambda_1}{2} + \frac{\lambda_1}{2} \cosh \frac{\lambda_1}{2} & \lambda_3 \sin h \frac{\lambda_3}{2} & \lambda_4 \sinh \frac{\lambda_4}{2} \\ (\lambda_1^2 - k^2) \cosh \frac{\lambda_1}{2} & 2\lambda_1 \cosh \frac{\lambda_1}{2} + \frac{\lambda_1^2 - k^2}{2} \sinh \frac{\lambda_1}{2} & (\lambda_3^2 - k^2) \cosh \frac{\lambda_3}{2} & (\lambda_4^4 - k^2) \cosh \frac{\lambda_4}{2} \\ [(\lambda_1^2 - k^2)^2 + Qk^2] \cosh \frac{\lambda_1}{2} & 4\lambda_1(\lambda_1^2 - k^2) \cosh \frac{\lambda_1}{2} + + [(\lambda_1^2 - k^2)^2 + Qk^2] \sinh \frac{\lambda_1}{2} & [(\lambda_3^2 - k^2)^2 + Qk^2] \cosh \frac{\lambda_3}{2} & [(\lambda_4^2 - k^2)^2 + Qk^2] \cosh \frac{\lambda_4}{2} \\ \lambda_1[(\lambda_1^2 - k^2)^2 + Qk^2] \sinh \frac{\lambda_1}{2} & (5\lambda_1^4 - 6k^2\lambda_1^2 + k^4 + Qk^2) \sinh \frac{\lambda_1}{2} + + \frac{\lambda_1}{2} [(\lambda_1^2 - k^2)^2 + Qk^2] \cosh \frac{\lambda_1}{2} & \lambda_3[(\lambda_3^2 - k^2)^2 + Qk^2] \sinh \frac{\lambda_3}{2} & \lambda_4[(\lambda_4^2 - k^2)^2 + Qk^2] \sinh \frac{\lambda_4}{2} \end{vmatrix} = 0.$$

(7.35) is written

(7.35)''
$$v = (A_1 - A_2) \cosh \lambda_1 x + A_2 (\lambda_2 - \lambda_1)\, x \times$$
$$\times \frac{\cosh \lambda_2 x - \cosh \lambda_1 x}{(\lambda_2 - \lambda_1)\, x} + A_3 \cosh \lambda_3 x + A_4 \cosh \lambda_4 x.$$

Putting $A_1' = A_1 - A_2$, $A_1' = A_2 (\lambda_2 - \lambda_1)$ and letting $\lambda_2 \to \lambda_1$ then (7.35)'' becomes (7.35)'. Equation (7.36)' is obtained from (7.36) in a similar way. Namely, let us subtract, in D_e, the first column from the second one, then devide the obtained column by $\lambda_2 - \lambda_1$ and finally let $\lambda_2 \to \lambda_1$. We find

(7.37)
$$D_{ec} = \left. \frac{\partial D_e}{\partial \lambda_2} \right|_{\lambda_2 = \lambda_1} .$$

Simple calculations show that $\left. \frac{\partial D_e}{\partial T} \right|_{\lambda_2 \to \lambda_1} = 0$. Since $\lambda_2 \to \lambda_1$ iff $T \to$ $\to 16a (3\sqrt{3})^{-1} Q \sqrt{Q}$ it follows that the equation of the NS for points belonging to C, denoted by NSC, is $D_e = 0$, $\left. \frac{\partial D_e}{\partial T} \right|_{T \to 16a (3\sqrt{3})^{-1} Q \sqrt{Q}} = 0$, $T = 16a (3\sqrt{3})^{-1} Q\sqrt{Q}$ consequently NSC $=$ NS $\cap$ C. Some extra detailed analysis show that NS is inside NSC and that NSC is the envelope of the NS equation.

The main advantage of method E consists in the fact that it provides the NS equation in a finite form $D_e = 0$ while the methods based on series expansions involve infinite determinants or series.

So far we had in view the derivation of NS. This equation will be solved numerically. The smallest amount of calculations is implied by the use of method E even if we take into account that this equation $D_e = 0$ is transcendental and to calculate the sinh's and cosh's which occur in D_e some series expansions must be done by the computer.

If the order of the given equation $Lw = 0$ is $2k \leqslant 8$ and $Lw = 0$ contains only even order derivatives then the characteristic equation can be solved exactly by means of the radicals. Correspondingly the NS can be written. If $2k > 8$ then generally $f(\lambda) = 0$ cannot be exactly solved. In this case the NS equation can be written formally but to derive the NS effectively the characteristic equation and the equations for NS must be solved simultaneously. This is not a difficulty for a computer.

Another advantage of method E is that the finite form of the NS allows an analytical study of the asymptotical behaviour of NS.

Finally, for problems with many physical parameters of actual interest method E leads to bounds for NS. These bounds are just the catastrophe (bifurcation) hypersurfaces of the characteristic equation.

REFERENCES

[1] Georgescu, A., *Variational formulation of some nonselfadjoint problems occurring in Bénard instability theory*, I, INCREST, Bucharest, Preprint Series in Mathematics No 35/1977.

[2] Harris, D. L., W. H. Reid, *On orthogonal functions which satisfy four boundary conditions*, The Astrophysical, J., **3**, 33, 429—453 (1958).

[3] Georgescu, A., *Catastrophe surfaces bounding the domain of linear hydromagnetic stability*, Central Institute of Physics, National Institute for Scientific and Technical Creation, Bucharest, Romania Preprint, FT-203—1981.

[4] Georgescu, A., A. Setelecan, *Methods to solve some eigenvalue problems which govern the linear stability of fluid flows*. I. (Romanian).

[5] Georgescu, A., I. Oprea, *On a direct method in linear hydrodynamic, hydromagnetic and elastic stability* (to be published).

[6] Georgescu, A., *The secular equations of some problems in thermal convection*, The 3rd Conference on naval constructions, Galatzi (Romania), 1981.

[7] Georgescu, A., V. Cardos, *Neutral stability curves for a thermal convection problem*, Acta Mechanica, 37 (1980) 165—168.

[8] Georgescu, A., O. Polotzka, *On a Bénard convection in the presence of dielectrophoretic forces*, J. Appl. Mech., 1981.

[9] Georgescu, A., *Exact solutions for some instability problems of Bénard type*, (to be published).

8. STABILITY OF NONSTATIONARY FLUID FLOWS

Linear case. Since modulation can induce dramatic modifications on the stability properties, a lot of papers dealing with linear stability of nonstationary basic fluid motions appeared after 1963. Mainly they look three directions: parallel shear flows, convective instabilities and centrifugal instabilities [1]. In order to illustrate the method used let us consider [2] a vertical plate oscillating vertically in a statically stably-stratified fluid. The problem which governs the linear stability of this fluid will be of the form

$$(8.1)\qquad 2\mathfrak{L}\varphi_t + i\alpha\,\mathrm{R}^{\delta_1}(\overline{W}\mathfrak{L}\varphi - \overline{W}_{xx}\varphi) = -\,2B^2\Gamma_{x}+ \mid 1 - B^2 \mid \mathfrak{L}^2\varphi,$$

$$(8.2)\qquad 2\Gamma_t + i\alpha\,\mathrm{R}^{\delta_1}(\overline{W}\Gamma - \bar{C}_x\varphi) = \mid 1 - B^2 \mid N_{Sc}^{-1}\mathfrak{L}\Gamma + 2\varphi_x,$$

$$(8.3)\qquad \varphi = \varphi_x = \Gamma_x = 0 \text{ at } x = 0,\ \beta_1$$

where $\mathfrak{L} = \dfrac{\partial^2}{\partial x^2} - \alpha^2$, $(0, 0, \overline{W}(x, t))$ is the velocity of the basic flow, $C(x, t)$ is the basic concentration, δ is the Stokes boundary layer thickness, $\delta_1 = \mid 1 - B^2 \mid^{-\frac{1}{2}}\delta$, $B = N/\omega$, N is the buoyancy frequency of the stratification, ω is the angular frequency of the oscillation of the plate, $\mathrm{R}^{\delta_1} = \mid 1 - B^2 \mid (W_0\delta_1/\nu)$, W_0 is the velocity amplitude of the oscillation of the plate, ν stands for the kinematic viscosity, and N_{Sc} is the Schmidt number. To obtain

(8.1)—(8.3) is has been supposed that the perturbation velocity and concentration $u = -\psi_z$, $w = \psi_x$ are normal modes i.e. $(\psi(x, z, t), C(x, z, t)) = (\varphi(x, t), \Gamma(x, t))\, e^{i\alpha z}$. Applying the Galerkin method the solutions of (8.1)—(8.3) are sought in the form

$$(8.4) \qquad \varphi(x, t) = \sum_{m=1}^{\infty} a_m(t)\, f_m(\xi), \quad \Gamma(x, t) = \frac{C_0(t)}{2} + \sum_{m=1}^{M-1} C_m(t) \cos m\pi\xi,$$

where $\xi = x/\beta_1$ and f_m are the functions C_m from Appendix 7 corresponding to the interval [0, 1]. Introducing (8.4) into (8.1), (8.2) we obtain the Galerkin system

$$(8.5) \qquad \dot{\mathbf{b}} = \mathbf{A}(t)\,\mathbf{b}, \quad \mathbf{A}(t + 2\pi) = \mathbf{A}(t)$$

where the transpose $\mathbf{b}^T$ of $\mathbf{b}$ is $\mathbf{b}^T = (a_1, a_2, \ldots, a_n, c_0, c_1, \ldots, c_{n-1})$. Then, by the Floquet theorem, a fundamental solution matrix $F(t)$ of (8.5) that satisfies $\mathbf{F}(0) = 1$ has the form $\mathbf{F}(t) = \mathbf{P}(t)\, e^{\mathbf{C}t}$, where $\mathbf{P}(t)$ is 2π -periodic and $\mathbf{C}$ is a constant matrix. If λ_1 is the eigenvalue of $\mathbf{C}$ having the largest real part then for $\mathcal{Re}\,\lambda_1 < 0$ there is stability, $\mathcal{Re}\,\lambda_1 > 0$ corresponds to instability and $\mathcal{Re}\,\lambda_1 = 0$ is the neutral case. To determine the neutral curve for (8.5) numerical integrations have to be performed.

Nonlinear case. Let us put the energy relation (1.1.15) in the form $\frac{dK}{d\tau} = -(J + \nu D)$ and define the variational problem

$$(8.6) \qquad f(\tau, \nu) = \max_N \frac{-J - \nu D}{K}$$

where N is the set of admissible functions which satisfy $\ln \frac{K(\tau)}{K(0)} \leqslant \int_0^\tau f(\eta, \nu)\, d\eta$ · A 2π-periodic motion is globally asymptotically stable in the mean if $\int_0^{2\pi} f(\eta, \nu)\, d\eta < 0$ and globally asymptotically stable in the strong sense if $f(\tau, \nu) < 0$ for $0 \leqslant \tau < \infty$ (i.e. (1.1.16)). Solving the Euler-Lagrange equations corresponding to the variational problem (8.6) by the Galerkin procedure as in the linear case, energy bounds for global stability may be found. For example in [3] the case of the modulated finite gap Taylor flow is analised.

REFERENCES

[1] Davis, S. H., *The stability of time-periodic flows*, Annual Review of Fluid Mech. vol. 8, Palo Alto, 1976, 57—74.

[2] Kerczek, Ch. von, S. H. Davis, *The instability of a stratified periodic boundary layer*, J. Fluid. Mech., **75**, 2, 287—303 (1976).

[3] Tustaniwskyi, J. I., S. Carmi, *Nonlinear stability of modulated finite gap Taylor flow*, The Physics of Fluids, **23**, 9 (1980) 1732—1739.

AFTERWORD

The Romanian version, which mainly contains the introduction, chs. 1, 2, 3, 5, the first part of Appendix 1 and Appendix 2 was written in 1974. Its revision was written in 1981 when chapter 4 and the rest of appendices were added. This supplimentary material concerns the new achievements and trends in hydrodynamic stability theory.

The aim of this book is to familiarized the reader with definitions, ideas, principles and modern or traditional methods on which hydrodynamic stability theory is based. This is why we did not mention all the contributions in the field (and we did not dare to assume such a task). Last decades more and more complex flows have been studied. We did not treat them because their analysis is not rigorous and no new idea is involved.

Three-dimensionality in linear stability. In ch. 1 we analysed the linear stability of some particular parallel or axi-symmetric incompressible flows against two-dimensional perturbations. Squire's method allows us to assert that for the stability of two-dimensional basic flows we have to consider only two-dimensional instead of three-dimensional perturbations. In practice this treatment is far from being sufficient. For instance, in the real flight of an aircraft, the effect of compressibility of the fluid, three-dimensionality of perturbations and slight nonparallelisms of the basic flow can amount to 40% [1]—[4]. In this case Squire's theorem no longer holds and crossflow profiles having inflection points lead (theorem 1.1.3, [5], [6]) to a crossflow instability. This type of instability was discoverd by Gray [7] in flight tests on aircraft with sweptback wings. This discovery gave impetus to studies of linear stability of compressible three-dimensional flows. For these flows three-dimensional perturbations must be considered. The slight nonparallelisms of the basic flows necessitated asymptotic methods, namely the method of multiple scales; while in the case of parallel basic flows other asymptotic approaches (the sadlle-point method) can be used. The theory of these perturbation methods may be found in [8], [9].

Last 15 years the linear stability of slightly nonparalel flows in channels recived considerable theoretical attention [10].

Spatial stability. In the case of nonparallel fluid flows of boundary layer type we must consider the spatial stability (i.e. with respect to disturbances of the boundary conditions at the initial section) together with temporal stability [11], [12]. In this case the single characteristic equation which relates the wavenumbers, phase, dimensionless numbers and other fluid flow characteristics which enter the eigenvalue problem which govern the linear stability are no longer sufficient to define these complex wavenumbers. To surpass this difficulty various relationships between the wavenumbers were postulated [13], [11]. These relationships involve the so-called group-velocity which is the derivative of the wavenumber which characterizes temporal instability with respect to the wavenumber which characterizes spatial instability. In the pure temporal stability only the velocity of wave propagation occurs.

Landau equation. In the linear theory of hydrodynamic stability the instability arives as an exponential growth of an initially very small periodic disturbance. This theory cannot state anything about the amplitude of the disturbances. But the finite amplitude of periodic disturbances in boundary layer flow are of exceptional practical importance in the initial stage of transition; therefore a nonlinear study has be carried out. In section 1.1.6 we presented the Landau equation and the Stuart approximate nonlinear theory which were successfull and intensively used. For their mathematic justification we quote [26] and [33] from Ch. 3. Nevertheless the Landau equation was obtained by Watson in the form of an infinite series with the restriction that the coefficient A in the Landau equation (1.1.58) is sufficiently small. The subsequent calculations shows only a qualitative agreement with experiments. This is why a reformulation of the method of Landau constants was carried out and better results were obtained [14].

We have to mention that the rigorous Hopf bifurcation cannot be established for the secondary boundary layer flows since the method of separation of variables cannot be applied in the case of boundary layer flows.

For the stability of a convective motion by perturbation methods we quote [15] where matched asymptotic expansions were used to construct time dependent solutions of the Boussinesq equations. In the course of this analysis the Landau equation is rigorously derived and a domain of stability of the convective state is determined. The rigorous justification of a perturbation method, similar to that introduced by Stuart and Watson was given.

Nonsteady basic flows. The stability of nonsteady basic flows was presented in Appendix 8. In the two cases mentioned above of boundary layer and convection secondary nonsteady flows bifurcating from steady basic motions were shown. In the second case an application of the extended Hopf theorem to particular fluid flows was given. For instance in [16] the bifurcation of the stationary Ekman flow into a stable periodic flow was proved. The bifurcation and stability for flows of fluids between rotating disks is at this time under very intensive investigation [17].

The Rayleigh-Synge criterion (Section 1.3.2) was extended to the case of general stationary and nonstationary basic flows between two coaxial cilinders [18].

Spherical solid boundaries. Fluid flows over spheres or between two spheres is of prime importance in applications [19]. The motion of the Earth's atmosphere belongs for instance to this class. The mathematical problem governing these motions was recently put into a generalized frame which allowed for a stability and branching analysis [20]. In [20] the first bifurcation of a steady flow into other steady waves was treated. The idea was advanced that the secondary bifurcation of these steady waves into time-periodic cyclone waves may be dealt with by the same method.

The stability studies rest on the analysis of the spectrum of the linearized problem while the branching needs in addition, the multiplicities of the eigenvalues.

The numerical calculations involved in secondary branching depend on the first bifurcating flow which generally can be described only numerically. This explains the small number of numerical studies on secondary bifurcation. At the same time the importance of global (with respect to time) studies for all values of the parameter is emphasized. In this line we mention the studies on the generic properties of the solutions of the Navier-Stokes equations, mainly done by Foiaș and Temam (§ 4.4).

The methods used in this book were predominantly analytical and the bifurcation was of local type. But, iu order to take into account higher-order bifurcation the geometric (topological) methods of so-called global analysis (i.e. the study of differential equations on manifolds and vector space bundles) must be applied. This is motivated by the fact that the Navier-Stokes equations are nonlinear and linear function spaces are a less satisfactory framework than the infinite dimensional manifold of maps. Last decade witnessed the penetration of this theory into the field of partial differential equations. For instances, the Liapunov-Schmidt Lemma can be rephrased in the geometric language of transversality. The key point in Foiaș and Temam studies is Smale's infinite dimensional version of Sard's theorem. In [3] (from Appendix 7) we succeded in bounding the neutral hypersurface by means of the catastrophe (branching) surface of the characteristic equation. A survey of applications of the geometric methods into bifurcation theory for fluid flows can be found in [21].

Exact solutions for the Navier-Stokes equations are highly desirable in stability as well as in branching theory; a large number of such solutions were derived by R. Berker [22]. Extending Serrin's first criterion (section 1.3.1) to the case of unbounded domains, he selected those solutions which are asymptotically stable in the mean among the set of exact solutions corresponding to a viscous fluid contained between two parallel plates rotating about the same axis.

Turbulence and chaos are two concepts which have been associated since the 1970s in connection with the explanation of the origin of turbulence by phenomenologic theories. This belongs to a strong line in dynamical systems lead by Ruelle, Bowen and Sinai and concerns the derivation of statistical properties of the attractors involved in phenomenological theories. In this respect the Lorenz attractor from fluid dynamics was widely investigated. The development of bifurcation theory and especially of global theories concerns the hydrodynamics case too; these studies treat the stages of transition to turbulence by repeated branching by loss of stability, symmetry breaking and pattern formation (Ch. 4).

A related topic is the analyticity of the solutions of the Navier-Stokes equations, successfully investigated by Foiaş and Temam by means of a finer splitting of the space $L^2(\Omega)$.

Synergetics [23]. The mathematical models involved in this book are rather simple. Some ideas concerning the dynamics of real fluids in supercritical regimes in the light of the stability and branching theory are presented in § 4.6. The further study of more complicated fluid flows, which include additional (thermal, chemical, electric, magnetic etc.) effects and which take place in the turbulent regime far from equilibrium, is no longer a matter of fluid dynamics alone. The analysis of these phenomena belong to the "interdisciplinary field of research called synergetics (i.e. cooperative effects) devoded to the global description of various dynamical systems", and their self-organizing behaviour which leads to the formation of structure and functionings. Special attention is paid to those phenomena where dramatic changes occur on a macroscopic scale when certain parameters are varied. These phenomena arise in systems which are composed of many subsystems.

According to Norbert Wiener a pattern is essentially an arrangement. Fluid dynamics perhaps presents the most examples of statical (i.e. stationary) paterns (Taylor vortices, Bénard cells, the von Kármán street, etc.) as well as dynamical patterns (nonstationary convective motions as for instance cloud streets, cyclone waves, etc.). Pattern formation, pattern recognition, chaos, turbulence are matters of synergetics. Stability and branching are tools to study these matters. The cooperative effects in fluid problems are briefly discussed in Sattinger's paper [23].

Synergetics reveals profound analogies between systems in different disciplines ranging from fluid mechanics to sociology such that it appears as an over-disciplinary field of knowledge belonging to natural philosophy.

Our book presents the mathematical model of the stability and branching (bifurcation) of fluid flows and the connection with the real world is made via synergetics. It is mainly devoted to fluid dynamicists. Since similar mathematical problems and interpretations are to be found in various other areas of science we think that this volume will be of profit to specialists in continuous media, physics, and chemistry too.

REFERENCES

[1] Mack, L. M. *Three-dimensional effects in boundary layer stability,* Proc. of the Twelfth Symposium on Naval Hydrodynamics, Naval Academy of Sciences, Washington, D. C., 1978, 63—76.

[2] Dunn, D. W., Lin C. C., *On the stability of the laminar boundary layer in a compressible fluid,* J. Aero. Sci., **22** (1955) 455—477.

[3] Mack, L. M., *Boundary layer stability theory,* Jet Propulsion Laboratory, Document 900—277 (Rev. A), 1969.

[4] Mack, L. M., *Transition prediction and linear stability theory,* AGARD Conference Proceedings No. 224, Laminar-Turbulent Transition Paper No. 1, 1977.

[5] Gregory, N., Stuart, J. T., Walker, W. S., *On the stability of three-dimensional boundary layers with applicaiion to the flow due to a rotating disk,* Phil. Trans. Roy. Soc. of London, **A248** (1955) 155—199.

[6] Owen, P. R., Randall, D. J., *Boundary layer transition on the swept back wing,* Royal Aircraft Establishment TM AERO, 277, 1952.

[7] Gray, W. E., *The effects of wing sweep on laminar flow,* Royal Aircraft Establishment TM Aero 255, 1952.

[8] Nayfeh, A. H., *Perturbation methods,* Wiley-Interscience, New York, 1973.

[9] Nayfeh, A. H., D. T. Mook, *Nonlinear oscillations,* New York, Chichester, John Wiley, 1979.

[10] Eagles, P. M., F. T. Smith, *The influence of nonparallelism in channel flow stability,* J. Engineering Mathematics, **14,** 3 (1980) 219—237.

[11] Gaster, M., *A note on the relation between temporally-increasing and spatially-increasing disturbances in hydrodynamic stability,* J. Fluid Mech., **14** (1962) 222—224.

[12] Nayfeh, A. H., Padhye, A., *The relation between temporal and spatial stability in three-dimensional flows,* AIAA Journal, **17** (1979) 1084—1090.

[13] Nayfeh, A. H. *Stability of three-dimensional boundary layers,* AIAA Journal, **18,** 4 (1980) 406—416.

[14] Herbert, Th., *On finite amplitudes of periodic disturbances of the boundary layer along a flat plate,* Proc. of the fourth Inst. Conf. on Numerical Methods in Fluid Dynamics, Boulder, Colorado, 1974.

[15] Gordon, N., F, C. Hoppensteadt, *An analysis of transient behaviour in the onest of convection,* Lecture Notes in Mathematics 594, Springer-Verlag, Berlin, 1977, 231—243.

[16] Iooss, G., H. B. Nielsen, H. True, *Bifurcation of the stationary Eckman flow into a stable periodic flow,* Arch. Rational Mech. Anal. **68,** 3 (1978) 227—256.

[17] Zandbergen, P. J., *New solutions of the Kármán problem for rotating flows,* Lecture Notes in Mathematics 771, Springer-Verlag, Berlin, 1980, 563—581.

[18] Kirchgässner, K., *Das Rayleigh-Syngesche Stabilitätskriterium für stationare und instationare zähe Strömungen,* ZAMM, **40** Sonderheft, (1960) T137—T139.

[19] Zierep, J., O. Sawatzki, *Three dimensional instabilities and vortices between two rotating spheres*, 8th Symposium on Naval Hydrodynamics, 1970.
[20] Herfort, P., H. True, *Bifurcation and nonlinear stability of solutions of thre quasi-geostrophic equation*, Gerlands Beitr. Geophysik, Leipzig, **86** (1977), 6, 425–442.
[21] Georgescu A., A. D. R. Choudary, *Bifurcation in fluid dynamics from the point of view of the differential topology*, Rev. Roum. Math. Pures et Appl. (to be published).
[22] Berker, R., *A new solution of the Navier-Stokes equation for the motion of a fluid contained between two parallel plates rotating about the same axis*, Archivum Mechaniki Stosowanej, **31**, 2 (1979) 265–280.
[23] Springer Series in Synergetics (vol. 1 *Synergetics. An Introduction* by H. Haken; vol 2 *Synergetics. A Workshop*, ed. H. Haken; vol. 3 *Synegetics. Far from Equilibrium*, eds. A. Pacault and C. Vidal; vol. **4** *Structural stability in Physics*, ed. W. Güttinger, H. Eikemeier; vol. 5 *Pattern formation by dynamic systems and pattern recognition*, ed. H. Haken; vol. 6 *Dynamics of Synergetic Systems*, ed. H. Haken; vol. 7 *Problems of biological physics* by L. A. Blumenfeld; vol. 8 *Chaos. The world of nonperiodic oscillations* by O. E. Rössler).

INDEX